Food Formulation

About the IFST Advances in Food Science Book Series

The Institute of Food Science and Technology (IFST) is the leading qualifying body for food professionals in Europe and the only professional organization in the UK concerned with all aspects of food science and technology. Its qualifications are internationally recognized as a sign of proficiency and integrity in the industry. Competence, integrity and serving the public benefit lie at the heart of the IFST philosophy. IFST values the many elements that contribute to the efficient and responsible supply, manufacture and distribution of safe, wholesome, nutritious and affordable foods, with due regard for the environment, animal welfare and the rights of consumers. IFST Advances in Food Science is a series of books dedicated to the most important and popular topics in food science and technology, highlighting major developments across all sectors of the global food industry. Each volume is a detailed and in-depth edited work, featuring contributions by recognized international experts and which focuses on new developments in the field. Taken together, the series forms a comprehensive library of the latest food science research and practice, and provides valuable insights into the food processing techniques that are essential to the understanding and development of this rapidly evolving industry. The IFST Advances series is edited by Dr Brijesh Tiwari, who is Senior Research Officer at Teagasc Food Research Centre in Ireland.

Forthcoming Titles in the IFST Series

Recent Advances in Micro- and Macroalgal Processing: Food and Health Perspectives edited by Gaurav Rajauria and Yvonne V. Yuan.

Oil and Oilseed Processing: Opportunities and Challenges edited by Tomás Lafarga, Gloria Bobo and Ingrid Aguiló.

Food Formulation

Novel Ingredients and Processing Techniques

Edited by

Shivani Pathania
Teagasc Food Research Centre, Ashtown, Dublin, Ireland

Brijesh K. Tiwari
Teagasc Food Research Centre, Ashtown, Dublin, Ireland

WILEY Blackwell

This edition first published 2021

Registered Offices
John Wiley & Sons, Inc., 111 River Street, Hoboken, NJ 07030, USA
John Wiley & Sons Ltd, The Atrium, Southern Gate, Chichester, West Sussex, PO19 8SQ, UK

Editorial Office
The Atrium, Southern Gate, Chichester, West Sussex, PO19 8SQ, UK

For details of our global editorial offices, customer services, and more information about Wiley products visit us at www.wiley.com.

Wiley also publishes its books in a variety of electronic formats and by print-on-demand. Some content that appears in standard print versions of this book may not be available in other formats.

Library of Congress Cataloging-in-Publication Data
Names: Pathania, Shivani, editor. | Tiwari, Brijesh K., editor.
Title: Food formulation : novel ingredients and processing techniques / edited by Shivani Pathania, Teagasc, Ashtown, Dublin, Ireland, Brijesh K. Tiwari, Teagasc, Ashtown, Dublin, Ireland.
Description: First edition. | Hoboken, NJ : Wiley Blackwell, 2020. | Series: About the IFST advances in food science book series | Includes bibliographical references and index.
Identifiers: LCCN 2020029618 (print) | LCCN 2020029619 (ebook) | ISBN 9781119614746 (hardback) | ISBN 9781119614753 (adobe pdf) | ISBN 9781119615248 (epub)
Subjects: LCSH: Food adulteration and inspection. | Food–Composition. | Food–Safety measures.
Classification: LCC TX531 .F564 2020 (print) | LCC TX531 (ebook) | DDC 363.19/264–dc23
LC record available at https://lccn.loc.gov/2020029618
LC ebook record available at https://lccn.loc.gov/2020029619

Cover Design: Wiley
Cover Image: Courtesy of Brijesh Tiwar

Set in 9.5/11.5pt TimesTenLTStd by SPi Global, Pondicherry, India.
Printed and bound by CPI Group (UK) Ltd, Croydon, CR0 4YY

C9781119614746_220221

Contents

List of Contributors

Maribel Abadias, Institute of Agrifood Research and Technology (IRTA), Lleida, Spain.

Ingrid Aguiló-Aguayo, Institute of Agrifood Research and Technology (IRTA), Lleida, Spain.

Carlos Álvarez, Teagasc Food Research Centre, Dublin, Ireland.

Pradip Behare, ICAR-National Dairy Research Institute, Karnal, Haryana, India. Department of Food Biosciences, Teagasc Food Research Centre, Moorepark, Cork, Ireland.

Bhesh Bhandari, School of Agriculture and Food Sciences, The University of Queensland, Brisbane, Queensland, Australia.

Cheenam Bhatia, Department of Clinical Nutrition, Sri Guru Ram Das University of Health Science & Research, Amritsar, Punjab, India.

Hanuman Bobade, Department of Food Science and Technology, Punjab Agricultural University, Ludhiana, Punjab, India.

Gloria Bobo, Institute of Agrifood Research and Technology (IRTA), Lleida, Spain.

B.N. Dar, Department of Food Technology, IUST-Kashmir, Jammu and Kashmir, India.

Arianna Dick, School of Agriculture and Food Sciences, The University of Queensland, Brisbane, Queensland, Australia.

Jesus M. Frias, Technological University Dublin, Environmental Sustainability and Health Institute, Dublin, Ireland.

Rajpreet Kaur Goraya, Department of Food Science & Technology, Punjab Agricultural University, Ludhiana, Punjab, India.

Ben Hillel, University of California, Hastings College of the Law, San Francisco (UC Hastings), USA

Shaik Abdul Hussain, ICAR-National Dairy Research Institute, Karnal, Haryana, India.
Department of Food Chemistry and Technology, Teagasc Food Research Centre, Moorepark, Cork, Ireland.

Tomás Lafarga, Institute of Agrifood Research and Technology (IRTA), Lleida, Spain.
Department of Chemical Engineering, University of Almería, Almería, Spain.

Olivia McAuliffe, Department of Food Biosciences, Teagasc Food Research Centre, Moorepark, Cork Ireland.
VistaMilk SFI Research Centre, Cork, Ireland.

Ciara McDonagh, Teagasc Food Research Centre, Dublin, Ireland.

Shabir Ahmad Mir, Department of Food Science and Technology, Government College for Women, Srinagar, Jammu and Kashmir, India.

Maneesha S. Mohan, Department of Wine, Food and Molecular Biosciences, Faculty of Agriculture and Life Sciences, Lincoln University, Lincoln, New Zealand.

Desirée Román Naranjo, Department of Food Biosciences, Teagasc Food Research Centre, Moorepark, Cork, Ireland.

Daniel Pando, Nanovex Biotechnologies, Parque Tecnológico de Asturias, Asturias, Spain.

Puneet Parmar, Animal and Grassland Research and Innovation Centre, Teagasc, Cork, Ireland.

Shivani Pathania, Teagasc Food Research Centre, Dublin, Ireland.

Camila A. Perussello, Technological University Dublin, Environmental Sustainability and Health Institute, Dublin, Ireland.

Milica Pojić, Institute of Food Technology, University of Novi Sad, Novi Sad, Serbia.

Sangeeta Prakash, School of Agriculture and Food Sciences, The University of Queensland, Brisbane, Queensland, Australia

Kai Purnhagen, University of Bayreuth, Campus Kulmbach, Chair for Food Law, Bayreuth, Germany.

Anna Vallverdú Queralt, Food Science and Gastronomy XaRTA, University of Barcelona, Institute of Nutrition and Food Safety (INSA-UB), Barcelona, Spain.
Consorcio CIBER, M.P. Fisiopatología de la Obseidad y Nutrición (CIBERObn), Instituto de Salud Carlos III (ISCIII), Madrid, Spain.

Manzoor Ahmad Shah, Department of Food Science & Technology, Government PG College for Women, Gandhi Nagar, Jammu, Jammu and Kashmir, India

Prateek Sharma, Department of Food Chemistry and Technology, Teagasc Food Research Centre, Moorepark, Cork, Ireland.
Department of Nutrition, Dietetics, and Food Sciences, Utah State University, USA.

Savita Sharma, Department of Food Science and Technology, Punjab Agricultural University, Ludhiana, Punjab, India.

Baljit Singh, Department of Food Science and Technology, Punjab Agricultural University, Ludhiana, Punjab, India.

Brijesh K. Tiwari, Teagasc Food Research Centre, Dublin, Ireland.

Jessica Vapnek, University of California, Hastings College of the Law, San Francisco (UC Hastings), USA.

Iuliana Vintilă, Dunarea de Jos University, Galati, Romania.

Preface

New product development and product reformulations have been implemented globally by food processors to develop novel foods. Food formulation is perhaps one of the most important components in new product development process and application of a new technology in food manufacturing business. Novel and healthy ingredients, along with increasing requirement of novel processing techniques for clean label food products as well as regulations and legislation to their use, is the premise of this book 'Food Formulation: Novel Ingredients and Processing Techniques'.

This book addresses three main areas of significant relevance to food formulation strategies required for food and nutraceutical sectors, namely: (i) Ingredients: This section covers the basis of a food formulation, i.e. ingredients. These chapters discuss the smart and functional ingredients from plant and animal sources, healthy ingredients such as salt, sugar and phosphate replacers, bioactives and peptides, macronutrient and micronutrient ingredient interactions. (ii) Processing techniques: This section includes the relevance and application of novel emerging technologies in processing of new food formulations. New techniques such as 3D printing, encapsulation and combination techniques have been discussed to improve the flavour and acceptability of functional food ingredients. Recent developments in greener technologies are discussed with emphasis on operational principles and inherent strengths and weaknesses of the technologies from food formulation viewpoint. (iii) New trends in food formulation: This section covers tailored food formulations for targeted populations, food targeting allergies and intolerance (gluten-free, lactose-free products), prebiotic and probiotic food formulations. Mathematical tools such as statistical modelling support new product development and food formulation process along with recent amendments/additions in regulations and legislations for novel food formulations are covered.

Editors of this book feel that this text will offer a reader a broad review of novel food ingredients and application of new processing techniques for advanced food formulations. This will serve as a comprehensive reference book for students, educators, researchers, food ingredients specialists and products developers providing an up-to-date insight into food formulation. The range of processes covered will provide food technologist, professionals, engineers and scientists working in food industry with a valuable resource for their work.

1

Food Formulation and Product Development

Shivani Pathania[1], Cheenam Bhatia[2], and Brijesh K. Tiwari[1]

[1] *Teagasc Food Research Centre, Ashtown, Dublin, Ireland*
[2] *Department of Clinical Nutrition, Sri Guru Ram Das University of Health Science & Research, Amritsar, Punjab, India*

Summary

Food formulation strategies are key for the product development process. Food formulation is critical in developing new food products along with rising consumer demand for products with key health and nutritional properties. As it is imperative to have a clear understanding of various ingredients, their roles, and judicious use of technologies available, this chapter provides an overview of the book and its relevance for various food formulations.

1.1 Introduction

Formulation is a common term used in various applications, with a fundamental meaning of placing a range of desired components in an appropriate manner according to a predefined formula. Food formulation can be defined as the activity of creating, designing or developing food products with an objective of providing some functionality. The functionality can range from providing additional nutritional benefits, to dietary supplementations over and above the basic function of the food. Food formulation can be perceived as a simple concept if it includes mixing ingredients, and it could be as complex and stringent as the pharmaceutical processes to regulate the concentration of highly functional ingredients such as bioactive peptides and fatty acids. Therefore, food formulation is perhaps one of the most important and critical steps involved in the new product development (NPD) process. Food formulation irrespective of the discipline, whether in food production, culinary research and development, or formulation of nutraceuticals/pharmaceuticals, requires careful consideration of ingredient selection and the application of technologies, considering consumer demand and legislative requirements. Food products are made by combining food

Food Formulation: Novel Ingredients and Processing Techniques, First Edition.
Edited by Shivani Pathania and Brijesh K. Tiwari.

ingredients in specific proportions in a formulation. Research on the effects of various formulations and processes on product qualities is common. However, research is focused on the effect of either a formulation or a novel processing technique on the product characteristics. Comprehensive, documented understanding of the trends in food formulation, its effect on the commercialisation of new technologies, and the impact of new technologies on formulations is limited.

1.2 New Product Development

NPD and product reformulations have been implemented globally by food processors to develop novel foods. Novel foods are newly developed, innovative foods produced using new technologies and production processes, as well as foods which are eaten outside of their traditional geographical boundary/region. Novel foods could be new sources of a major food component (e.g. protein, carbohydrate, fibre) or minor food component (e.g. vitamins, minerals, enzymes), or extracts from existing food or food derived from new production processes such as UV-treated foods (milk, bread, mushrooms, and yeast). New food formulations and processing techniques have been adopted commercially for better nutrient retention, quality improvement, and shelf life enhancement, and to support sustainability. Moreover, in recent times, consumers' attention, interest, and awareness of the health impact of various ingredients, as well as the manufacturing processes (e.g. thermal treatment) used in formulations, have grown tremendously. Health-conscious consumers have driven the increasing consumption of minimally and alternatively processed clean label products, which have been shown to fulfil tailored consumer needs for high-protein, low-fat, low-calorie products. From flavour enhancement and blocking to formulating for an ageing populace to carbohydrate systems for weight management, the future focus is on ingredients that promote better overall health. Thus, there is a requirement for detailed industrially relevant information concerning novel food ingredients and their application in the formulation of novel and health-promoting food products. In addition, industrial adoption of novel processing techniques is in its infancy; however, some technologies have seen commercial application.

1.3 Challenges in Food Formulations

Food formulation is a systematic, critical, complex, and commercially-oriented research process, and a great number of challenges are faced by the formulators in finalizing a formulation. Key components associated with the food formulation include: (i) the identification and selection of raw materials, which is generally based on the target end product and its use; (ii) the interaction of raw materials at the ingredient level and molecular level, i.e. protein–protein, protein–lipid, etc.; and (iii) the application of technology depending on food components, target consumer, quality, shelf life, and cost (Figure 1.1). Cost is one of the most important factors in food formulation. From the price of the raw materials to the cost of manufacturing, everything needs special consideration. It is important to formulate a product that follows certain requirements, especially in terms of its pricing. Therefore, a clear process is required to facilitate the sharing of information, from the raw materials needed to the manufacturing technique to the packaging material.

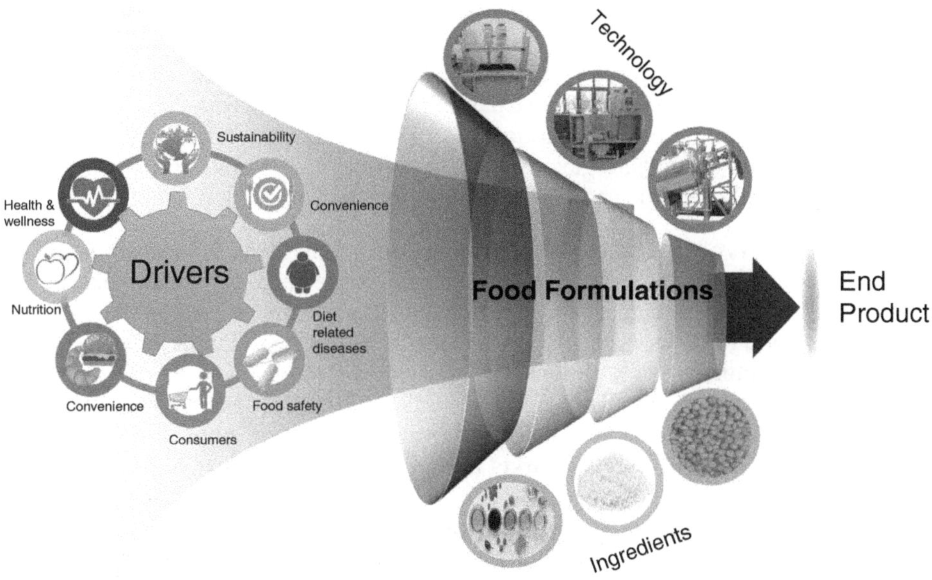

Figure 1.1 Factors impacting new food formulation process.

1.4 Relevance of the Book and Objectives

Applications of new and innovative technologies, either individually or in combination, and their resulting effects on ingredients and product characteristics are always of great interest to academic, industrial, nutrition, and health professionals. With the advancements in novel ingredients, including novel sources of raw material (proteins, starch), an understanding of the effects of food sources and their processing, as well as the interactions between ingredients, is critical. This book provides a comprehensive overview of novel food ingredients, the role of emerging processing techniques on product and health benefits, and customized food formulations for targeted populations. It provides a comprehensive assessment of food ingredients, their interaction, and the application of emerging technologies to formulate new food products. The book addresses three main areas of significant relevance to food and nutraceutical sectors, namely:

i. *Ingredients and their interactions* (Chapters 2–5): The first section of the book consists of a number of chapters covering the basis of a food formulation, i.e. ingredients. Smart and functional ingredients from plant and animal sources; healthy ingredients such as replacements for salt, sugar, and phosphate, bioactives, and peptides; macronutrient and micronutrient ingredient interactions.
ii. *Processing techniques* (Chapters 6–9): This section outlines the application of emerging technologies in the processing of new food formulations. New techniques such as 3D printing for tailor-made, demand-driven production and products with geometric intricacy is covered (Chapter 6). The role of encapsulation techniques (Chapter 7), extrusion (Chapter 8), and combination techniques to improve the flavour and acceptability of functional food ingredients and recent developments in greener technologies including high pressure, microwave

heating, ultrasound, supercritical fluid extraction, pulsed electric fields, ohmic heating, dielectric heating and other emerging technologies are discussed, with emphasis on the operational principles and inherent strengths and weaknesses of the technologies (Chapters 8 and 9) from a food formulation viewpoint.

iii. *New trends in food formulation* (Chapters 11–12): This section describes recent trends in the development of food for specific challenges for targeted populations, food targeting allergies and intolerances (gluten-free and lactose-free products), and prebiotic and probiotic food formulation for gut health.

The final chapters of the book focus on mathematical tools identified to support NPD and the food formulation process. Chapter 13 covers a wide range of mathematical tools, ranging from experimental design to consumer studies for the assessment of nutrient bioavailability, and outlines various examples of studies having different objectives (e.g. the effect of ingredients on a formulation vs consumer studies and panel studies vs nutrient delivery) and the mathematical tools utilized in each of them. One of the key hurdles associated with NPD and any food formulation is the route to market. Key issues associated with the route to market are regulatory and legislative issues around the use of ingredients and novel technologies. Therefore, novel and healthy ingredients, along with the increasing requirement for novel processing techniques for clean label food products, as well as regulations and legislation are covered in Chapter 14. Hitherto, this book provides a comprehensive text covering recent advances in food formulation, and offers the reader a broad review of novel food ingredients and the application of new processing techniques for advanced food formulations.

1.5 Conclusions

The novel food industry has made significant steps in the last several years, and is predicted to grow exponentially with product formulations and their nutrition, health, and wellness applications. This book provides a comprehensive assessment of the ingredients, their interactions, and the technologies available for developing new products.

2
Smart Functional Ingredients

Milica Pojić
Institute of Food Technology, University of Novi Sad, Novi Sad, Serbia

Summary

The increasing global population, climate change, the depletion of natural resources, and new consumer requirements tending to consume more natural, less processed, more sustainable and clean label food, impose the need to find alternative raw materials and technologies to meet these challenges. Therefore, new food product development more and more requires the utilization of non-traditional raw materials and ingredients that can be utilized either as alternative sources of proteins and/or as alternative sources of carbohydrates. These novel sources of nutrients, together with the specificity of their production, environmental benefits, and sustainability and safety aspects are reviewed in this chapter.

2.1 Introduction

Changes in the preferences of consumers, their increasing awareness about the significance of a balanced diet for maintaining good health, and changed dietary patterns have led to innovations and developments in the food industry encompassing both ingredients and food products. In new food product development, the research focus has been shifted from concentrated extracts of natural active ingredients from different origins to the utilization of non-traditional raw materials and ingredients (Calabrese and Ferranti 2019). Globalization, migration, and multiculturalism contribute to this trend as well (Fernández-Segovia et al. 2018).

On the other hand, the increase in global population, reaching 10 billion by 2050, climate change, the need of sustainability assurance in both agricultural and industrial production, and food security have necessitated research on alternative food ingredients, alternative processing technologies, and other solutions that will satisfy these requirements (Fasolin

Food Formulation: Novel Ingredients and Processing Techniques, First Edition.
Edited by Shivani Pathania and Brijesh K. Tiwari.

et al. 2019). By employing cutting-edge science to change the composition of food and make it healthier for people and the environment, different types of functional food products, encapsulated food and genetically modified food have been proposed (Fasolin et al. 2019).

The acceptability of new foods by consumers is a complex and challenging issue influenced by many factors, sensory preferences, and personal factors. The fear of eating new or unfamiliar food, known as neophobia, may hinder the consumption of novel food and ingredients (de Beukelaar et al. 2019).

2.2 *Smart* Protein Ingredients

The increasing demands for food proteins caused by the increasing world population can be met by the utilization of proteins from novel sources, which include alternative plants, aquatic photosynthetic organisms, microorganisms, insects, and waste streams from food processing (Pojić et al. 2018). Novel sources of protein could have many advantages compared to traditional sources – they offer opportunities to valorize biomass commonly treated as waste, do not compete with food crops for land and water, and have a year-round harvesting capability with high biomass yield, the ability to be cultivated on non-arable land utilizing non-irrigation water, a high protein yield, and resistance to pest and diseases (Bleakley and Hayes 2017; van der Spiegel et al. 2013).

2.2.1 Novel Sources of Proteins of Plant Origin

The emerging trends of vegetarianism and veganism and the negative environmental impacts of meat, dairy, and fish processing have initiated the growing demand for novel sources of plant protein (Pojić et al. 2018). The occurrence of allergies and intolerances associated with the consumption of milk and egg proteins, with the prevalence of egg allergy of 0.5–2.5% and milk allergy of 1.2–17% of young children in the Western world also contribute to the growing interest in plant proteins (Dang et al. 2016; Leonard 2016). Apart from health and environmental concerns, the ongoing replacement of animal with plant proteins in the diet has also been motivated by social image and prices (Vainio et al. 2016). Although a plant-based diet is generally considered healthier, safer, and more sustainable than animal-based diet, the provision of sufficient quantities of plant-based food for the growing global population is estimated at 60% more than current food production by 2050, will be difficult but feasible (Calabrese and Ferranti 2019). Therefore, several types of plant-based protein sources have emerged. The most valuable protein sources of plant origin are those that are cereals and legumes (Luigia 2019). Beside them, the potential of minor legumes (e.g. chickpea, faba bean, lentil, lupin, pea), pseudocereals (e.g. quinoa, chia, amaranth), and nontraditional plants, such as rape, industrial hemp, and cotton seeds, has recently been proven not only because they represent valuable sources of plant proteins providing substantial amounts of the essential amino acid lysine, but also because of the environmental sustainability of their cultivation (Dapčević-Hadnađev et al. 2019; López et al. 2018; Luigia 2019; Pojić et al. 2015) (Figure 2.1). Their additional advantage is that they are gluten-free, so in addition to being a protein source, they can be utilized in the production of special food products for consumers with special needs (Dapčević-Hadnađev et al. 2014; López et al. 2018). Increasing attention has been paid to the bioactive peptides arising from the enzymatic digestion of proteins from plant seeds and legumes (López et al. 2019), as well as from industrial by-products due to their beneficial health effects.

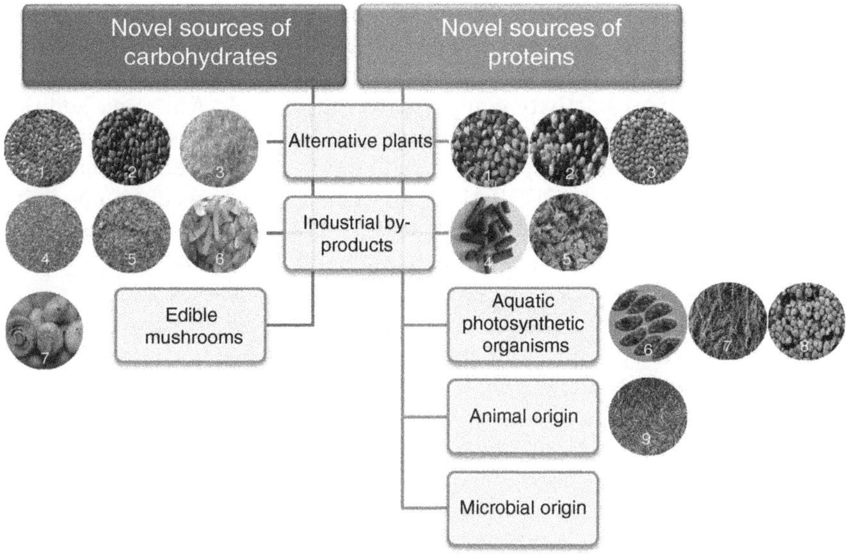

Figure 2.1 Alternative sources of **carbohydrates** (1-flaxseed, 2-chia seed, 3-psyllium husk, 4-apple pomace, 5-cereal brans, 6-potato peel, 7-edible mushrooms) and **proteins** (1-buckwheat, 2-quinoa, 3-hemp seed, 4-hempseed press cake, 5-brewers' spent grains, 6-microalgae, 7-macroalgae, 8-duckweed, 9-insect larvae). Source: Adapted from Heimann and Huerlimann (2015).

2.2.1.1 Alternative Plants as Novel Sources of Protein

2.2.1.1.1 Pseudocereals With its protein content in the range of 12–23%, quinoa is characterized with a higher protein and lower carbohydrate content then those of major cereals (e.g. wheat, barley, maize, and rice) (Dakhili et al. 2019). The isolation procedures of quinoa, chia, and amaranth proteins, their composition, and their structural, physico-chemical, and functional properties were comprehensively reviewed by Dakhili et al. (2019) and López et al. (2018). Although representing a simple procedure comprising the treatment of defatted flour with alcohol or dilute acid to solubilize carbohydrates and obtain a protein concentrate, no attempts to obtain amaranth, quinoa, or chia protein concentrates have yet been reported in the literature, unlike the obtainment of protein isolates by alkaline solubilization followed by isoelectric precipitation (Chim-Chi et al. 2018; López et al. 2018). A dry fractionation procedure to obtain protein-rich fractions from defatted chia flour has been reported, whereby two fractions were obtained: the fibre-rich fraction and the protein-rich fraction (Vázquez-Ovando et al. 2010). The protein-rich fraction contained 44.62% protein, consisting mainly of globulins (64.86%) and glutelins (20.21%), with an *in vitro* digestibility of 77.53% and high levels of essential sulphur amino acids, and non-essential amino acids, while tryptophan and lysine were deficient (Vázquez-Ovando et al. 2010). Quinoa contains saponins, which can be decreased by abrasive milling/pearling processes, from 557 to 36 mg/100 g by dry weight (DW) (Mattila et al. 2018). The antioxidant capacity of Mexican chia (*Salvia hispanica* L.) protein hydrolysates was confirmed (Chim-Chi et al. 2018), as well as the inhibiting activity of the protein hydrolysates of quinoa and chia seeds towards angiotensin I-converting enzyme (ACE), all in *in vitro* studies (Segura-Campos et al. 2013). On the other hand, the inhibiting activity of amaranth protein hydrolysates towards didpeptidyl peptidase IV was confirmed in *in vitro* studies. Segura-Campos et al. (2013) demonstrated the possibility of incorporating the chia protein hydrolysates as ingredients in functional food products (e.g. white bread and carrot cream), which resulted in products with improved ACE-inhibitory activity compared to the controls, while the addition of hydrolysates did not affect the antioxidant activity of white bread and only slightly that of the carrot cream, due to processing conditions that caused hydrolysis of the ACE-inhibitory peptides. García Fillería and Tironi (2015) showed the possibility of incorporating amaranth peptides in restructured fish products with reduced salt content to act as binders and antioxidants to inhibit lipidoxidation.

2.2.1.1.2 Lupin The significance of this novel high yielding protein crop – lupin – is rising, especially in Europe, since this crop represents a native European legume that is an alternative to soya bean (Fasolin et al. 2019). The significance of lupin is due to its high quality protein (up to 46%), mainly consisting of storage proteins – α-conglutin and β-conglutin (up to 80%) – as well as smaller protein fractions – γ-conglutin and δ-conglutin (Czubinski and Feder 2019; Cerletti et al. 1978).

When utilizing lupin as a source of protein for food, several aspects must be taken into account: its undesirable alkaloid content, the conglutin-γ protein fraction being of interest in insulin resistance and diabetes control, and the interactions between lupin proteins and the phenolic compounds, which affect the digestibility of the proteins and the bioavailability of the phenolic compounds (Czubinski and Feder 2019; Terruzzi et al. 2011). However, conglutins –α, –β, –γ, and –δ are designated as lupin allergens. Although lupin allergy is rare, due to its cross-reactivity with peanut, lupin has been included in the allergen list of the EU directive on labelling (Commission Directive 2006/142/EC). Lupin contains a considerable quantity of undesirable alkaloids that are bitter and toxic, requiring a debittering process, in which hydration is the first stage for facilitating the alkaloid extraction (Miano et al. 2019). Miano et al. (2019) demonstrated an enhancement of this process by the utilization of ultrasound technology to improve alkaloid extraction, in which an increase of 21% more extracted alkaloids was achieved in comparison to the conventional process.

Lupin is processed into flour, protein isolates, and concentrates for bakery and gluten-free products (Hoehnel et al. 2019). Palanisamy et al. (2019a) demonstrated the production of texturized lupin protein meat analogues from a lupin protein concentrate and isolate (50:50) mixture by high moisture extrusion. When spirulina powder is incorporated with the lupin protein and processed by high-moisture extrusion, meat analogues with improved physico-chemical and nutritional properties were obtained, reflected in improved protein digestibility and antioxidant properties (Palanisamy et al. 2019b).

2.2.1.1.3 Industrial Industrial hemp is a multiple-use crop, nowadays utilized as a seed oil crop, fibre crop, or dual-purpose crop (House et al. 2010). Due to its macronutrient composition and the quality of the protein of hemp seed, it is considered a valuable novel source of proteins. The main storage proteins of hemp seed are the salt-soluble globulins – edestin (~75%) – and the water-soluble – albumin (~25%) (Aluko 2016). The popularity of hemp proteins is also due to their high arginine content, being beneficial for heart health, as it serves as a precursor of the vasodilatory agent, nitric oxide (Aluko 2016; Malomo and Aluko 2015).

Hemp seeds and hemp seed meal remaining after cold-oil extraction (hemp seeds contain approximately 33–35% oil) are ingredients of interest. The remaining seed cake or meal, containing approximately 10% residual oil, has a high protein content, typically 30–50%, which is after grinding and fractionation marketed as a source of vegetable protein in the form of hemp protein powder or hemp flour, which can be utilized in a wide range of different food products, from an ingredient for the production of meat substitutes, edible films, or gluten-free bakery products, to the material for micro- and nano-encapsulation (House et al. 2010; Korus et al. 2017; Mamone et al. 2019; Pojić et al. 2014, 2015, 2018).

Pojić et al. (2015) described how, by applying a simple dry fractionation procedure – sieving – the concentration of target valuable compounds from hemp meal can be achieved. The concentration of proteins was increased from 27.9 ± 0.12% in wholemeal up to 44.4 ± 0.02%. Thereafter, the protein isolates from hemp seed meal were obtained by alkaline extraction/isoelectric precipitation and the micellization process, both yielding protein content higher than 90% (Hadnađev et al. 2018). The type of protein extraction procedure did not affect the electrophoretic mobility, subunit composition, and amino acid composition, unlike the physico-chemical properties and functionality of the proteins. It was shown that the proteins extracted by alkaline extraction/isoelectric precipitation were partially denatured with confirmed changes in secondary structure and conformational changes that affected their gelling behaviour and rheological and structural properties (Dapčević-Hadnađev et al. 2019; Hadnađev et al. 2018).

Malomo and Aluko (2015) described a procedure for the conversion of a low-protein hemp seed meal having 37% protein content into a functional protein concentrate having 74% protein content, by the utilization of the enzymatic digestion of fibre and membrane ultrafiltration. This procedure allows protein isolates to be obtained with superior functionality, solubility, and digestibility, in comparison to the traditional method of protein isolation by isoelectric precipitation.

Hemp proteins form high-quality emulsions with oil droplet sizes similar to those of milk-based emulsions (Aluko 2016). Wang et al. (2018) demonstrated a process for obtaining hemp milk, a plant-based milk alternative, in which physically and oxidatively stable hemp milk was obtained by applying high-pressure homogenization (HPH) combined with pH shift treatment. Thus, the structure of the protein was modified and the emulsifying capacity of hemp protein was improved to prevent flocculation, coalescence, and creaming (Wang et al. 2018). Moreover, it was also shown that the emulsifying capacity of hemp protein is dependent on the isolation technique applied, being different for alkaline extraction/isoelectric precipitation and micellization, due to their denaturation state after isolation (Dapčević-Hadnađev et al. 2019; Hadnađev et al. 2018).

The significance of hemp as a protein source can be ascribed to high digestibility and low allergenicity, as determined *in vitro* using a static model of gastrointestinal digestion (Mamone et al. 2019). Hemp flour as an ingredient that can enhance the physico-chemical and antioxidant properties of wheat bread was demonstrated by Mikulec et al. (2019) and Pojić et al. (2015). Apart from being a protein-enriching ingredient, due to their surface-active and foaming properties, hemp proteins can be utilized as structure-forming agents in gluten-free breads, thus allowing retention and stabilization of the gas bubbles in the dough and the bread (Korus et al. 2017). The defatted and whole powders of hemp can be used as valuable ingredients for the formulation of energy bars (Norajit et al. 2011).

2.2.2 Aquatic Photosynthetic Organisms as Novel Sources of Protein

2.2.2.1 Duckweed Duckweed represents small floating aquatic plants, and has recently attracted the attention of the scientific community due to its high amount of high-quality protein (35–43% of dry matter) with a more favourable essential amino acid composition than that of many other plant proteins, close to the WHO recommendations (Calabrese and Ferranti 2019; de Beukelaar et al. 2019). Duckweed protein contains 4.8% lysine, 2.7% methionine and cysteine, and 7.7% phenylalanine and tyrosine (Appenroth et al. 2017). Its importance as an alternative protein source is mostly due to its environmental adaptability, easy cultivation, rapid growth, and the presence of easily digestible walls that are naturally low in lignin (Calabrese and Ferranti 2019; Sowinski et al. 2019). Kaplan et al. (2018) revealed the bioavailability of the protein of *Wolffia globosa* duckweed is similar to that of well-established animal and plant protein sources. Its ability to concentrate and remove pollutants such as heavy metals, phenols, pesticides, dioxins, and pathogens from contaminated aquatic environments, which so far has been exploited for sustainability purposes, may hinder its utilization in human consumption due to safety issues (Calabrese and Ferranti 2019; Hadnađev et al. 2017; van der Spiegel et al. 2013).

Duckweed has been used traditionally for human consumption in some parts of Southeast Asia, including Laos, Thailand, and Myanmar, but for Western consumers duckweed represents a relatively unknown food (van der Spiegel et al. 2013).

The procedure for obtaining protein concentrates from duckweed (*Fanruily Lenrnnaceae*) by precipitation with acid and organic solvents (acetone, isobutanol, and isopropanol) was demonstrated by Ismail (1998). Ismail (1998) also revealed a low level of antinutritional factors in *Spirodela polyrrhiza* and *Lemna perpusilla*. The average content of antinutritional factors was higher in the protein concentrates than in fresh duckweed, while the content of trypsin inhibitor was higher in the fresh duckweed.

Extraction methods for duckweed protein, duckweed protein gel, and duckweed protein powder have been reported (Zhangli 2015) but the products made from duckweed protein have not yet been studied, except that which was related to the enrichment of dough with duckweed (in the form of a whole plant, pieces of the plant, or a powder of the whole plant) to form various types of dough suitable for the production of dough-based food products (Elituv 2013).

The green and degreened protein powders of duckeed (whole *Lemnaceae*) are in accordance with food-grade product specifications and are considered to be Generally Recognized As Safe when consumed as a nutritional ingredient in commercial food products at a maximum level of 201 g per person per day. When protein is extracted from ambient-dried duckweed, a higher purity is achieved (67.8%) than that of fresh and frozen duckweed, as the initial protein content of duckweed (*Spirodela polyrhiza*) is 34.5%. In relation to techno-functional properties, duckweed proteins showed a low stability of their emulsion and poor emulsifying properties, being more hydrophobic than soy proteins at the same pH value, which suggested their application – the improvement of the water resistance of protein-based adhesives (Yu et al. 2011).

2.2.2.2 Micro- and Macroalgae Algae are classified as microalgae, representing single-celled organisms, and seaweed, representing complex multicellular organisms. Micro- and macroalgae have been proven to be a versatile source of different bioactive compounds that can be used as (i) preservatives and pigments (e.g. colourants, antimicrobials), (ii) improvers of the nutritional profile of foods (e.g. antioxidant bioactive compounds, polyunsaturated fatty acids, proteins and essential amino acids, etc.), and (iii) improvers of the technological characteristics of the products (e.g. emulsifiers, hydrocolloids, etc.) (Barba 2017).

2.2.2.2.1 Microalgae Owning to a highly valuable nutritional composition – high protein content with a balanced amino acid profile, carotenoids, fatty acids, vitamins, polysaccharides, sterols, phycobilins, and other biologically active compounds – microalgae can be utilized for the enhancement of the nutritional value of conventional food and feed products more effectively than traditional raw materials (Calabrese and Ferranti 2019). Microalgae represent a huge and diverse group comprising approximately 200000 species, with *Bacillariophyta* (diatoms), *Chlorophyta* (green algae), *Chrysophyta* (golden algae), and *Cyanophyta* (blue-green algae) as the most abundant divisions (Bleakley and Hayes 2017), of which several species are currently exploited for a variety of applications, including cosmetics, animal feed, fatty acids, alginates extraction, wastewater treatment, and biofuel. So far the greatest importance in food application is held by *Arthrospira platensis* (Spirulina) and *Chlorella vulgaris* (Chlorella), due to their high protein content (51–58% DW for Chlorella) and favourable essential amino acid composition (Bleakley and Hayes 2017). As they represent a huge biodiversity, both in nature and within the different culture collections, their usefulness should be determined in terms of safety, biochemical composition, and digestibility, thus contributing to their wider inclusion in the human diet (Niccolai et al. 2019).

There are several challenges related to the application of microalgae as food ingredients, where the demonstration of non-toxicity in the concentrations at which microalgae is intended for use is most important. Another obstacle to the utilization of microalgae is their limited digestibility, determined to be at most 89% due to the robust cell walls of certain strains, which prevents the action of the digestive enzymes (Fasolin et al. 2019; Niccolai et al. 2019). Therefore, the utilization of microalgae requires the assistance of effective processing treatments to disrupt the cell walls.

Marti-Quijal et al. (2019) demonstrated that the addition of proteins derived from Chlorella and Spirulina microalgae into fresh pork sausages as alternatives to soy proteins affected the colour and texture of the sausages, but provided a nutritionally favourable profile – they increased the total amino acid content and the ratio of essential/non-essential amino acids. Although the utilization of alternative proteins in food formulation sometimes affects more favourable amino acid profiles, it also affects the texture, rheological properties and colour, leading to different sensory properties of the final product. Therefore, the substitution of animal proteins with alternative sources represents a challenging task in new product development (Marti-Quijal et al. 2019).

2.2.2.2.2 Macroalgae Seaweed (macroalgae) has traditionally been utilized for the extraction of techno-functional polysaccharides used as stabilizing and thickening agents, and only recently as a source of proteins (Harnedy and FitzGerald 2013; Kadam et al. 2017). The utilization of seaweed proteins as a component of animal feed has been demonstrated (Angell et al. 2016; Magnusson et al. 2019; Makkar et al. 2016). Ramin et al. (2019) indicated that by replacing a part of grass silage with extracted seaweed proteins, it is possible to increase the level of utilizable crude protein and decrease the level of methane (CH_4) production. However, apart from its direct consumption, the utilization of seaweed for the extraction and fractionation of proteins, peptides, and amino acids to be used as a food ingredient is gaining importance. However, there are several related concerns: the high variability in protein content (3–47%) and its dependence on species, geographical location, and growing season; the protein extraction has been demonstrated only at laboratory scale; off-flavours; and the accumulation of different hazardous compounds, e.g. heavy metals, polychlorinated dibenzo-p-dioxins (PCDDs), polychlorinated dibenzofurans (PCDFs), and other organic micropollutants (Calabrese and Ferranti 2019; Kadam et al. 2017; Laohakunjit et al. 2014). Moreover, the content of certain essential amino acids such as tryptophan, methionine, and leucine is limited (Vieira et al. 2018).

Vieira et al. (2018) determined that red and green macroalgae species are superior protein sources to brown macroalgae species, with protein contents of 19.1–28.2% and 20.5–23.3% DW respectively, compared with 6.90–19.5% DW, with high concentrations of lysine, especially, in red (2.71–3.85% of protein) and green (2.84–4.24% of protein) seaweeds. Brown seaweeds contain a lower content of essential amino acids, but a higher content of non-essential and free amino acids.

In addition to their high nutritional value, fresh, dehydrated or powdered seaweeds can be directly utilized as a food ingredients, and due to the their functional properties, such as swelling capacity and oil and water holding capacity, they can act as texture improvers and salt and synthetic additive replacers (Fernández-Segovia et al. 2018), which is in line with a clean label approach.

2.2.3 By-products of Plant Origin as Renewable Protein Sources

An increasing need for the assurance of sustainable food production initiated research on different industrial food by-products as sources of plant proteins. Hence, maximum utilization of the resources at our disposal and effectiveness of production are achieved (Calabrese and Ferranti 2019).

Several sources of proteins originating from different industrial processes, where they are regarded as by-products, can be identified. They are most commonly used as animal feed livestock, but numerous recent studies have shown that that material can be redirected from animal to human consumption, provided that existing problems related to protein allergenicity, toxicity, off-flavour, and the presence of high amounts of fibres and antinutritional factors (e.g. tannins, glucosinolates, phytates, erucic acid, trypsin inhibitors) are overcome (Pojić et al. 2018). Antinutritional factors are reported to be quite high in certain types of food by-products, especially in the meal remaining after oil extraction (Arntfield 2011). Depending on the type of technology used for protein isolation/concentration, the antinutritional factor content can be managed. By utilizing air classification technology to obtain protein-rich fractions, the resulting protein concentrates will contain antinutritional factors. On the other hand, antinutritional factors are reduced in protein isolation – by utilizing ultrafiltration, as well as precipitation, an effective reduction in antinutritional factors can be achieved (Arntfield 2011; Arntfield and Maskus 2011). Moreover, reductions in antinutritional factors can be achieved by the utilization of different conventional (e.g. heating, enzyme technology, soaking, irradiation, fermentation) and novel processing methods (e.g. high pressure processing, microwave heating, and extrusion) (Linsberger-Martin et al. 2013; Nikmaram et al. 2017; Zarei 2019).

Good sources of proteins from industrial by-products that are of low cost and readily available are:

- rice bran, the protein of which appears to be comparable to casein – gluten-free, easily digestible, and nutritionally superior to soybean, corn and wheat proteins, with a favourable protein efficiency ratio (1.6–1.9) (Han et al. 2015; Watanabe et al. 2019);
- brewers' spent grain (He et al. 2019; Ibbett et al. 2019);
- pulse by-products (broken grains, coarse and fine fractions, and husk) (Mullen et al. 2015);

- oilseed cakes, in the form of defatted meal derived from the oil industry, such as sunflower, canola, rapeseed, palm, and peanuts. (Mullen et al. 2015; Pojić et al. 2014, 2018).

Rice bran, with 12–16% protein, has attracted interest in the food industry due to its hypoallergenicity and high nutritional quality, represented by a high content of threonine, valine, lysine, histidine, and tryptophan, more than that of other vegetable proteins. Due to its hypoallergenicity, it represents a suitable ingredient for infant formulas for children with a cow's milk allergy. Rice bran proteins have been compared to casein and have been reported to have anti-cancer activity (Cozzano Ferreira et al. 2019).

2.2.4 Novel Sources of Protein of Animal Origin: Insects

Approximately two billion people are already using insects in their traditional diets, whereby more than 1900 species are used as food, mainly in Asia, Africa, Latin America, and Australia (Van Huis et al. 2013, Murefu et al. 2019). The most commonly consumed insects are beetles (*Coleoptera*) (31%); caterpillars (*Lepidoptera*) (18%); bees, wasps and ants (*Hymenoptera*) (14%); grasshoppers, locusts and crickets (*Orthoptera*) (13%); and cicadas, leafhoppers, planthoppers, scale insects and true bugs (*Hemiptera*) (10% percent) (Van Huis et al. 2013). Recently, the acceptance of insect consumption has increased in the countries of the Western World due to their provision of nutritional and environmental benefits, but disgust and food neophobia are still barriers to insect consumption (Orsi et al. 2019). Insects represent a smart food choice for several reasons. Their requirement for natural resources (e.g. water, feed, land) is less than that of conventional livestock. Their production is more sustainable, due to lower greenhouse gas and ammonia emissions. Insects are characterized by the efficient biotransformation of organic matter into biomass, quick growth, high rates of reproduction, short lifespans, and high survival rates (Orsi et al. 2019; Patel et al. 2019b). Edible insects are characterized by a high amount of protein (70% or more by dry weight), with a favourable amino acid composition comprising histidine, isoleucine, leucine, lysine, phenylalanine, threonine, and valine in adequate amounts. Only the concentrations of methionine and tryptophan are limited (de Castro et al. 2018). Montowska et al. (2019) revealed the nutritional profile of cricket powder obtained from adult crickets (*Acheta domesticus*), which was characterized with significant amounts of protein (55–60%), fat (24–29%), fibre (3.5–7%), and minerals (Ca, K, Mg, Na, Cu, Fe, Mn, and Zn). It must be noted that the nutritional value of edible insects is highly variable depending on the insect species, the metamorphic stage of the insect, its habitat, and diet.

In order to be fully implemented as food ingredients, edible insects must meet consumer acceptance, clear legislation to regulate their use must exist, and their toxicity and allergenicity must be determined (de Castro et al. 2018; Patel et al. 2019b).

Haber et al. (2019) demonstrated the potential of grasshopper (*Schistocerca gregaria*) powder and defatted grasshopper powder as ingredients for the enrichment of white bread at levels of 10 and 20%. Grasshopper powder exhibited a higher water- and oil-holding capacity in comparison to that of wheat flour, which consequently resulted in a softer bread crumb texture contributing to the preservation of sensory properties during storage. The inclusion of grasshopper powder in the formulation of bread decreased the specific volume of grasshopper-based bread, which was expected due to gluten dilution. It was found that the bread with 10% enrichment was sensory acceptable, whilst the 20% supplement level resulted in a distinctive odour of the bread. Roncolini et al. (2019) used mealworm (*Tenebrio molitor* L.) powder for the enrichment of bread doughs at 5 and 10% substitution levels. The addition of mealworm powder (MP) did not negatively affect the technical features of the dough, and exhibited unaltered leavening ability, which consequently had a positive effect on bread volume and crumb texture, especially at the 5% substitution level.

Da Rosa Machado and Thys (2019) demonstrated the potential of cricket powder for the enrichment of gluten-free bread. Apart from the nutritional improvements of gluten-free bread, the enriched bread had higher hardness and chewiness of crumb than those of the control bread.

Zielińska et al. (2018) indicated that whole insects and isolated insect protein can be suitable for the development of new food formulations, due to good water- and oil-holding capacity, foaming capacity, foam stability, and emulsion stability.

Duda et al. (2019) demonstrated that cricket powder appeared to be a good protein enrichment ingredient for durum wheat pasta, and when used at the level of 5%, decreases in cooking weight and cooking loss were achieved. The enriched pasta was characterized by higher firmness and high sensory acceptance, resembling the properties of wholemeal pasta in the perception of consumers.

A knowledge gap in terms of food safety hazards and antinutritional factors associated with edible insects still exists. Chakravorty et al. (2016) reported the presence of phytic acid and tannins in *Oecophylla smaragdina* (Hymenoptera: Formicidae) and *Odontotermes* sp. (Isoptera: Termitidae) being much lower than that of some common foods of plant origin.

The true prevalence of food allergy to edible insects is still unknown (Pali-Schöll et al. 2019a). Proteins found in insects, such as tropomyosin, arginine kinase, larvalcuticle protein, myosin light and heavy chain isoforms, and troponin, are responsible for causing allergic reactions and cross-reactivity with other allergenic species (e.g. house dust mite, shrimp, prawn, and crab), so that persons allergic to these species are potentially at risk when consuming insects (de Gier and Verhoeckx 2018). Chitin, an insect component, may be an allergen, too (Gallo 2019). Pali-Schöll et al. (2019b) revealed that crustacean-, house dust mite- and stable fly-allergic patients exhibited immune reactions to desert locust and house cricket proteins, while crustacean-allergic patients exhibited reactions to fly proteins. Insect protein allergenicity can be decreased, increased, or unchanged by thermal processing, depending on the heat treatment that is applied, which causes a difference in protein solubility. Moreover, the cross-reactivity and allergenicity of insect proteins can be modulated by enzyme hydrolysis. However, knowledge of the impact of process technologies on the integrity of insect proteins, and their immunoreactivity and allergenicity, is still limited and requires further investigation (Pali-Schöll et al. 2019b).

2.2.5 Novel Sources of Proteins of Microbiological Origin: Single-cell Protein

Single-cell protein means protein obtained by fermentation and extracted from pure or mixed microbial culture biomass – yeast, fungi, or bacteria – which can be used in animal and human nutrition, as an alternative for agricultural proteins

(Chuck-Hernández et al. 2019; Spalvins et al. 2018). As in the case of insect and algal proteins, the production of single-cell proteins meets the principles of sustainable production in terms of environmental friendliness, with lower water consumption and smaller land requirements. One of the most important aspects of single-cell protein technology is its cultivation on biodegradable industrial wastes, by which reductions in production costs, the need for waste treatment, and negative environmental impact are achieved (Spalvins et al. 2018). The utilization of orange waste from orange juice production (Zhou et al. 2019) and cheese whey (Yadav et al. 2016) for single-cell protein production have been demonstrated.

Although the production of yeast is commercially viable, its use as a source of protein for food is still limited, while protein isolates and protein concentrates from single-cell proteins are not listed in the Novel Food list in the EU (Chuck-Hernández et al. 2019).

Buitrago Mora et al. (2019) demonstrated the incorporation of *Candida utilis* single-cell protein extract into a fermented milk beverage, for the partial substitution of milk. It was found that a substitution of 30% vol/vol did not cause changes in the microstructure of the beverage, except for a viscosity reduction.

By examining different biodegradable industrial by-products as substrates for the production of single-cell proteins (SCP), it was found that carbon compounds, especially waste gases and glycerol, have the highest potential for SCP production. This was confirmed by Khoshnevisan et al. (2019), who valorized urban biowaste by coupling anaerobic digestion and SCP production in a process which involved feeding a methanotroph mixed culture with raw and upgraded biogas. The SCP obtained was rich in essential amino acids, with aspartic acid, glutamic acid, glycine, and lysine constituting more than 50% of the total amino acid profile, ranking the single-cell protein that was produced alongside the other protein sources (Khoshnevisan et al. 2019).

2.3 *Smart* Carbohydrate Ingredients

Unlike bioactive peptides from other food protein sources, which have been subjected to intense scientific research in recent decades, bioactive carbohydrates – polysaccharides and oligosaccharides – are still insufficiently explored (Hayes and Tiwari 2015). They encompass insoluble dietary fibre (cellulose, high molar mass glucan, lignin, partially insoluble fructan, and structurally different heteropolysaccharides) and soluble dietary fibre (glucan, pectin, fructan, and heteropolysaccharides from gums, mucilages, and hemicelluloses) (Cruz-Rubio et al. 2018). Polysaccharides can be classified as homopolysaccharides, in which the main-chain polymer consists of a single monosaccharide, and heteropolysaccharides, in which the main-chain polymer consists of more than one monosaccharide (Lynch et al. 2018).

The attention that has increasingly been given to the research of bioactive polysaccharides is due to their potential health benefits – anticancer, antioxidant, and antidiabetic (Kaur and Sharma 2019). Apart from their health benefits, they are significant from an industrial point of view, due to their techno-functional properties – solubility, viscosity, and gelation properties – which is why they are utilized as thickeners, stabilizers, emulsifiers, and textural and gelation agents in wide range of different food products (e.g. milk and milk products, bakery, meat, and extruded products) (Kaur and Sharma 2019).

They can be isolated from different natural sources – marine seaweeds, terrestrial plants, mushrooms, and dairy sources. Moreover, the emerging sources of polysaccharides and oligosaccharides are cereal-based, fruit-based and legume-based by-products originating from the corresponding food processing phases – milling, juice extraction, and de-hulling (Cheung 2013; Fernandes et al. 2019). Fructan-containing plants such as asparagus, onion, artichoke, legumes (e.g. beans, peas, and lentils), wholegrain cereals, fruits, diverse seeds, and succulents contain high levels of soluble fibre (Cruz-Rubio et al. 2018). Recent research has emphasized the significance of microbial over plant-derived polysaccharides due to their defined and reproducible production parameters, which are independent of environmental conditions, the high yield and high quality of the final product (Zeng et al. 2019). Almost all alternative protein sources discussed above can be sources of bioactive carbohydrates which can be recovered in downstream processing (Hayes and Tiwari 2015).

The extraction of polysaccharides is simple, and commonly requires only hot water. In the last decade, several novel extraction technologies have emerged to facilitate and assist not only the extraction of polysaccharides, but also that of other bioactives (Trygg et al. 2019). However, the antioxidant activity of plant polysaccharides extracted by hot water was stronger than that of polysaccharides extracted by ultrasonic-assisted extraction, enzyme extraction, and microwave-assisted extraction, possibly due to the destruction of the secondary structures of polysaccharides responsible for their bioactivity, caused by the assisting technologies (Cheng et al. 2013; Trygg et al. 2019).

Research on alternative sources of non-starchy polysaccharides is urgently needed at the moment, due to the very intensive production of fibre-enriched food products intended to prevent or improve the status of bowel disease and/or other digestive disorders. Therefore, the market for fibre-enriched food products is becoming highly competitive, and these products are more and more required by consumers (Cheung 2013; Cruz-Rubio et al. 2018).

2.3.1 Alternative Plants as Novel Sources of Polysaccharides

Nowadays, structurally different heteropolysaccharides from the mucilages of alternative and non-traditional plants (e.g. psyllium, flax, and chia) are gaining importance as useful sources of dietary fibre with prebiotic characteristics (Cruz-Rubio et al. 2018).

Mucilage from chia seeds has been demonstrated as a fat and/or egg substitute in baking products (e.g. bread, cakes, and cookies) (Felisberto et al. 2015; Fernandes and Salas-Mellado 2017; Punia and Dhull 2019). Punia and Dhull (2019) revealed that chia seed mucilage can replace up to 30% of fat without affecting the sensory properties of cookies, while the replacement of up to 50% of fat did not affect the technological and physical characteristics of breads and chocolate cakes (Fernandes and Salas-Mellado 2017).

The utilization of psyllium (*Plantago ovata* Forssk) husk and seeds as an alternative source of polysaccharides has also been reported (Ji et al. 2019; Nie et al. 2018; Patel et al. 2018, 2019a; Zhang et al. 2019). Psyllium polysaccharides comprise D-xylose, L-arabinose, D-glucose, D-galactose, and L-rhamnose (Patel et al. 2019a). Depending on the type of seed fraction used as the raw material for polysaccharide extraction, the extracted polysaccharides exhibited different techno-functional properties: the polysaccharide fraction from the husk exhibited higher viscosity compared to the polysaccharide fraction from the psyllium seed. Hydrogel-like behaviour was observed for crude polysaccharides, unlike purified fractions, which exhibited the greatest antioxidant and scavenging activities (Rashid et al. 2019).

The heteropolysaccharide fraction extracted and purified from flaxseeds – flax gum – consists of xylose, arabinose, galactose, rhamnose, fructose, and glucose. The techno-functional properties of flax gum are characterized by better water holding ability, foaming capacity, swelling index, and foaming and emulsion stability than those of guar and xanthan gums, which are in widespread commercial use (Rashid et al. 2019).

2.3.2 By-products as a Renewable Polysaccharide Source

The most common, abundant, and therefore commercially interesting sources of polysaccharides are the brans of different cereals – by-products originating from the cereal milling process (Cruz-Rubio et al. 2018; Dapčević-Hadnađev et al. 2018). In that sense, the widespread utilization of oat as a source of β-glucans and wheat bran as a source of arabinoxylans means that oat β-glucan and wheat arabinoxylan preparations are already commercially viable. However, those preparations are still not sufficiently pure, containing residues of starch and protein and residuals of other active substances, due to which an enzyme methodology for the removal of residual starch and protein to obtain purified oat-β glucan isolate has been suggested (Harasym et al. 2019).

Winemaking by-products, residues from fermentation (Campestrini et al. 2020), potato peel, residues from the potato processing industry (Xu et al. 2019), bamboo shoot processing by-products (Chen et al. 2018), and seedless pea pod by-products (Belghith-Fendri et al. 2018) have been demonstrated as potential sources of polysaccharides.

Campestrini et al. (2020) demonstrated the efficacy of different procedures (aqueous, acidic, and alkaline) for the extraction of polysaccharides from winemaking by-products. The heterogeneous profiles of polysaccharides were found regardless of the applied conditions. Aqueous extracts contained a high content of uronic acid, thus indicating the presence of pectin, while acidic extracts contained pectin and hemicellulose. Alkaline extracts contained a complex mixture of monosaccharides, mainly xylose, and only this extract exhibited anti-inflammatory potential.

Polysaccharides can be used as food ingredients with techno-functional properties or as prebiotics with health-functional properties that selectively stimulate the growth and/or activity of beneficial bacteria (probiotics) in the colon. In the case of polysaccharides isolated from bamboo shoots (*Chimonobambusa quadrangularis*), it was shown that their prebiotic activity was dependent on the type of extraction processes applied (Chen et al. 2019). Between hot water-, accelerated solvent-, ultrasound-, microwave-, and enzyme-assisted extraction processes, ultrasound- and enzyme-assisted extraction appeared to be the procedures that yielded polysaccharides with better prebiotic activity, exhibiting more proliferation of probiotic bacteria and higher production of short chain fatty acids than polysaccharides obtained by other applied extraction procedures. Moreover, they also exhibited higher total carbohydrate contents, lower molecular weights, and smaller particle sizes. However, the type of applied extraction procedure did not affect the digestibility of the polysaccharide fractions that were obtained (Chen et al. 2019).

Fernandes et al. (2019) demonstrated the procedure of polysaccharide extraction with hot water (at pH 7) from apple pomace. The polymeric material obtained comprised 65% polysaccharides, mainly arabinose (58 mol%), galacturonic acid (16 mol%), and glucose (10 mol%). Moreover, by adjusting the pH to 3, an additional fraction was obtained representing a further 4% of the polymeric material, which contained 53% polysaccharides, comprising galacturonic acid (66 mol%).

2.3.3 Edible Mushrooms as a Source of Polysaccharides

Edible mushrooms are considered a novel source of polysaccharides, of which their content and composition vary depending on their morphological stages – the fruit body, the mycelium, and the sclerotium, which contains the highest level of non-starch polysaccharides (Cheung 2013). The extraction of the polysaccharides from mushroom by-products (e.g. cups and stalks) was also reported (Aguiló-Aguayo et al. 2017).

Fu et al. (2019) applied ultrasound and hot water to the extraction of water-soluble 1,3-β-glucans from edible mushrooms (*Dictyophora indusiata*). The maximum extraction yields of water-soluble 1,3-β-glucans (1.20%) and total polysaccharides (5.41%) were achieved at an ultrasound amplitude of 56%, ultrasound extraction time of 15 minutes, and a ratio of liquid to raw material of 22 ml/g.

Chou et al. (2013) demonstrated the extraction of polysaccharides from inexpensive mushroom wastes – bases or stipes of mushrooms (e.g. *Lentinula edodes* stipe, *Pleurotus eryngii* base, and *Flammulina velutipes* base) – using hot water, trichloroacetic acid precipitation, and ethanol precipitation, and its incorporation into a fermented milk system. It was shown that a relatively low concentration (0.1–0.5%) of polysaccharides can enhance the survival rate of *Lactobacillus acidophilus*, *Lactobacillus casei*, and *Bifidobacterium longum* subsp. *longum* during cold storage. The synergistic effects with the peptides and amino acids from a yoghurt culture were observed to be responsible for maintaining probiotics above 10^7 CFU/ml during cold storage.

In order to increase the hot water extraction yield of polysaccharides from the fungal cell wall of *Agaricus bisporus*, an ethanol–acid pretreatment was employed, which resulted in an increase in the yield of polysaccharides by 46%, an increase in the content of glucan by 10%, and decreased biological activities of the extracted polysaccharides (Trygg et al. 2019).

By applying a pulsed electric field (PEF) (38.4 kV/cm field intensity for 272 μs and a treatment temperature of 85 °C) as a pretreatment for the extraction of polysaccharides from white button mushrooms, a 98% extraction yield of polysaccharides was obtained, compared with 56% for polysaccharides obtained by conventional extraction, due to the permeabilization of the cell membrane caused by electric pulses. However, it must be noted that the extraction performance of PEF is not solely due to the electric pulses, but also due to a synergistic effect between the electric pulses and the temperature generated during processing (Xue and Farid 2015).

Cheng et al. (2018) demonstrated the possibility of the large-scale production of sulphated polysaccharides from *Antrodia cinnamomia* that was cultivated in a fermentor. Different fractions of sulphated polysaccharides were obtained with molecular weights ranging from 0.51 to 523.48 kDa; each exhibited different bioactivity. Another procedure scalable to an industrial level was demonstrated by Morales et al. (2019), who demonstrated the process for obtaining β-D-glucan-rich extracts from

Shiitake mushrooms (*L. edodes*), which involved the combination of boiling water extraction, cross-flow membrane clarification, and the reverse osmosis/nanofiltration concentration of β-D-glucans.

2.4 Conclusion and Future Considerations

The emergence of different alternative and non-traditional ingredients for food formulation has been triggered mainly by the global need to reduce and eliminate poverty, hunger, and malnutrition by 2030, as well as to preserve the environment, as set by the Food and Agriculture Organization of the United Nations (FAO)'s Sustainable Development Goals. To achieve the set goals, a variety of innovation and developments in food science and technology has been proposed, encompassing ingredients, technological processes, and food products. Although emerging ingredients are characterized by valuable nutritional composition, sustainability, and environmental benefits, the food safety aspects of these novel food ingredient sources are still not well known. Potential hazards may include a range of contaminants, pathogens, antinutritional factors, allergens, and also the hazards associated with the production methods and processing conditions. The legislation is still insufficiently clear on several issues regarding the use of novel ingredient sources for food and feed solutions.

The demonstration of the extraction of different ingredients from novel sources has been done mainly at laboratory scale, and it is still unclear if an upscaled process would be sufficiently cost-effective and economically viable. Moreover, the discomfort related to the eating of new or untraditional food and ingredients may hinder their commercial exploitation and consumption. Therefore, increasing the likelihood of acceptance of new foods by consumers is a challenging task. Without any doubt, the utilization of novel and alternative food ingredients offers the possibility to develop new food products that would create a new market niche. Due to the abundance of natural bioactive compounds and nutrients, their utilization may effect the reduced use of synthetic ingredients and additives, which is in accordance with current consumer requirements for the use of natural ingredients and clean labelling.

Acknowledgement

The author would like to acknowledge the financial support of the Ministry of Education, Science and Technological Development of the Republic of Serbia (No. 451 – 03 – 68/2020 – 14/200222).

References

Aguiló-Aguayo, I., Walton, J., Viñas, I., and Tiwari, B.K. (2017). Ultrasound assisted extraction of polysaccharides from mushroom by-products. *LWT - Food Science and Technology* 77: 92–99.

Aluko, R.E. (2016). Hemp seed (*Cannabis sativa* L.) proteins: composition, structure, enzymatic modification, and functional or bioactive properties. In: *Sustainable Protein Sources* (eds. S.R. Nadathur, J.P.D. Wanasundara and L. Scanlin), 121–132. Academic Press.

Angell, A.R., Angell, S.F., de Nys, R., and Paul, N.A. (2016). Seaweed as a protein source for mono-gastric livestock. *Trends in Food Science & Technology* 54: 74–84.

Appenroth, K.-J., Sree, K.S., Böhm, V. et al. (2017). Nutritional value of duckweeds (*Lemnaceae*) as human food. *Food Chemistry* 217: 266–273.

Arntfield, S.D. (2011). Canola and other oilseed proteins. In: *Handbook of Food Proteins* (eds. G.O. Phillips and P.A. Williams), 289–315. Woodhead Publishing.

Arntfield, S.D. and Maskus, H.D. (2011). Peas and other legume proteins. In: *Handbook of Food Proteins* (eds. G.O. Phillips and P.A. Williams), 233–266. Woodhead Publishing.

Barba, F.J. (2017). Microalgae and seaweeds for food applications: challenges and perspectives. *Food Research International* 99: 969–970.

Belghith-Fendri, L., Chaari, F., Jeddou, K.B. et al. (2018). Identification of polysaccharides extracted from pea pod by-products and evaluation of their biological and functional properties. *International Journal of Biological Macromolecules* 116: 947–954.

de Beukelaar, M.F.A., Zeinstra, G.G., Mes, J.J., and Fischer, A.R.H. (2019). Duckweed as human food. The influence of meal context and information on duckweed acceptability of Dutch consumers. *Food Quality and Preference* 71: 76–86.

Bleakley, S. and Hayes, M. (2017). Algal proteins: extraction, application, and challenges concerning production. *Food* 6 (5): 33.

Buitrago Mora, H.M., Piñeros, M.A., Espinosa Moreno, D. et al. (2019). Multiscale design of a dairy beverage model composed of *Candida utilis* single cell protein supplemented with oleic acid. *Journal of Dairy Science* 102 (11): 9749–9762.

Calabrese, M.G. and Ferranti, P. (2019). Novel foods: new food sources. In: *Encyclopedia of Food Security and Sustainability* (eds. P. Ferranti, E.M. Berry and J.R. Anderson), 271–275. Elsevier.

Campestrini, L.H., Rasera, G.B., de Camargo, A.C. et al. (2020). Alkaline conditions better extract anti-inflammatory polysaccharides from winemaking by-products. *Food Research International* 131: 108532.

de Castro, R.J.S., Ohara, A., Aguilar, J.G.d.S., and Domingues, M.A.F. (2018). Nutritional, functional and biological properties of insect proteins: processes for obtaining, consumption and future challenges. *Trends in Food Science & Technology* 76: 82–89.

Cerletti, P., Fumagalli, A., and Venturin, D. (1978). Protein composition of seeds of *Lupinus albus*. *Journal of Food Science* 43: 1409–1414.

Chakravorty, J., Ghosh, S., Megu, K. et al. (2016). Nutritional and anti-nutritional composition of *Oecophylla smaragdina* (Hymenoptera: Formicidae) and *Odontotermes* sp. (Isoptera: Termitidae): two preferred edible insects of Arunachal Pradesh, India. *Journal of Asia-Pacific Entomology* 19 (3): 711–720.

Chen, G., Chen, K., Zhang, R. et al. (2018). Polysaccharides from bamboo shoots processing by-products: new insight into extraction and characterization. *Food Chemistry* 245: 1113–1123.

Chen, G., Chen, X., Yang, B. et al. (2019). New insight into bamboo shoot (*Chimonobambusa quadrangularis*) polysaccharides: impact of extraction processes on its prebiotic activity. *Food Hydrocolloids* 95: 367–377.

Cheng, H., Feng, S., Jia, X. et al. (2013). Structural characterization and antioxidant activities of polysaccharides extracted from *Epimedium acuminatum*. *Carbohydrate Polymers* 92 (1): 63–68.

Cheng, J.-J., Chao, C.-H., and Lu, M.-K. (2018). Large-scale preparation of sulfated polysaccharides with anti-angionenic and anti-inflammatory properties from *Antrodia cinnamomia*. *International Journal of Biological Macromolecules* 113: 1198–1205.

Cheung, P.C.K. (2013). Mini-review on edible mushrooms as source of dietary fiber: preparation and health benefits. *Food Science and Human Wellness* 2 (3): 162–166.

Chim-Chi, Y., Gallegos-Tintoré, S., Jiménez-Martínez, C. et al. (2018). Antioxidant capacity of Mexican chia (*Salvia hispanica* L.) protein hydrolyzates. *Journal of Food Measurement and Characterization* 12 (1): 323–331.

Chou, W.-T., Sheih, I.-C., and Fang, T.J. (2013). The applications of polysaccharides from various mushroom wastes as prebiotics in different systems. *Journal of Food Science* 78 (7): M1041–M1048.

Chuck-Hernández, C., Baigts Allende, D.K., and Mahlknecht, J. (2019). Novel food ingredients for food security. In: *Encyclopedia of Food Security and Sustainability* (eds. P. Ferranti, E.M. Berry and J.R. Anderson), 369–375. Elsevier.

Commission Directive 2006/142/EC of 22 December 2006 amending Annex IIIa of Directive 2000/13/EC of the European Parliament and of the Council listing the ingredients which must under all circumstances appear on the labelling of foodstuffs. Official Journal of the European Union L368: 110–111.

Cozzano Ferreira, S., Fernández, A.M., del Castillo Bilbao, M.D., and Medrano Fernández, A. (2019). New functional ingredients from agroindustrial by-products for the development of healthy foods. In: *Encyclopedia of Food Security and Sustainability* (eds. P. Ferranti, E.M. Berry and J.R. Anderson), 351–359. Elsevier.

Cruz-Rubio, M.J., Loeppert, R., Viernstein, H., and Praznik, W. (2018). Trends in the use of plant non-starch polysaccharides within food, dietary supplements, and pharmaceuticals: beneficial effects on regulation and wellbeing of the intestinal tract. *Scientia Pharmaceutica* 86: 49.

Czubinski, J. and Feder, S. (2019). Lupin seeds storage protein composition and their interactions with native flavonoids. *Journal of the Science of Food and Agriculture* 99 (8): 4011–4018.

Dakhili, S., Abdolalizadeh, L., Hosseini, S.M. et al. (2019). Quinoa protein: composition, structure and functional properties. *Food Chemistry* 299: 125161.

Dang, T.D., Peters, R.L., and Allen, K.J. (2016). Debates in allergy medicine: baked egg and milk do not accelerate tolerance to egg and milk. *World Allergy Organization Journal* 9: 2.

Dapčević-Hadnađev, T., Hadnađev, M., and Pojić, M. (2014). 'OSA starches as improvers in gluten-free bread from hemp seed meal', *Proceedings of 2nd International Congress "Food Technology, Quality and Safety"*, Institute of Food Technology, Novi Sad, Serbia, pp. 166–170.

Dapčević-Hadnađev, T., Hadnađev, M., and Pojić, M. (2018). The healthy components of cereal by-products and their functional properties. In: *Sustainable Recovery and Reutilization of Cereal Processing By-Products* (ed. C. Galanakis), 27–61. Woodhead Publishing. DOI 10.1016/B978-0-08-102162-0.00002-2, ISBN 9780081021620.

Dapčević-Hadnađev, T., Dizdar, M., Pojić, M. et al. (2019). Emulsifying properties of hemp proteins: effect of isolation technique. *Food Hydrocolloids* 89: 912–920.

Duda, A., Adamczak, J., Chełminska, P. et al. (2019). Quality and nutritional/textural properties of durum wheat pasta enriched with cricket powder. *Food* 8: 46.

Elituv, E. (2013). Dough comprising plant material from duckweeds and food products prepared therefrom. US Patent US2015/0216187 A1, filed July 31, 2013 and issued August 6, 2015.

Fasolin, L.H., Pereira, R.N., Pinheiro, A.C. et al. (2019). Emergent food proteins – towards sustainability, health and innovation. *Food Research International* 125: 108586.

Felisberto, M.H.F., Wahanik, A.L., Gomes-Ruffi, C.R. et al. (2015). Use of chia (*Salvia hispanica* L.) mucilage gel to reduce fat in pound cakes. *LWT - Food Science and Technology* 63 (2): 1049–1055.

Fernandes, S.S. and Salas-Mellado, M.d.l.M. (2017). Addition of chia seed mucilage for reduction of fat content in bread and cakes. *Food Chemistry* 227: 237–244.

Fernandes, P.A.R., Silva, A.M.S., Evtuguin, D.V. et al. (2019). The hydrophobic polysaccharides of apple pomace. *Carbohydrate Polymers* 223: 115132.

Fernández-Segovia, I., Lerma-García, M.J., Fuentes, A., and Barat, J.M. (2018). Characterization of Spanish powdered seaweeds: composition, antioxidant capacity and technological properties. *Food Research International* 111: 212–219.

Fu, Y., Lin, S., Lu, M. et al. (2019). Quantitative evaluation of ultrasound-assisted extraction of 1,3-ß-glucans from *Dictyophora indusiata* using an improved fluorometric assay. *Polymers* 11 (5): 864.

Gallo, M. (2019). Novel foods: insects – safety issues. In: *Encyclopedia of Food Security and Sustainability* (eds. P. Ferranti, E.M. Berry and J.R. Anderson), 294–299. Elsevier.

García Fillería, S.F. and Tironi, V.A. (2015). Application of amaranth protein isolate and hydrolysate on a reduced salt fish restructured product: antioxidant properties, textural and microbiological effects. *International Journal of Food Science & Technology* 50 (6): 1452–1460.

de Gier, S. and Verhoeckx, K. (2018). Insect (food) allergy and allergens. *Molecular Immunology* 100: 82–106.

Haber, M., Mishyna, M., Martinez, J.J.I., and Benjamin, O. (2019). The influence of grasshopper (*Schistocerca gregaria*) powder enrichment on bread nutritional and sensorial properties. *LWT - Food Science and Technology* 115: 108395.

Hadnađev, M., Dapčević-Hadnađev, T., Pojić, M. et al. (2017). Progress in vegetable proteins isolation techniques: a review. *Food and Feed Research* 44 (1): 11–21.

Hadnađev, M., Dapčević-Hadnađev, T., Lazaridou, A. et al. (2018). Hempseed meal protein isolates prepared by different isolation techniques. Part I. physicochemical properties. *Food Hydrocolloids* 79: 526–533.

Han, S.-W., Chee, K.-M., and Cho, S.-J. (2015). Nutritional quality of rice bran protein in comparison to animal and vegetable protein. *Food Chemistry* 172: 766–769.

Harasym, J., Żyła, E., Dziendzikowska, K., and Gromadzka-Ostrowska, J. (2019). Proteinaceous residue removal from oat β-glucan extracts obtained by alkaline water extraction. *Molecules* 24 (9): 1729.

Harnedy, P.A. and FitzGerald, R.J. (2013). Extraction of protein from the macroalga *Palmaria palmata*. *LWT - Food Science and Technology* 51 (1): 375–382.

Hayes, M. and Tiwari, B.K. (2015). Bioactive carbohydrates and peptides in foods: an overview of sources, downstream processing steps and associated bioactivities. *International Journal of Molecular Sciences* 16 (9): 22485–22508.

He, Y., Kuhn, D.D., Ogejo, J.A. et al. (2019). Wet fractionation process to produce high protein and high fiber products from brewer's spent grain. *Food and Bioproducts Processing* 117: 266–274.

Heimann, K. and Huerlimann, R. (2015). Microalgal classification: major classes and genera of commercial microalgal species. In: *Handbook of Marine Microalgae* (ed. S.-K. Kim), 25–41. Academic Press https://doi.org/10.1016/B978-0-12-800776-1.00003-0.

Hoehnel, A., Axel, C., Bez, J. et al. (2019). Comparative analysis of plant-based high-protein ingredients and their impact on quality of high-protein bread. *Journal of Cereal Science* 89: 102816.

House, J.D., Neufeld, J., and Leson, G. (2010). Evaluating the quality of protein from hemp seed (*Cannabis sativa* L.) products through the use of the protein digestibility-corrected amino acid score method. *Journal of Agricultural and Food Chemistry* 58 (22): 11801–11807.

Ibbett, R., White, R., Tucker, G., and Foster, T. (2019). Hydro-mechanical processing of brewer's spent grain as a novel route for separation of protein products with differentiated techno-functional properties. *Innovative Food Science & Emerging Technologies* 56: 102184.

Ismail, M. (1998). Chemical characterization of protein contentrates of duckweed (*Fanruily Lenrnnaceae*). *Pertanika Journal of Science & Technology* 6 (1): 7–21.

Ji, X., Hou, C., and Guo, X. (2019). Physicochemical properties, structures, bioactivities and future prospective for polysaccharides from *Plantago* L. (*Plantaginaceae*): a review. *International Journal of Biological Macromolecules* 135: 637–646.

Kadam, S.U., Álvarez, C., Tiwari, B.K., and O'Donnell, C.P. (2017). Extraction and characterization of protein from Irish brown seaweed *Ascophyllum nodosum*. *Food Research International* 99: 1021–1027.

Kaplan, A., Zelicha, H., Tsaban, G. et al. (2018). Protein bioavailability of *Wolffia globosa* duckweed, a novel aquatic plant–a randomized controlled trial. *Clinical Nutrition* 38 (6): 2576–2582.

Kaur, R. and Sharma, M. (2019). Cereal polysaccharides as sources of functional ingredient for reformulation of meat products: a review. *Journal of Functional Foods* 62: 103527.

Khoshnevisan, B., Tsapekos, P., Zhang, Y. et al. (2019). Urban biowaste valorization by coupling anaerobic digestion and single cell protein production. *Bioresource Technology* 290: 121743.

Korus, J., Witczak, M., Ziobro, R., and Juszczak, L. (2017). Hemp (*Cannabis sativa* subsp. *sativa*) flour and protein preparation as natural nutrients and structure forming agents in starch based gluten-free bread. *LWT - Food Science and Technology* 84: 143–150.

Laohakunjit, N., Selamassakul, O., and Kerdchoechuen, O. (2014). Seafood-like flavour obtained from the enzymatic hydrolysis of the protein by-products of seaweed (*Gracilaria* sp.). *Food Chemistry* 158: 162–170.

Leonard, S.A. (2016). Debates in allergy medicine: baked milk and egg ingestion accelerates resolution of milk and egg allergy. *World Allergy Organization Journal* 9: 1.

Linsberger-Martin, G., Weiglhofer, K., Thi Phuong, T.P., and Berghofer, E. (2013). High hydrostatic pressure influences antinutritional factors and *in vitro* protein digestibility of split peas and whole white beans. *LWT - Food Science and Technology* 51 (1): 331–336.

López, D.N., Galante, M., Robson, M. et al. (2018). Amaranth, quinoa and chia protein isolates: physicochemical and structural properties. *International Journal of Biological Macromolecules* 109: 152–159.

López, D.N., Galante, M., Raimundo, G. et al. (2019). Functional properties of amaranth, quinoa and chia proteins and the biological activities of their hydrolyzates. *Food Research International* 116: 419–429.

Luigia, D.S. (2019). New protein sources: novel foods. In: *Encyclopedia of Food Security and Sustainability* (eds. P. Ferranti, E.M. Berry and J.R. Anderson), 276–279. Elsevier.

Lynch, K.M., Coffey, A., and Arendt, E.K. (2018). Exopolysaccharide producing lactic acid bacteria: their techno-functional role and potential application in gluten-free bread products. *Food Research International* 110: 52–61.

Magnusson, M., Glasson, C.R.K., Vucko, M.J. et al. (2019). Enrichment processes for the production of high-protein feed from the green seaweed *Ulva ohnoi*. *Algal Research* 41: 101555.

Makkar, H.P.S., Tran, G., Heuzé, V. et al. (2016). Seaweeds for livestock diets: a review. *Animal Feed Science and Technology* 212: 1–17.

Malomo, S.A. and Aluko, R.E. (2015). Conversion of a low protein hemp seed meal into a functional protein concentrate through enzymatic digestion of fibre coupled with membrane ultrafiltration. *Innovative Food Science & Emerging Technologies* 31: 151–159.

Mamone, G., Picariello, G., Ramondo, A. et al. (2019). Production, digestibility and allergenicity of hemp (*Cannabis sativa* L.) protein isolates. *Food Research International* 115: 562–571.

Marti-Quijal, F.J., Zamuz, S., Tomašević, I. et al. (2019). Influence of different sources of vegetable, whey and microalgae proteins on the physicochemical properties and amino acid profile of fresh pork sausages. *LWT - Food Science and Technology* 110: 316–323.

Mattila, P.H., Pihlava, J.-M., Hellström, J. et al. (2018). Contents of phytochemicals and antinutritional factors in commercial protein-rich plant products. *Food Quality and Safety* 2 (4): 213–219.

Miano, A.C., Rojas, M.L., and Augusto, P.E.D. (2019). Using ultrasound for improving hydration and debittering of Andean lupin grains. *Journal of Food Process Engineering* 42: 13170.

Mikulec, A., Kowalski, S., Sabat, R. et al. (2019). Hemp flour as a valuable component for enriching physicochemical and antioxidant properties of wheat bread. *LWT - Food Science and Technology* 102: 164–172.

Montowska, M., Kowalczewski, P.Ł., Rybicka, I., and Fornal, E. (2019). Nutritional value, protein and peptide composition of edible cricket powders. *Food Chemistry* 289: 130–138.

Morales, D., Smiderle, F.R., Piris, A.J. et al. (2019). Production of a ß-d-glucan-rich extract from Shiitake mushrooms (*Lentinula edodes*) by an extraction/microfiltration/reverse osmosis (nanofiltration) process. *Innovative Food Science & Emerging Technologies* 51: 80–90.

Mullen, A.M., Álvarez, C., Pojić, M. et al. (2015). Classification and target compounds. In: *Food Waste Recovery* (ed. C.M. Galanakis), 25–57. Academic Press.

Murefu, T.R., Macheka, L., Musundire, R., and Manditsera, F.A. (2019). Safety of wild harvested and reared edible insects: a review. *Food Control* 101: 209–224.

Niccolai, A., Chini Zittelli, G., Rodolfi, L. et al. (2019). Microalgae of interest as food source: biochemical composition and digestibility. *Algal Research* 42: 101617.

Nie, S., Cui, S.W., and Xie, M. (2018). Psyllium polysaccharide. In: *Bioactive Polysaccharides* (eds. S. Nie, S.W. Cui and M. Xie), 395–443. Academic Press.

Nikmaram, N., Leong, S.Y., Koubaa, M. et al. (2017). Effect of extrusion on the anti-nutritional factors of food products: an overview. *Food Control* 79: 62–73.

Norajit, K., Gu, B.-J., and Ryu, G.-H. (2011). Effects of the addition of hemp powder on the physicochemical properties and energy bar qualities of extruded rice. *Food Chemistry* 129 (4): 1919–1925.

Orsi, L., Voege, L.L., and Stranieri, S. (2019). Eating edible insects as sustainable food? Exploring the determinants of consumer acceptance in Germany. *Food Research International* 125: 108573.

Palanisamy, M., Franke, K., Berger, R.G. et al. (2019a). High moisture extrusion of lupin protein: influence of extrusion parameters on extruder responses and product properties. *Journal of the Science of Food and Agriculture* 99 (5): 2175–2185.

Palanisamy, M., Töpfl, S., Berger, R.G., and Hertel, C. (2019b). Physico-chemical and nutritional properties of meat analogues based on Spirulina/lupin protein mixtures. *European Food Research and Technology* 245 (9): 1889–1898.

Pali-Schöll, I., Verhoeckx, K., Mafra, I. et al. (2019a). Allergenic and novel food proteins: state of the art and challenges in the allergenicity assessment. *Trends in Food Science & Technology* 84: 45–48.

Pali-Schöll, I., Meinlschmidt, P., Larenas-Linnemann, D. et al. (2019b). Edible insects: cross-recognition of IgE from crustacean- and house dust mite allergic patients, and reduction of allergenicity by food processing. *World Allergy Organization Journal* 12 (1): 100006.

Patel, M.K., Tanna, B., Mishra, A., and Jha, B. (2018). Physicochemical characterization, antioxidant and anti-proliferative activities of a polysaccharide extracted from psyllium (*P. ovata*) leaves. *International Journal of Biological Macromolecules* 118: 976–987.

Patel, M.K., Tanna, B., Gupta, H. et al. (2019a). Physicochemical, scavenging and anti-proliferative analyses of polysaccharides extracted from psyllium (*Plantago ovata* Forssk) husk and seeds. *International Journal of Biological Macromolecules* 133: 190–201.

Patel, S., Suleria, H.A.R., and Rauf, A. (2019b). Edible insects as innovative foods: nutritional and functional assessments. *Trends in Food Science & Technology* 86: 352–359.

Pojić, M., Mišan, A., Sakač, M. et al. (2014). Characterization of byproducts originating from hemp oil processing. *Journal of Agricultural and Food Chemistry* 62 (51): 12436–12442.

Pojić, M., Dapčević Hadnađev, T., Hadnađev, M. et al. (2015). Bread supplementation with hemp seed cake: a by-product of hemp oil processing. *Journal of Food Quality* 38 (6): 431–440.

Pojić, M., Hadnađev, M., Dapčević Hadnađev, T., and Mišan, A. (2018). 'Versatile utilization of by-products from hemp (*Canabis sativa L.*) oil processing', *5th International ISEKI_Food Conference, Book of abstracts, 3–5 July 2018, Stuttgart, Germany*, p. 90. ISBN 978-3-900932-57-2.

Pojić, M., Mišan, A., and Tiwari, B. (2018). Eco-innovative technologies for extraction of proteins for human consumption from renewable protein sources of plant origin. *Trends in Food Science & Technology* 75: 93–104.

Punia, S. and Dhull, S.B. (2019). Chia seed (*Salvia hispanica* L.) mucilage (a heteropolysaccharide): functional, thermal, rheological behaviour and its utilization. *International Journal of Biological Macromolecules* 140: 1084–1090.

Ramin, M., Franco, M., Roleda, M.Y. et al. (2019). *In vitro* evaluation of utilisable crude protein and methane production for a diet in which grass silage was replaced by different levels and fractions of extracted seaweed proteins. *Animal Feed Science and Technology* 255: 114225.

Rashid, F., Ahmed, Z., Hussain, S. et al. (2019). *Linum usitatissimum* L. seeds: flax gum extraction, physicochemical and functional characterization. *Carbohydrate Polymers* 215: 29–38.

Roncolini, A., Milanović, V., Cardinali, F. et al. (2019). Protein fortification with mealworm (*Tenebrio molitor* L.) powder: effect on textural, microbiological, nutritional and sensory features of bread. *PLoS One* 14 (2): e0211747.

da Rosa Machado, C. and Thys, R.C.S. (2019). Cricket powder (*Gryllus assimilis*) as a new alternative protein source for gluten-free breads. *Innovative Food Science & Emerging Technologies* 56: 102180.

Segura-Campos, M.R., Salazar-Vega, I.M., Chel-Guerrero, L.A., and Betancur-Ancona, D.A. (2013). Biological potential of chia (*Salvia hispanica* L.) protein hydrolysates and their incorporation into functional foods. *LWT - Food Science and Technology* 50 (2): 723–731.

Sowinski, E.E., Gilbert, S., Lam, E., and Carpita, N.C. (2019). Linkage structure of cell-wall polysaccharides from three duckweed species. *Carbohydrate Polymers* 223: 115119.

Spalvins, K., Zihare, L., and Blumberga, D. (2018). Single cell protein production from waste biomass: comparison of various industrial by-products. *Energy Procedia* 147: 409–418.

van der Spiegel, M., Noordam, M.Y., and van der Fels-Klerx, H.J. (2013). Safety of novel protein sources (insects, microalgae, seaweed, duckweed, and rapeseed) and legislative aspects for their application in food and feed production. *Comprehensive Reviews in Food Science and Food Safety* 12 (6): 662–678.

Terruzzi, I., Senesi, P., Magni, C. et al. (2011). Insulin-mimetic action of conglutin-gamma, a lupin seed protein, in mouse myoblasts. *Nutrition, Metabolism and Cardiovascular Diseases* 21: 197–205.

Trygg, J., Beltrame, G., and Yang, B. (2019). Rupturing fungal cell walls for higher yield of polysaccharides: acid treatment of the basidiomycete prior to extraction. *Innovative Food Science & Emerging Technologies* 57: 102206.

Vainio, A., Niva, M., Jallinoja, P., and Latvala, T. (2016). From beef to beans: eating motives and the replacement of animal proteins with plant proteins among Finnish consumers. *Appetite* 106: 92–100.

Van Huis, A., Van Itterbeeck, J., Klunder, H. et al. (2013). *Edible insects: future prospects for food and feed security (FAO Forestry Report No. 171)*. Rome: Food and Agriculture Organization of the United Nations.

Vázquez-Ovando, J.A., Rosado-Rubio, J.G., Chel-Guerrero, L.A., and Betancur-Ancona, D.A. (2010). Dry processing of chía (*Salvia hispanica* L.) flour: chemical characterization of fiber and protein. *CyTA Journal of Food* 8 (2): 117–127.

Vieira, E.F., Soares, C., Machado, S. et al. (2018). Seaweeds from the Portuguese coast as a source of proteinaceous material: total and free amino acid composition profile. *Food Chemistry* 269: 264–275.

Wang, Q., Jiang, J., and Xiong, Y.L. (2018). High pressure homogenization combined with pH shift treatment: a process to produce physically and oxidatively stable hemp milk. *Food Research International* 106: 487–494.

Watanabe, M., Yamada, C., Maeda, I. et al. (2019). Evaluating of quality of rice bran protein concentrate prepared by a combination of isoelectronic precipitation and electrolyzed water treatment. *LWT - Food Science and Technology* 99: 262–267.

Xu, G.-Y., Liao, A.-M., Huang, J.-H. et al. (2019). The rheological properties of differentially extracted polysaccharides from potatoes peels. *International Journal of Biological Macromolecules* 137: 1–7.

Xue, D. and Farid, M.M. (2015). Pulsed electric field extraction of valuable compounds from white button mushroom (*Agaricus bisporus*). *Innovative Food Science & Emerging Technologies* 29: 178–186.

Yadav, J.S.S., Yan, S., Ajila, C.M. et al. (2016). Food-grade single-cell protein production, characterization and ultrafiltration recovery of residual fermented whey proteins from whey. *Food and Bioproducts Processing* 99: 156–165.

Yu, G., Liu, H., Venkateshan, K. et al. (2011). Functional, physiochemical, and rheological properties of duckweed (*Spirodela polyrhiza*) protein. *Transactions of the ASABE* 54 (2): 555–561.

Zarei, M. (2019). Effects of using radiation processing in nutrition science and their restriction: a review. *International Journal of Advanced Biological and Biomedical Research* 7 (1): 19–29.

Zeng, Y.-J., Yang, H.-R., Ou, X.-Y. et al. (2019). Fungal polysaccharide similar with host *Dendrobium officinale* polysaccharide: preparation, structure characteristics and biological activities. *International Journal of Biological Macromolecules* 141: 460–470.

Zhang, J., Wen, C., Zhang, H., and Duan, Y. (2019). Review of isolation, structural properties, chain conformation, and bioactivities of psyllium polysaccharides. *International Journal of Biological Macromolecules* 139: 409–420.

Zhangli, L. (2015). Extraction method for duckweed protein, duckweed protein gel, duckweed protein powder and duckweed protein gel foodstuff. Chinese Patent CN105166318A, filed August 18, 2015 and issued January 1, 2019)

Zhou, Y.-M., Chen, Y.-P., Guo, J.-S. et al. (2019). Recycling of orange waste for single cell protein production and the synergistic and antagonistic effects on production quality. *Journal of Cleaner Production* 213: 384–392.

Zielińska, E., Karaś, M., and Baraniak, B. (2018). Comparison of functional properties of edible insects and protein preparations thereof. *LWT - Food Science and Technology* 91: 168–174.

3 Healthy Ingredients

Ciara McDonagh
Food Industry Development Department, Teagasc Food Research Centre, Ashtown, Dublin, Ireland

Summary

This chapter looks at the growing requirement for healthy ingredients, driven by consumer demands and legislative approaches to improve nutrition. Replacements for sugar, salt, and phosphates are considered in terms of their applications in the food industry and their implications for human health. The use of bioactives, peptides, and other ingredients with potential health benefits in the development of food products is also discussed.

3.1 Introduction

Non-communicable diseases (NCDs), such as heart disease, stroke, cancer, chronic respiratory diseases, and diabetes, are the leading cause of mortality in the world. People of all ages, genders, and socio-economic classes are vulnerable to the risk factors contributing to NCDs, whether from unhealthy diets, physical inactivity, exposure to tobacco smoke or the harmful use of alcohol. NCDs have a huge economic impact. Health at a Glance (OECD 2016) reported that the economic loss due to premature death from NCDs amounts to 115 billion euros annually. The development of NCDs has prompted a reaction worldwide and national ministries of health are developing and implementing policies and interventions to prevent and control the further rise of these diseases. These preventative actions include strategies to reduce the levels of fat, sugar, and salt in food. For food manufacturers, this has resulted in pressure to replace these ingredients with healthier alternatives whilst ensuring the same quality, taste, and shelf life that the consumer expects.

The growing awareness around the direct role of diet and the foods we consume on human health and well-being has resulted in a rising demand for functional foods, or

Food Formulation: Novel Ingredients and Processing Techniques, First Edition.
Edited by Shivani Pathania and Brijesh K. Tiwari.

foods that provide health benefits beyond the provision of essential nutrients. Consumers today perceive food products not just as a means to combat hunger, but also to provide essential nutrients to the body, to be able to prevent nutritional deficiency, and also improve physical and mental well-being. Therefore, in recent years, a significant body of research has focused on the identification and development of health-promoting ingredients for incorporation into food. However, there still remains a large body of scientific work to substantiate innovative health claims in relation to these ingredients. These claims are primarily based on human efficacy studies as the health benefits frequently identified are credence attributes, or qualities that cannot be observed by a consumer.

3.2 Need for Healthy Ingredients

With health and wellness trends continuing to rise and consumer preferences moving towards more natural and functional food products, offering healthy ingredients is a hot topic for the food industry. Manufacturers are under increasing pressure to alter the composition of processed foods and improve the nutritional profile of their products whilst maintaining the taste and quality. This can be done through the reduction and elimination of ingredients such as sugar and salt that cause concern or the addition of healthy ingredients to create nutritionally-enhanced or functional foods. The key requirement for healthy ingredients is to improve diets and help reduce the prevalence of diet-related disease in the population.

On average, the global population consumes too much salt and sugar. The Guideline on Sodium Intake published by the World Health Organization (2012) reported that current levels of salt consumption are contributing to high blood pressure, which in turn is increasing the risk of heart disease and stroke. They published a recommendation that reducing salt intake to less than 5 g per day could prevent 1.7 million deaths each year. People are often unaware of the amount of hidden salt they consume in processed foods such as cheese, meat, and bread and tend to only consider the amount added during cooking or at the point of food consumption. The Guideline on Sugars Intake for Adults and Children published by the World Health Organization (2015a) recommended that the intake of free sugars should be reduced to less than 10% of total energy intake and that a reduction to less than 5% of total energy intake would provide additional health benefits. High sugar consumption has been linked to the development of NCDs such as obesity, type 2 diabetes, and dental caries. Evidence also shows that free sugars influence blood pressure and serum lipids, and suggests that a reduction in intake reduces risk factors for cardiovascular diseases (Te Morenga et al. 2014). The food industry's response to increasing consumer awareness and the growing demand for healthier food products has been to reduce or replace the sugar and salt content using clean label alternatives/replacers, which will be discussed in detail in this chapter. To ensure a competitive edge, food manufacturers are also seeking new opportunities to fortify and enrich their products, whereby existing nutrient contents are increased and/or ingredients with potential beneficial health effects are added to enhance food products.

Selecting the most appropriate healthy ingredients is highly dependent on what is motivating the product reformulation. For example, is it to reduce levels of unhealthy ingredients or is it the development of functional foods, which provide the consumer with specific health benefits, or is it creating products for people with specific dietary needs (e.g. the avoidance of particular allergens)?

A key driving force in the application and inclusion of new and exciting healthy ingredients by manufacturers is the steady stream of emerging innovations from ingredient suppliers. Understanding, characterizing, and optimizing the functionality of these ingredients is essential to deliver enhanced, healthier products with broad consumer appeal.

3.3 Salt Replacers

Salt is a crystalline compound, consisting of about 0.4 g of sodium and 0.6 g of chloride, commonly used as a food ingredient. The words salt and sodium are often used interchangeably but consumers tend to understand and use the term salt rather than sodium. It was traditionally added to foods to prevent spoilage, inhibiting the growth and survival of undesirable microorganisms. Whilst the development of refrigerated storage, along with more recent advances in packaging techniques and distribution capabilities, has diminished this role, salt still remains in widespread use for assisting in the extension of food product shelf life. Salt also plays a key role in the production of fermented foods such as pickles, sauerkraut, cheeses, and fermented sausages, creating an environment that is conducive towards the growth of good lactic acid bacteria, whilst inhibiting the growth of undesirable spoilage bacteria and fungi naturally present in these foods. The development of particular physical properties of some foods relies on the use of salt. In certain baked goods, the addition of salt is used to control the stickiness of the dough. In meats, cheeses, and extruded snack products (e.g. cheese balls, shaped potato snacks), salt can help develop the characteristic texture expected by consumers. For example, in cheeses, salt acts to remove excess water, creating a firmer texture and, in some cases, a rind (Guinee and Fox 2004). Salt also contributes to characteristics such as meltability, shredding, stretching, and flow in processed cheese products (Reddy and Marth 1991).

International health related organizations have issued recommendations to limit salt intake to less than 5 or 6 g per day, with lower recommendations for children. Adequate intakes of salt in adults have been estimated to be approximately 3.8 g salt/day; however, mean daily intakes of salt around the world far exceed this minimum physiological need. In the majority of European countries, the range of intake is 7–12 g of salt per day based on the European Commission's survey on Member States' Implementation of the EU Salt Reduction Framework. High salt intake can lead to high blood pressure, which is one of the major factors contributing to the development of cardiovascular disease. The World Health Organization reports that Ischaemic heart disease and stroke are the highest causes of death globally and have remained the leading causes for the past 15 years. Nutritionists predict that a reduction in salt consumption to 6 g per day would lead to a reduction of 2.5 million deaths worldwide per year (He et al. 2008; Macgregor and Sever 2003). WHO Member States have agreed to reduce the global population's intake of salt by a relative 30% by 2025.

The food industry is therefore under increasing pressure to reduce the amount of salt, i.e. sodium chloride, in food products without any resulting compromising effects on the flavour, texture, appearance, and safe shelf life of those products. Several large international food companies have already changed their product portfolios to promote healthier diets. A review by Kloss et al. (2015), on 'Sodium intake and its reduction by food reformulation in the European Union' reported that at that time, Nestlé had removed almost 7500 tons of sodium from their products since 2005.

Kellogg's reported a sodium reduction of 38% in many cereals over that time period and Unilever had screened more than 16000 of their products and subsequent food reformulation resulted in a reduction of over 3000 tons of sodium. Under EU legislation, manufacturers can only claim that the salt/sodium content has been reduced in a food product where the reduction in content is at least 25% compared to a similar product (Council Directive 90/496/EEC 1990). Currently, the salt reduction strategies being applied include incremental reductions to added salt quantities, an increased use of salt replacers, and the addition of taste enhancers. In recent years, the industry has focused their attention on the development of salt replacers, which is a term applied to both: (i) salts with reduced sodium content relative to NaCl (on a weight basis); or (ii) fully sodium-free (0% Na) salt substitutes. Within the EU, the term 'salt with reduced sodium content' is a legally permitted category term for declaration of such products on the market with a minimum of 25% reduced NaCl content. However, Petit et al. (2018) rightly stated that 'One of the disadvantages of salt substitutes is that they do not encourage the consumer to accept a lower perceived saltiness.'

Potassium Chloride (KCl) is one of the most commonly used salt replacers in processed foods, as it imparts the salty flavour that consumers expect. It is a naturally occurring mineral salt, which is extracted from rock and sea salts, in a similar way to sodium chloride. However, KCl is characterized by relatively offensive acrid, metallic, and bitter aftertastes (Sinopoli and Lawless 2012). To successfully formulate KCl-based salt replacers, numerous taste-improving agents and formulation concepts have been developed. These include food acids, amino acids and their nutritionally acceptable salts; food polymers; umami ingredients; simple carbohydrates, sugar alcohols and sugar substitutes; herbs and spices and their extracts. Modified potassium chloride products have also been developed, to exhibit the same functional properties as pure potassium chloride without the bitter/metallic taint often experienced. These salt replacers are produced by modifying the size of the crystal structure of potassium chloride. Both Pasin et al. (2008) and Stanley et al. (2017) showed that the replacement of salt with modified potassium chloride in pork sausage patties was well accepted ($p<0.001$) in sensory trials, suggesting this could be a positive replacement product, avoiding the bitter aftertaste imparted by standard potassium chloride. They also found that, statistically, the effect of substitution had no effect on moisture, protein, fat, textural properties, lipid oxidation, or colour (Figure 3.1).

Potassium chloride has gained regulatory acceptance for use in food products in the United States, European Union and numerous other international scientific bodies and regulatory authorities. It is the most promising salt replacer from a consumer, production and, potentially, also from a health point of view. Other mineral salts such as ammonium chloride, calcium chloride, and magnesium sulphate are also used as salt replacers but can deliver unwanted flavours and insignificant nutritional benefits. More positively, dietary intake of potassium is associated with a decreased risk of hypertension, the opposite effect to sodium (Aburto et al. 2013; Binia et al. 2015). Accumulating evidence has shown an association between potassium intake and blood pressure reduction in adults and, consequently, coronary heart disease (Weaver 2013). Whilst the average global intake of sodium is too high, most people are not meeting the recommended WHO guidelines for potassium intake (at least 3510mg of potassium daily). VanBuren et al. (2016) examined the dietary impact of adding potassium chloride to foods and concluded that the replacement of sodium chloride by KCl results in better compliance to the WHO potassium intake guideline without exceeding European Food Safety Authority (EFSA) safety recommendations, thereby posing little risk of adverse effects in the general healthy adult population.

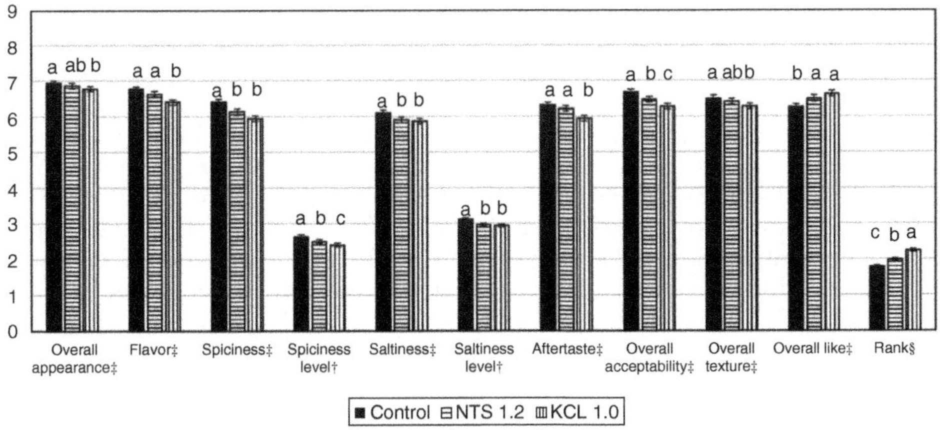

Figure 3.1 Effect of sodium reduction and replacement with standard (KCl 1.0) or modified potassium chloride (NTS 1.2) on sensory characteristics of sausage patties. Error bars indicate standard error of means. ‡ Evaluated on a 9-point hedonic scale with 1 = 'dislike extremely' and 9 = 'like extremely' used as anchors. † Evaluated on a 5-point JAR scale with 1 = 'not spicy/salty', 3 = 'just about right' and 5 = 'too spicy/salty.' § Ranking from most preferred to least preferred. a–c Means within a trait with a common superscript are similar ($P > 0.05$). *Source*: Stanley et al. (2017), Figure 1, p. 40. CC BY 4.0.

Ammonium chloride is a colourless crystal or white granular powder that dissolves easily in water. It is odourless and has a salty and bitter taste. It is commonly used in the baked goods industry, as a functional salt and as a dough improver for yeasts. It gives the dough good elasticity, extensibility, and machinability, producing breads with good quality characteristics in terms of volume, colour, aroma, texture, and elasticity. Ammonium chloride is considered safe in small amounts. The US Food and Drug Administration (FDA) granted ammonium hydroxide status as a GRAS, or Generally Recognized as Safe, substance in 1974 and it is also recognised as a food additive in the EU. Raseta et al. (2018) substituted sodium chloride with potassium chloride (33 and 50%) and ammonium chloride (33 and 50%) in cooked pea and pork burgers. Their study highlighted the importance of the product matrix before introducing sodium chloride substitution. Cooked pea samples, substituted with 33% of ammonium chloride substitution, were scored lowest on the sensory scale. Ammonium chloride is known to enhance the perceived intensity of the 'salt taste', therefore, one-third substitution of sodium chloride with ammonium chloride was not accepted well by the panellists in the cooked pea product. However, for the cooked pork product, this effect, for the similar substitution level, was not observed.

Calcium chloride is an approved food additive and salt replacer worldwide. It has an extremely salty taste and can also impart a bitterness, but has other beneficial functional properties across a wide range of food products, including pickled products, tofu, jellies and jams, meat products and cheese. Calcium chloride provides firmness when applied to products such as canned vegetables and tofu. It is commonly used as a salt replacer in pickled products because of its strong salty flavour. When used in brewing beer, it can be used to correct mineral deficiencies in the brewing water as well improving the flavour. In cheese-making, calcium chloride is sometimes added to pasteurized milk to restore the natural balance between calcium and protein in casein. However, Grummer et al. (2012) reported that when applied to Cheddar-style cheese, calcium chloride resulted in flavour differences that would not be desirable in a high-quality cheese.

Magnesium sulphate is commonly known as Epsom salts. It is similar in appearance to salt/sodium chloride but can be bitter and unpalatable. Common applications within the food industry include flour, cheese, and confectionary products. As well as being used as a salt replacer, because of the salty flavour it imparts, it provides additional functional contributions in terms of the nutrient properties and product texture. Magnesium sulphate is used to enhance the flavour and firmness of canned vegetables. It is also used as a fermentation aid in brewing beer and malt beverages, enhancing enzymatic activity and promoting the desired flavour profile.

Petit et al. (2019) compiled a comprehensive review on salt intake from processed meat products and compiled a table synthesizing the available substitutes to reduce sodium in processed meat products (Table 3.1).

It is evident that further research on the use of these salt replacers, effective combination approaches, and their inclusion levels, is required to address the successful reduction of salt/sodium content. The growing demand for healthier food and the requirement for reduced salt content in processed foods means that work in this area will continue and new and emerging formulations will contribute to reductions in dietary sodium intake going forward.

3.4 Sugar Replacers

Sugars are sweet tasting soluble carbohydrates that provide energy to the body in the form of glucose, which is required for the human brain, muscles, and major organs to function properly. Sucrose is the scientific name for commonly used table sugar and is made up of glucose and fructose. Although sugar/sucrose is mainly used for its sweet taste and its contribution to the flavour profile of foods, it has a number of additional technological functions. It provides a distinctive, desirable colour to browned foods such as baked goods, chocolate, coffee, and meat, following the chemical Maillard reaction between amino acids and reducing sugars. Sugar is also used as a bulking agent impacting the mouthfeel and texture of many foods. It contributes to the fermentation process in the production of products such as yoghurts, vinegar, sour cream, wine, beer, bread, cheese, soy sauce, and sauerkraut. Because of its hygroscopic nature, sugar plays a crucial role in reducing the water activity in foods and thus contributes to shelf life extension. Sugar also prevents baked goods from becoming dry or stale and preserves the colour of frozen fruits and jellies.

High sugar consumption has been linked to the development of NCDs such as obesity, type 2 diabetes, and dental caries. Sugar provides energy but no specific nutrient contribution and has therefore been the target for reduction by several organisations worldwide. The World Health Organization (2015a) recommended a reduced intake of free sugars throughout the life course of both adults and children, to less than 10% of total energy intake. In recent years there has been a global interest in the reduction of sugar in processed foods and consumer attitudes towards added sugar has changed drastically. Following the publication of the WHO's Global Action Plan for the Prevention and Control of NCDs 2013–2020 (WHO 2015b), governmental bodies worldwide have introduced sugar taxation intended to reduce obesity and rising levels

Table 3.1 Synthesis of available substitutes to reduce sodium in processed meat products.

Substitute	Nature	Interest	Effect	Example of trade name	References
KCl	Metal salt	Technological and sensory	Variable, depends on the product and the substitution rate. Stronger ham cohesion and less cooking exudation.	Pansalt®/ AlsoSalt®	Inguglia et al. (2017), Mäki (2000)
CaCl2			Microbiological interest. Technological impact very variable.		
MgCl2			Microbiological interest. Technological impact very variable.		
MgS04			Not currently authorized as an additive by the European Union. Increases the hydration of meat proteins	Pansalt®	Mäki (2000)
Potassium lactate (E326) Sodium lactate (E325) Sodium acetate & diacetate (E262) Sodium citrate (E331)	Organic acid	Flavour enhancer	Not currently authorized as an additive by the European Union. Increases the hydration of meat proteins Bacteriostatic effect + increase salty perception, enhances the flavour of certain products + optimization of the required brining process (lactate ions diffusion coefficient < Na +, Cl- and K +) + increase in firmness + increase in the juiciness of pork cooked. Potassium lactate less cohesive than sodium.	Purasal® P, Purasal S, Purasal OptiForm®, Ksalt®	Martin (1940), Meijer et al. (2007), Kroon et al. (2010)

(Continued)

Table 3.1 (Continued)

Substitute	Nature	Interest	Effect	Example of trade name	References
Sodium glutamate and derivatives Glutamic acid (E620), Monosodium glutamate (E621), Monopotassium glutamate E622, Calcium diglutamate (E623), Ammonium glutamate (E624), Magnesium diglutamate (E625),	Glumatic additive		Taste enhancers the best known and most used in the world. Additives with mandatory labelling, use rate limited to 10 g/kg. Technological impact little studied because effect on the texture or the colour is marginal.		
Guanylic acid (E626), disodium guanylate (E627), dipotassium guanylate (E628), calcium guanylate (E629), inosinic acid (E630), disodium inosinate (E631), dipotassium inosinate (E632), calcium inosinate (E633), disodium ribonucleotide (E634), calcium ribonucleotide (E635)	Nucleotide		Known to be partially responsible for the development of flavour of meat products, flavour enhancer, and umami flavour enhancement. Technological effect very little studied, except significant increase in the sensory score attributed to the texture, increase in the hardness and chewability of the products. No change in colour, pH, or water retention capacity shown to date.		
Arginine, taurine, lysine, glycine	Amino acid		Taste enhancer (glutamic acid in particular), particularly interesting in the problem of salt substitution. Acids basic amines (lysine, arginine) increase saline perception. No technological change demonstrated to date.	Pansalt, AlsoSalt®	Inguglia et al. (2017), Mäki (2000)

Trehalose	Reducing sugar	Masking agent	Taste enhancer, bitterness inhibitor, masking agent without specific taste but able to mask other substances. No technological effect reported or documented.	Ascend®, SaltWise®	Sahurkar and Karadbhajne (2018)
Adenosine monophosphate (AMP)	Ribonucleotide			Betrasalt®	McGregor (2007)
Sodium gluconate	Sodium salt			Sub4salt®	Sahurkar and Karadbhajne (2018)
Whey and derivatives	Ingredient	Aroma	Few studies have examined milk-based substitutes, let alone their mechanisms of action and their effects, both technological and sensory.	Savoury Powder®/ LactoSalt Opti Taste®	Inguglia et al. (2017)
Soy sauce and derivatives			Changes in technological characteristics have been reported, which vary considerably depending on the meat products: Texture changes (decrease or increase in hardness depending on the products, decrease or increase in cohesion and chewability), colour (decrease in L^*, a^* or b^*).	Maggi®	Olofsson et al. (1980)
Yeast extracts			From a technological point of view, the products that include the extracts have their pH drop more rapidly during storage, and the final Aw of the products is significantly lower.	Koji-Aji®	Chiba et al. (2011)

(Continued)

Table 3.1 (Continued)

Substitute	Nature	Interest	Effect	Example of trade name	References
Hydrolyzed vegetable protein			Additional ingredient in the product recipe		
Seaweed extracts			(High levels of sodium glutamate). Strong influence on cooking efficiency thanks to fibre algae content (better water retention and fat in products after cooking)	Algysalt®, Seagreens®	Strauss (2010), Triki et al. (2017)
Tomato extracts			(High levels of sodium glutamate)	SANTE®	Locke and Fielding (1990)
Mushroom extract			(High levels of sodium glutamate)	Mycoscent®	Kilcast and Den Ridder (2007)
Black pepper, ginger, basil, curry, garlic, onion, cumin, coriander, paprika, mint, lemon, vinegar	Herb/spice		Additional ingredient in the product recipe		
Carrot fibre, β-glucan	Plant fibre	Functional agent	Water retention		
Potato starch, carrageenans	Polysaccharide		Additional ingredient in the product recipe		
Transglutaminases, polyphenoloxydases	Enzyme		Additional ingredient in the product recipe		

Source: Petit et al. (2019), Table 2, p. 1464–1465. Reproduced with permission of John Wiley & Sons.

Table 3.2 Physiologic properties and metabolism of sugar alcohols and polydextrose.

Sweetener/ Name of Compound	% Relative Sweetness vs. Sucrose (normal sugar)	Impact on Blood Sugar and Insulin Secretion	Calorie Value kcal/g	Derived From
Mannitol	50–70	Low	1.6	Fructose
Sorbitol	50–70	Low	2.6	Glucose
Sorbitol Syrup	25–50 (depending on sorbitol content)	Low	3	Corn, Wheat or Potato Starch
Xylitol	100	Low	3	D-xylose
Maltitol	90	Low	3	High Maltose Corn Syrup
Maltitol Syrup	25–50 (depending on maltitol content)	Low	3	Corn, Wheat or Potato Starch
Lactitol	30–40	Low	2	Lactose
Isomalt	45–65	Low	2	Sucrose
Erythritol	60–80	Low	0.2	Glucose
Polydextrose	0	Low	1	Dextrose (Glucose), Sorbitol and Citric or Phosphoric Acid

Source: Health Canada (2005).

of diabetes. This has resulted in significant pressure on food manufacturers and ingredient companies to find alternatives to sugar. There are a range of technological solutions available to replace sugar, all with benefits and limitations that will be outlined here.

High Intensity Sweeteners (HIS), both artificial and natural, have been commonly applied as sugar replacers, particularly in beverages. When used in solid food products, a bulking agent is also required. Examples of artificial HISs include aspartame, sucralose, cyclamate, neotame, altame, and acesulfame-k. They deliver sweetness ranging from 200 to 13 000 times the sweetness of sugar and their use is highly regulated in food products. These sweeteners are most frequently used in an attempt to reduce calories and their long-term effect on health is controvertial. Trends are therefore moving towards natural HISs, such as Stevia (steviol glycosides) or Monk Fruit (Mogroside V). However, they tend to exert a high degree of bitterness. This can be a problem in solid products such as confectionary but not to the same degree in beverages on account of their high solubility rate in liquid products. HIS production is currently moving towards the use of biotransformation technologies to meet the needs for commercial-scale production of natural sweeteners (Philippe et al. 2014).

Sugar alcohols or polyols (Table 3.2) are energy-reduced hydrogenated sugars that are used as a bulk sweetener in a wide range of products. They are not as sweet as sugar and therefore are commonly combined with HISs to deliver desirable flavour intensity as well as an acceptable mouthfeel. Examples include erythritol, maltitol, mannitol, sorbitol, xylitol, hydrogenated starch hydrolysates (HSH), and isomalt. Interactions between polyols and macronutrients tend to reduce postprandial glycaemia (Livesey 2003), indicating that they have potential for sugar replacement in diabetic foods. The low rate of absorption of polyols by the small intestine results in lower calorific intake; however, because they are not completely absorbed, they can cause gas, bloating, and diarrhoea. Products containing more than 10% of polyols have to carry a warning on the package stating that excessive eating may produce a laxative effect.

There is a growing trend currently in the food industry to look at sweet enhancers or alternative sugars such as allulose or tagatose, which are both low in calories. Sweet enhancers amplify the intensity and flavour of dietary sugars and are therefore classed as additives. They are added in low quantities and therefore do not affect the calorific value of a food product or the process by which it is manufactured. Most flavour/sweet enhancers are artificial so this remains an active area of research in terms of developing naturally produced versions. Alternative sugars, when incorporated into food products, can provide a palatable sweet taste without the after taste associated with natural and artificial HIS. Low caloric alternative sugars include allulose and tagatose. Allulose provides approximately 70% of the sweetness of sucrose and has the clean taste, bulk, and functionality of sugar. It can be used effectively to replace sugar in a wide range of products. Tagatose is often found in dairy products, and is similar in texture to sucrose with 92% the sweetness. Both allulose and tagatose have a minimal effect on blood glucose and insulin levels; however, the long-term toxicological effects remain to be determined. Isomaltulose and trehalose are examples of alternative sugars that have a similar caloric content to sucrose. Isomaltulose is present naturally in honey and sugarcane extracts. Commercial isomaltulose is produced from sucrose by enzymatic rearrangement and has been used as a sugar in Japan since 1985, in the EU since 2005, and in the US since 2006. Isomaltulose is gaining interest as an alternative sweetener to sucrose, primarily because of its low glycaemic index (GI) properties. Low GI has been implicated in the prevention and management of chronic diseases such as cardio-metabolic diseases and cancers (Shyam et al. 2018). Trehalose is a naturally occurring sugar that functions in many organisms as an energy source or a protectant against the effects of freezing or dehydration. It is considered a novel food in the EU and classified as GRAS by the US FDA. Trehalose exhibits the same technological properties as sucrose with a relative sweetness of 40–45% and has uses across a wide range of food products as a flavouring, stabilizer, cryoprotectant, and for its rehydration abilities.

Maltodextrins are polysaccharides used commonly in the food industry as thickeners, sweeteners, and stabilizers. They are industrially produced by the hydrolysis of starch from corn, rice, potato starch, or wheat generally. Their use is widespread as they are considered safe, inexpensive, and extremely water-soluble. They are relatively bland in flavour and have a low sweetness when compared to sucrose, but have similar calorie content and are also high on the GI. However, research has shown that digestion-resistant maltodextrin helps to relieve constipation and support regular bowel function (Watanabe et al. 2018). New trends are looking towards replacing widely used maltodextrins with dietary fibres to confer added health benefits. Dietary fibre is a term that is used for plant-based carbohydrates that, unlike other carbohydrates (i.e. sugars and starch), are not digested in the small intestine. Dietary fibres contain approximately half the calories of sucrose and reduction of calorie intake is one of the main benefits of substitution. Dietary fibre also holds all the characteristics required to be considered as an important ingredient in the formulation of food products. Some types of soluble fibres, such as pectins, inulin, guar gum, and carboxymethyl-cellulose, are utilized as functional ingredients in milk products, whilst insoluble fibres, from cereal sources, vegetables, and pulses, are used as sugar replacers in cereal-based products such as biscuits and wafers.

Replacing or reducing sugar with alternatives, while still meeting consumer expectations, remains a challenging task for the food industry. Future growth in the sugar replacement market will lie in the development of alternatives that deliver taste, functionality, and offer the health benefits required by consumers and regulators alike.

3.5 Phosphate Replacers

Phosphates are essential nutrients and a form of phosphorous which is required by the human body. They occur naturally in protein-rich foods such as meats, poultry, fish, nuts, beans, and dairy products. However, when we discuss replacement, it is in relation to the group of authorized food additives referred to as 'phosphates'. These phosphates are added to a wide range of foods for 'technological' functions, including pH stabilization, metal cation sequestration, emulsification, leavening, hydration, and bactericidal actions (Molins 1991). Their use is widespread in the food and drink industries, as they have a diverse range of functions and applications. For example, phosphates are incorporated to affect the chemical leavening of cake and biscuit products; they are used to maintain the structure and texture of processed meat products; they contribute to the fluidity of evaporated milk, the distinctive flavour of cola beverages, and the free flow of spice blends. Within this category of additives, there are many types of phosphates, and each form has unique characteristics and optimum conditions for use that determine the food product applications that they are most suited to.

Phosphate additives have been authorized at EU level in 65 different food categories, including processed cheese, UHT milk to bread, cereal-based baby food, and some alcoholic drinks. In the United States, food-grade phosphates are GRAS, but usage levels have been defined for some foods by the Code of Federal Regulations. Molins reported in 'Phosphates in Food', 1991, that over 40 million pounds of phosphorus additives were used annually in the US during the 1970s by the meat industry alone. Baked goods at that time were estimated to contain nearly 10-fold higher amounts of phosphorus additives because of the key role that phosphorus acids play as a dough leavening agent. In 2019, the EFSA set out an acceptable daily intake or ADI for phosphates, which covers both phosphates occurring in food naturally and those used as food additives. This ADI of 40 mg per kg of body weight per day equates to an intake of 2.8 g of phosphorus per day for an average adult weighing 70 kg. This ADI does not apply to people with moderate to severe reduction in kidney function, who are unable to excrete phosphorous to the same extent, resulting in high serum phosphorous levels (hyperphosphataemia) associated with increased morbidity and mortality in those with Chronic Kidney Disease. There has been some evidence to suggest a relationship between high serum phosphorus levels and increased risk of cardiovascular disease (Ritz et al. 2012), but a cause and effect relationship has not been demonstrated to date.

The growth of clean-label processing has led to a growing demand for food manufacturers to remove phosphates from their products. Global health initiatives and the implementation of legislation or voluntary guidelines to reduce sodium intake in many countries will impact on the use of sodium phosphates in food products. The general demand for healthy foods and beverages by consumer health organizations drives the research and development agenda around phosphate alternatives to provide the same technological function within food systems.

In bakery products, the major function of phosphates is in the chemical leavening of the products, causing the doughs and batters to rise by the release of gases. This leavening action is responsible for good volume, improved eating quality, and a uniform cell structure. It can be achieved using other methods such as yeast fermentation, the mechanical incorporation of air by mixing and creaming, or the formation of water vapour during baking. However, phosphates are difficult to replace in baked goods because of their unique ability to delay or control the release of leavening gas until a specific point in the baking process. Fat-coated microcapsules of organic acids or soda can be effective replacements for slow-acting phosphate leavening agents, but most organic acids are too rapid to deliver the right timing of leavening gas.

Phosphates are widely used in the meat processing industry for several reasons, such as changing and/or stabilizing the pH value, increasing water holding capacity in order to lead to higher yields, decreasing losses of weight in cooking, improving texture and sensory properties (tenderness, juiciness, colour, and flavour), and extending shelf life. The most common categories of products that use phosphates are cooked sausages, hams, and other whole-muscle products where moisture retention is important. In meat products where cured meat colour is important but yield is not a concern, the use of phosphates to reduce the pH slightly can increase important curing reactions. Clean label phosphate replacement in meat products has predominantly focused on optimizing the water-holding capacity of the products. Native starches, such as rice starch, have proved effective in poultry products – they have a neutral colour and flavour and the small granule size allows the rice starch to activate at lower cooking temperatures. Citrus and vegetable-based fibre sources bind water effectively without heating due to the small hollow-tube structure of alpha cellulose. Powell et al. (2019) showed that the replacement of sodium tripolyphosphate with citrus fibre did not significantly alter the physical, chemical, or sensory characteristics of Bologna sausages during refrigerated storage. A 2015 review of plum ingredients reported that the use of higher amounts of plum can affect the water-holding capacity of meat products, but at lower levels can achieve similar marinade pickup as compared with industry standard marinades containing phosphates (Jarvis et al. 2015). Winter mushroom powder has been shown to effectively replace phosphates in meat products. Addition levels above 1.0% increased the pH of sausage emulsions and efficiently inhibited the exudation of fat from the products. Below 2%, no adverse effects were observed with respect to the colour and sensory properties of the sausages containing this mushroom powder (Choe et al. 2018). Proteins from meat, dairy, and plant sources are also effective phosphate replacers in terms of improving meat yields. The high solubility, clean flavour, and affordability of isolated soy proteins and soy protein concentrates mean that they are used in many processed meats. A study by Bae et al. (2017) investigated the application of natural calcium powders to replace phosphate in cooked ground pork products. The results of this work suggested that the combined use of egg shell calcium and oyster shell calcium could be a potentially effective method for replacing synthetic phosphate in these meat products. Combinations of the above ingredients are now being used commercially in conjunction with gums and buffering agents to replace phosphates in meat products; however, research continues to determine the optimum replacer for this multifunctional food additive.

Phosphates are used in processed cheeses for emulsification and they act as creaming and buffering agents. Their role during processing is to sequester calcium in the natural cheese, to solubilize protein and increase its hydration and swelling, to facilitate the emulsification of fat, and to adjust and stabilize the pH. Their inclusion in the formulation produces a cheese that is less likely to separate and split during heating and cooling. Stabilization of the pH ensures better stretch and melt characteristics, and prevents a brittle or crumbly texture. Citrates and potassium-based salts have been explored as phosphate replacers in processed cheeses. Chavhan et al. (2015) reported the successful use of potassium-based emulsifying salts and KCl in processed Mozzarella cheese up to approximately 71% without jeopardizing its sensory, textural, and functional properties. A study by Chen and Liu (2012) showed that processed cheeses made with trisodium citrate had similar hardness, pH, and degree of casein dissociation, as those processed using sodium tripolyphosphate (STPP).

As alluded to previously, the major challenge in replacing phosphates in food products is the fact that they are multifunctional. Whilst alternative clean-label emulsifiers, moisture retention, and fortification ingredients have been developed, a significant amount of formulation work is required to find the correct combination to replace a single phosphate ingredient. So, whilst consumer demand may dictate an eventual move away from the use of phosphates in manufactured products, food producers continue to use them for now.

3.6 Bioactives

Bioactive compounds are components of food that are capable of modulating metabolic processes, resulting in the promotion of better health. They are distinct from nutrients because bioactive compounds are not essential and, currently, there are no recommended daily intake values. Bioactive compounds can be derived from milk, eggs, meat, fish, fruits, vegetables, whole grains, and more. There has been considerable interest and research committed to the discovery of bioactive compounds, their production, and the impact they can have on human health in the prevention of certain NCDs.

Phytochemicals are bioactive compounds produced by plants including fruits, vegetables, grains, and beans. They can be classified into major categories, such as carotenoids and polyphenols, which include bioactive compounds such as flavonoids, phenolic acids, and lignans, amongst others. Carotenoids are yellow, orange, and red pigments synthesized by plants. The most prevalent carotenoids in the human diet are α-carotene, β-carotene, β-cryptoxanthin, lycopene, lutein, and zeaxanthin. The bioavailability of carotenoids from most foods is low; however, this can be increased through processes such as chopping, homogenizing, and cooking, that disrupt the plant matrix. A growing body of research-based evidence exists regarding the effects of carotenoids on chronic diseases in humans. Studies have been conducted to determine the effects of carotenoids on certain cancers, cardiovascular diseases, photosensitivity diseases, cataracts, and age-related macular degeneration (Mayne 1996). Despite the body of research conducted to date, there is insufficient evidence to substantiate the role of carotenoids in the prevention of chronic diseases and to date no health claims have been approved for these bioactives.

Polyphenols are beneficial compounds found in many plant foods that can be grouped into flavonoids, phenolic acid, lignans, and other polyphenols. Interest in polyphenols has grown in recent years due to their antioxidant properties, their great abundance in the human diet, and their potential role in the prevention of various diseases associated with oxidative stress, such as cancer and cardiovascular and neurodegenerative diseases (Scalbert et al. 2005). It has been suggested that polyphenols may also impact genes and gene expression and influence our gut bacteria. Foods contain complex mixtures of polyphenols but the most abundant of these bioactives are not always the most active within the body. This can be on account of low intrinsic activity, poor absorption from the intestine, or that they are highly metabolized, or rapidly eliminated on ingestion.

Flavonoids are part of the polyphenol class of phytochemicals. The main dietary sources of flavonoids include tea, citrus fruit, citrus fruit juices, berries, red wine, apples, and legumes, but they are found in almost all fruit and vegetables. There are several significant groups of flavonoids, including anthocyanidins, flavanols, flavones, flavonols, flavonones, and isoflavones. Each of these subgroups and each type of flavonoid carries its own distinct set of actions, benefits, and originating foods. Flavonoids are associated with a broad spectrum of health-promoting effects. Because of their antioxidant and anti-inflammatory behaviours, flavonoids have been associated with cardiovascular disease prevention. A significant cross-cultural study was carried out over a 25-year period to determine whether flavonoid intake can explain differences in mortality rates from chronic diseases between populations. Hertog et al. (1995) concluded that average flavonoid intake may partly contribute to differences in coronary heart disease mortality across populations. A more recent review of existing cohort studies of flavonoid intake and cardiovascular disease risk in the US and Europe by Peterson et al. (2012) concluded that there is intriguing, but not yet compelling, evidence that relatively small amounts of certain dietary flavonoids may lower the risk of coronary heart disease mortality. The authors stated that more research is needed to establish that cardioprotective relationships exist and, if they prove to be protective, what consumption levels may be required to achieve health benefits. Whilst more recent studies have also demonstrated that flavonoids can have anticarcinogenic properties against different types of cancers (Darband et al. 2018; Sudhadkaran et al. 2019), an updated review by Rodríguez-García et al. (2019) concluded that further studies are needed to investigate and confirm this hypothesis that a healthy diet can help decrease the incidence of different types of cancer. There is ongoing research into the potential health benefits of individual flavonoids; however, to date, neither the FDA nor the EFSA has approved any health claim for these bioactive components.

Phenolic acids are found in a variety of plant-based foods, the seeds and skins of fruits, and the leaves of vegetables contain the highest concentrations. Little is known about the bioactivity of phenolic acid *in vivo* and the mechanisms by which they may contribute towards disease prevention. There is accumulating evidence suggesting that phenolic acids are rapidly metabolized in the human body (Rechner et al. 2002; Scalbert and Williamson 2000). Whilst foods rich in antioxidant phenolic acids may be beneficial for human health, it proves difficult to determine if the benefit is due to the phenolic acids, or to inherent nutrients and other phytochemicals also found in those foods.

Lignans are present in a wide variety of plant foods, including seeds, whole grains, bran, beans, fruit, and vegetables. Flaxseed is the richest dietary source of lignans, and bioavailability can be improved by crushing or milling flaxseed. A variety of factors may affect the lignan contents of plants, including geographical location, climate, maturity, and storage conditions. Lignan-rich diets have been consistently associated with reductions in the risk of cardiovascular disease (Peterson et al. 2010; Zanwar et al. 2014). However, these cardioprotective effects may also be attributed to the varied nutrients and other phytochemicals present in lignan-containing foods. Experimental evidence in animals has shown clear anticarcinogenic effects of flaxseed or pure lignans on many types of cancer (Calado et al. 2018; Demark-Wahnefried et al. 2008; Ezzat et al. 2018). The results are promising, but much work is still needed. Corrêa et al. (2018) recently reported that lignans possess anti-ageing properties.

Discoveries in bioactive components and their potential for improving human health in terms of preventing chronic disease are escalating. However, evidence-based criteria to allow for public health statements, or for food manufacturers to make incorporated product health claims, is lagging behind. It is a challenging area with further research and human *in vivo* studies are required.

3.7 Peptides

Bioactive peptides are isolated small fragments of proteins that exert a beneficial effect on body functions. They are commonly extracted from animal proteins such as egg, milk and meat proteins, or from plant sources including soy, pulses, oat, wheat, hemp, and flaxseed. Proteins from marine sources can also be used. Bioactive peptides can be produced by enzymatic hydrolysis of the food proteins, using proteolytic enzymes from either plants or microbes; by hydrolysis with digestive enzymes; or by fermentation using starter cultures. The activity of peptides is based on their inherent amino acid composition and sequence. Bioactive peptides from food proteins offer major potential for incorporation into functional foods and nutraceuticals.

Bioactive peptides from milk were amongst the first food-derived peptides studied. Milk contains approximately 3.5% protein, of which 80% are casein and 20% whey proteins. Milk peptides are derived from these proteins by enzymatic breakdown by digestive enzymes or by the proteinase enzymes produced by lactobacilli during the fermentation of milk. These bioactive peptides are usually comprised of 2–20 amino acids and they exert multifunctional properties including antimicrobial (Mohanty et al. 2014), antioxidant (Tonolo et al. 2019), immune modulatory (Gill et al. 2000), antihypertensive (Korhonen and Pihlanto 2007), and opioid activities (Meisel and FitzGerald 2000).

Meat and meat products are promising food sources for bioactive compounds that can promote health and/or complement the effect of nutrients. Examples include carnosine, choline, L-carnitine, conjugated linoleic acid, glutathione, taurine, and creatine, which have been studied for their physiological properties. The most widely used method for extracting bioactive peptides from meat is enzymatic hydrolysis, often conducted using digestive enzymes such as pepsin, trypsin, and chymotrypsin. Alternatively, proteases of bacterial, animal, and plant origin have been used to generate bioactive peptides from meat sources. Meat products produced using drying, curing, ripening, and fermentation processes can also result in the release of bioactive peptides from parent proteins. Many studies have been conducted to date to evaluate the beneficial health effects of meat bioactive compounds. Bhat et al. (2018) reported that a remarkable increase has been observed in the documentation of peptides from animal proteins, such as meat, with demonstrated *in vitro* and *in vivo* antihypertensive properties. A review by Liu et al. (2016) identified that peptides in meat muscle protein and by-products exhibit strong antioxidant capacity. Meat-derived peptides have also demonstrated antimicrobial and cytotoxic physiological activities (Jang et al. 2008).

Fish, shellfish, seaweed, and algae are important sources of unique bioactive peptides. These marine organisms constitute half of the worldwide biodiversity and there is growing interest in identifying compounds from this natural and plentiful resource.

Similar to peptides derived from milk and meat muscle sources, these bioactive peptides are indicated to have biological functions such antioxidant, antithrombotic, antihypertension, immunomodulation, anticancer, and antimicrobial effects. A review by Kang et al. (2015) presented the structures and anti-infective activity of peptides isolated from marine sources (sponges, algae, bacteria, fungi, and fish) and reported that these peptides have been found to exhibit antimicrobial, antifungal, antimalarial, antiprotozoal, anti-tuberculosis, and antiviral activities. Cheung et al. (2015) documented the bioactivity and applications of a wide range of marine bioactive peptides including antihypertensive peptides, antioxidant peptides, cardiovascular protective peptides, immune-modulatory peptides, antidiabetic peptides, analgesic peptides, and more.

To date, a wide range of food-derived bioactive peptides have been identified and their associated health benefits researched. However, there is currently a lack of sufficient evidence in terms of clinical trials for the authorization of many of the health claims associated with peptides derived from milk, meat, or marine sources. The commercial application of these bioactive peptides has also been slow on account of the lack of scalable production methods.

3.8 Conclusions and Future Trends

Health and wellness show an accelerating trend that has been a fundamental driver of change in the food and beverage industry worldwide. Equally, the growing consumer requirement for transparency is driving the development of new and innovative healthy ingredients. Consumers want to know and understand what ingredients are going into the food products they eat and are spending more time reading product labels and interpreting claims. Clean labels are increasingly important, particularly to a growing segment of consumers with special dietary needs, which means there is a preference for natural ingredients over artificial. Consumers are also starting to query heavily fortified products and are looking to replace them with food products that are inherently nutrient-rich or contain healthy bioactive components.

Sugar and salt have been primary nutrition targets in food and beverages for a number of years. Recommendations from public health authorities and government agencies worldwide, as well as the introduction of penalty taxes, will continue to drive the reduction in sugar and salt content across all categories of food. Unfortunately for food manufacturers there is no simple solution in terms of reduction and replacement without affecting the integrity of the product and, therefore, significant research and development work will continue in this area. However, trends are moving towards more natural replacers and alternatives, with consumers expecting additional health and functional advantages from the ingredients used.

Consumer demand for foods with benefits beyond basic nutrition has created a category of functional foods containing bioactive ingredients becoming more prominent in the marketplace. This growing market is one of the most promising in recent times; however, there is a need for 'evidence-based' functional foods. Research on bioactives currently forms a significant portion of the project portfolios of governments, universities, and food manufacturers. This work will continue to uncover novel bioactive components, developing suitable technologies for their isolation and extraction. However, reliable, science-based information is lacking for the health benefits of bioactive foods and constituents. Therefore, screening methods need to be developed for the measurement of the long-term beneficial effects, *in vivo*, in order to confirm indisputably that they can promote good health.

To conclude, the development of healthy ingredients and replacers is dominating the food industry and claims around the functional advantages of food products are increasingly expected by consumers who desire to feel benefits from the food they eat. The future in terms of ingredient development and application is exciting, but not without significant challenges for ingredient companies and food manufacturers.

References

Aburto, N.J., Hanson, S., Gutierrez, H. et al. (2013). Effect of increased potassium intake on cardiovascular risk factors and disease: systematic review and meta-analyses. *British Medical Journal* 346: 1378.

Bae, S.M., Cho, M.G., and Jeong, J.Y. (2017). Effects of various calcium powders as replacers for synthetic phosphate on the quality properties of ground pork meat products. *Korean Journal for Food Science of Animal Resources* 37 (3): 456.

Bhat, Z.F., Mason, S., Morton, J.D. et al. (2018). Antihypertensive peptides from animal proteins. In: *Bioactive Molecules in Food*, Reference Series in Phytochemistry (eds. J.M. Mérillon and K. Ramawat). Cham, Switzerland: Springer.

Binia, A., Jaeger, J., Hu, Y. et al. (2015). Daily potassium intake and sodium-to-potassium ratio in the reduction of blood pressure: a meta-analysis of randomized controlled trials. *Journal of Hypertension* 33: 1509.

Calado, A., Neves, P.M., Santos, T., and Ravasco, P. (2018). The effect of flaxseed in breast cancer: a literature review. *Frontiers in Nutrition* 5: 4.

Chavhan, G.B., Kanawjia, S.K., Khetra, Y., and Puri, R. (2015). Effect of potassium-based emulsifying salts on sensory, textural, and functional attributes of low-sodium processed mozzarella cheese. *Dairy Science & Technology* 95 (3): 265.

Chen, L. and Liu, H. (2012). Effect of emulsifying salts on the physicochemical properties of processed cheese made from mozzarella. *Journal of Dairy Science* 95 (9): 4823.

Cheung, R.C.F., Ng, T.B., and Wong, J.H. (2015). Marine peptides: bioactivities and applications. *Marine Drugs* 13 (7): 4006.

Chiba, S., Saegusa, T., and Ishii, M. (2011). Composition for low-salt food or beverage. US Patent US20120135110A1, filed December 21, 2011.

Choe, J., Lee, J., Jo, K. et al. (2018). Application of winter mushroom powder as an alternative to phosphates in emulsion-type sausages. *Meat Science* 143: 114.

Corrêa, R.C.G., Peralta, R.M., Haminiuk, C.W.I. et al. (2018). New phytochemicals as potential human anti-aging compounds: reality, promise, and challenges. *Critical Reviews in Food Science and Nutrition* 58 (6): 942.

Council Directive 90/496/EEC of 24 September 1990 on nutrition labelling for foodstuffs. *Official Journal of the European Communities*. N L 276/40. 6.10.90.

Darband, S.G., Kaviani, M., Yousefi, B. et al. (2018). Quercetin: a functional dietary flavonoid with potential chemo-preventive properties in colourectal cancer. *Journal of Cellular Pysiology* 233 (9): 6544.

Demark-Wahnefried, W., Polascik, T.J., George, S.L. et al. (2008). Flaxseed supplementation (not dietary fat restriction) reduces prostate cancer proliferation rates in men presurgery. *Cancer Epidemiology, Biomarkers and Prevention* 17 (12): 3577.

Ezzat, S.M., Shouman, S.S., Elkhoely, A. et al. (2018). Anticancer potentiality of lignin rich fraction of six flaxseed cultivars. *Nature Scientific Reports* 8: 544.

Gill, H.S., Doull, F., Rutherfurd, K.J., and Cross, M.L. (2000). Immunoregulatory peptides in bovine milk. *British Journal of Nutrition* 84: 111.

Grummer, J., Karalus, M., Zhang, K. et al. (2012). Manufacture of reduced-sodium Cheddar-style cheese with mineral salt replacers. *Journal of Dairy Science* 95 (6): 2830.

Guinee, T.P. and Fox, P.F. (2004). *Cheese: Chemistry, Physics and Microbiology*. San Diego, CA: Elsevier, Academic Press.

He, K., Zhao, L., Daviglus, M.L. et al. (2008). Association of mono-sodium glutamate intake with overweight in Chinese adults: the INTERMAP study. *Obesity* 16 (8): 1875.

Hertog, M.G., Kromhout, D., Aravanis, C. et al. (1995). Flavonoid intake and long-term risk of coronary heart disease and cancer in the seven countries study. *Archives of Internal Medicine* 155 (4): 381.

Health Canada (2005). Sugar Alcohols (Polyols) and Polydextrose Used as Sweeteners in Foods. https://www.canada.ca/en/health-canada/services/food-nutrition/food-safety/food-additives/sugar-substitutes/sugar-alcohols-polyols-polydextrose-used-sweeteners-foods-food-safety.html

Inguglia, E.S., Zhang, Z., Tiwari, B.K. et al. (2017). Salt reduction strategies in processed meat products–a review. *Trends in Food Science & Technology* 59: 70.

Jang, A., Jo, C., Kang, K.S., and Lee, M. (2008). Antimicrobial and human cancer cell cytotoxic effect of synthetic angiotensin-converting enzyme (ACE) inhibitory peptides. *Food Chemistry* 107: 327.

Jarvis, N., O'Bryan, C.A., Ricke, S.C., and Crandall, P.G. (2015). The functionality of plum ingredients in meat products: a review. *Meat Science* 102: 41.

Kang, H.K., Seo, C.H., and Park, Y. (2015). Marine peptides and their anti-infective activities. *Marine Drugs* 13 (1): 618.

Kilcast, D. and Den Ridder, C. (2007). Sensory issues in reducing salt in food products. In: *Reducing Salt in Foods* (eds. D. Kilcast and F. Angus), 201–220. Cambridge, UK: Woodhead Publishing.

Kloss, K., Meyer, J., and Detter, W. (2015). Sodium intake and its reduction by food reformulation in the European Union — a review. *NFS Journal* 1: 9.

Korhonen, H. and Pihlanto, A. (2007). Technological options for the production of health-promoting proteins and peptides derived from milk and colostrum. *Current Pharmaceutical Design* 13: 829.

Kroon, H., Ide, S., Galligan III, W.J., and Akimoto, N. (2010). Use of alkali metal lactate in curing of seafood surimi products. US Patent US20120045541A1, filed March 10, 2010.

Liu, R., Xing, L., Fu, Q. et al. (2016). A review of antioxidant peptides derived from meat muscle and by-products. *Antioxidants* 5 (3): 32.

Livesey, G. (2003). Health potential of polyols as sugar replacers, with emphasis on low glycaemic properties. *Nutrition Research Reviews* 16 (2): 163.

Locke, K.W. and Fielding, S. (1990). Salt taste enhancer. US Patent US4997672A, filed July 25, 1990 and granted March 5, 1991.

Macgregor, G.A. and Sever, P.S. (2003). Salt – overwhelming evidence but still no action: can a consensus be reached with the food industry? *British Medical Journal* 312 (7041): 1287.

McGregor, R. (2007). The use of bitter blockers to replace salt in food products. In: *Reducing Salt in Foods* (eds. D. Kilcast and F. Angus), 221–230. Cambridge, UK: Woodhead Publishing.

Mäki, J. (2000). Physiological food salt product. US Patent US6787169B1, filed January 12, 2000 and granted September 7, 2004.

Martin, C.H. (1940). Salt tablet containing a calcium salt and process for making the same. US Patent US2333873A, filed December 2, 1940 and granted November 9, 1943.

Mayne, S.T. (1996). Beta-carotene, carotenoids, and disease prevention in humans. *The FASEB Journal* 10 (7): 690.

Meijer, F.R., Houdijk, C., Muijlwijk, C., and Van, N.C.O. (2007). Reduced-salt dairy product with improved taste. US Patent US20100047391A1, filed December 17, 2007.

Meisel, H. and FitzGerald, R.J. (2000). Opioid peptides encrypted in milk protein sequences. *British Journal of Nutrition* 84: 27.

Mohanty, D.P., Tripathy, P., Mohapatra, S., and Samantaray, D.P. (2014). Bioactive potential assessment of antibacterial peptide produced by Lactobacillus isolated from milk and milk products. *International Journal of Current Microbiology and Applied Sciences* 3: 72.

Molins, R.A. (1991). *Phosphates in Food*. Boca Raton: CRC Press, Inc.

OECD (2016). Health at a Glance: Europe 2016. State of Health in the EU Cycle. Paris: OECD Publishing.

Olofsson, M., Buhler, M. and Wood, R. (1980). Process for the preparation of a purified protein hydrolysate. US Patent US4293571A, filed February 28, 1980 and granted October 6, 1981.

Pasin, G., O'Mahony, M., York, G. et al. (2008). Replacement of sodium chloride by modified potassium chloride (co crystallized disodium-5′-inosinate and disodium-5′-guanylate with potassium chloride) in fresh pork sausages: acceptability testing using signal detection measures. *Journal of Food Science* 54 (3): 553.

Peterson, J.J., Dwyer, J.T., Adlercreutz, H. et al. (2010). Dietary lignans: physiology and potential for cardiovascular disease risk reduction. *Nutrition Reviews* 68 (10): 571.

Peterson, J.J., Dwyer, J.T., Jacques, P.F., and McCullough, M.L. (2012). Do flavonoids reduce cardiovascular disease incidence or mortality in us and European populations? *Nutrition Reviews* 70 (9): 491.

Petit, G., Jury, V., de Lamballerie, M. et al. (2019). Salt intake from processed meat products: benefits, risks and evolving practices. *Comprehensive Reviews in Food Science and Food Safety* 18 (5): 1453–1473.

Philippe, R.N., De Mey, M., Anderson, J., and Ajikumar, P.K. (2014). Biotechnological production of natural zero-calorie sweeteners. *Current Opinion in Biotechnology* 26: 155.

Powell, M.K.J., Prusa, K.J., Sebranek, J.G., and Tarte, R. (2019). Evaluation of citrus fiber as a natural replacer of sodium phosphate in alternatively-cured all-pork Bologna sausage. *Meat Science* 157: 107883.

Raseta, M., Lazic, I.B., Lilic, S. et al. (2018). Partial replacement of sodium chloride with potassium chloride or ammonium chloride in a prepared meal-cooked peas with pork burger. *Scientific Journal Meat Technology* 59 (2): 114–119.

Rechner, A.R., Kuhnle, G., Bremner, P. et al. (2002). The metabolic fate of dietary polyphenols in humans. *Free Radical Biology and Medicine* 33: 220.

Reddy, K.A. and Marth, E.H. (1991). Reducing the sodium content of foods: a review. *Journal of Food Protection* 54 (2): 138.

Ritz, E., Hahn, K., Ketteler, M. et al. (2012). Phosphate additives in food-a health risk. *Deutsches Ärzteblatt International* 109: 4955.

Rodríguez-García, C., Sánchez-Quesada, C., and Gaforio, J.J. (2019). Dietary flavonoids as cancer chemopreventive agents: an updated review of human studies. *Antioxidants* 8 (5): 137.

Sahurkar, M.R. and Karadbhajne, S.V. (2018). Replacement of fat, salt, and sugar: need of an hour for food industries. *International Journal of Food Science and Nutrition* 3 (4): 200.

Scalbert, A. and Williamson, G. (2000). Dietary intake and bioavailability of polyphenols. *Journal of Nutrition* 130: 2073.

Scalbert, A., Manach, C., Morand, C. et al. (2005). Dietary polyphenols and the prevention of diseases. *Critical Reviews in Food Science and Nutrition* 45 (4): 287.

Shyam, S., Ramadas, A., and Chang, S.K. (2018). Isomaltulose: recent evidence for health benefits. *Journal of Functional Foods* 48: 173.

Sinopoli, D.A. and Lawless, H.T. (2012). Taste properties of potassium chloride alone and in mixtures with sodium chloride using a check-all-that-apply method. *Journal of Food Science* 77 (9): 319.

Stanley, R.E., Bower, C.G., and Sullivan, G.A. (2017). Influence of sodium chloride reduction and replacement with potassium chloride based salts on the sensory and physicochemical characteristics of pork sausage patties. *Meat Science* 133: 36.

Strauss, S. (2010). Parse the salt, please. *Nature Medicine* 16: 841–843.

Sudhadkaran, M., Sardesai, S., and Doseff, A.I. (2019). Flavonoids: new frontier for immuno-regulation and breast cancer control. *Antioxidants* 8 (4): 103.

Te Morenga, L.A., Howatson, A., Jones, R.M., and Mann, J. (2014). Dietary sugars and cardiometabolic risk: systematic review and meta-analyses of randomized controlled trials of the effects on blood pressure and lipids. *American Journal of Clinical Nutrition* 100 (1): 65.

Tonolo, F., Moretto, L., Ferro, S. et al. (2019). Insight into antioxidant properties of milk-derived bioactive peptides *in vitro;* and in a cellular model. *Journal of Peptide Science* 25 (5): 3162.

Triki, M., Khemakhem, I., Trigui, I. et al. (2017). Free-sodium salts mixture and AlgySalt® use as NaCl substitutes in fresh and cooked meat products intended for the hypertensive population. *Meat Science* 133: 194.

VanBuren, L., Dötsch-Klerk, M., Seewi, G., and Newson, R.S. (2016). Dietary impact of adding potassium chloride to foods as a sodium reduction technique. *Nutrients* 8 (4): 235.

Watanabe, N., Suzuki, M., Yamaguchi, Y., and Egashira, Y. (2018). Effects of resistant maltodextrin on bowel movements: a systematic review and meta-analysis. *Clinical and Experimental Gastroenterology* 11: 85.

Weaver, C.M. (2013). Potassium and health. *Advances in Nutrition* 4 (3): 368S.

World Health Organisation (2012). *Guideline: Sodium Intake for Adults and Children*. Geneva: World Health Organization.

World Health Organisation (2015a). *Guideline: Sugars Intake for Adults and Children*. Geneva: World Health Organization.

World Health Organization (2015b). Global Action Plan for the Prevention and Control of Noncommunicable Diseases 2013–2020. Geneva: WHO Press.

Zanwar, A.A., Hegde, M.V., and Bodhankar, S.L. (2014). Flax Lignan in the prevention of atherosclerotic cardiovascular diseases. In: *Polyphenols in Human Health and Disease* (eds. R.R. Watson, V.R. Preedy and S. Zibadi), 915–921. San Diego, CA, USA: Academic Press.

4

Macromolecules Interactions in Food Formulations

Puneet Parmar[1], Rajpreet Kaur Goraya[2], and Shivani Pathania[3]

[1] *Animal and Grassland Research and Innovation Centre, Teagasc, Cork, Ireland*
[2] *Department of Food Science & Technology, Punjab Agricultural University, Ludhiana, Punjab, India*
[3] *Teagasc Food Research Centre, Dublin, Ireland*

Summary

Different ingredients present in food systems interact with each other and in formulations based on their affinity or aversion with other ingredients present in the system. Physical and chemical changes such as protein–lipid interactions, protein–carbohydrate interactions, etc. directly or indirectly influence the various characteristics of food formulation along with their microstructure and particle size distribution. This chapter outlines the intra- as well as inter-macromolecular interactions which take place in food systems and how these interactions contribute to the physical properties as well as flavour of popular foods as we know them today.

4.1 Introduction

Food, in raw as well as processed form, is generally a complex system comprising numerous molecules and compounds of varying kinds. Foods of both plant and animal origin contain macronutrients like carbohydrates, proteins, and fats, together with micronutrients like vitamins and minerals. Besides these macro- and micronutrients, many foods contain minor compounds such as phytochemicals, bioactives, toxins, anti-nutritional factors, and many others. Being complex, heterogeneous, and non-uniform in nature, foods are made up of a combination of these compounds. No food in its original form consists of only one of these compounds. The occurrence of these compounds in food depends on many factors such as origin, variety, breed, cultivation, or rearing practises, followed by atmospheric conditions amongst many other factors. As the human body

---

Food Formulation: Novel Ingredients and Processing Techniques, First Edition.
Edited by Shivani Pathania and Brijesh K. Tiwari.

needs many nutrients for its normal working, the presence of as many compounds as possible (except toxins and non-nutrients) in our food is considered desirable. For example, milk, a diverse food, contains lactose, casein, whey proteins, fats, vitamins, minerals, acids and other compounds, is considered as an almost complete food. However, these constituents in foods do not remain in isolation and interact in various ways with themselves as well as with each other and so further affect the functional properties of food systems. The interactions of these macromolecules and formation of complexes is of great importance in deciding the quality of food and food products. For instance, the protein–polysaccharide complexation is essential in the formation of complex gels such as alginate-gelatine, alginate-casein gels, etc. and the protein-polysaccharide complexes are often applied as emulsion stabilizers due to their high surface activity and ability to form thick gel-like layers. For example, they are added in ice-cream mixes (Tolstoguzov 1996). As stated earlier, food may contain macro-components, i.e. carbohydrates, proteins, lipids, water and micro-components such as vitamins, minerals, enzymes, etc. This chapter provides a comprehensive overview of interactions between macromolecules present in key food ingredients required for food formulations.

4.2 Role of Macromolecular Interactions

The molecular and colloidal dispersions of biopolymers and their complexes make up the structure of most foods. The structure, rheology, and other physico-chemical properties of foods, as well as their sensory characteristics, are fundamentally governed by the major biopolymers in foodstuff. Proteins occur as structure and texture-macromolecular modifiers in foods, where multiple interactions with the rest of the biopolymers can take place (Lopes da Silva and Rao 2007). Appelqvist and Debet (1997) also mentioned that the structure, rheological, and other physico-chemical properties of many processed and convenience foods are determined to a large extent by the behaviour of the protein and polysaccharide components. The authors also found the interaction between these biopolymers within the food system to be of great interest.

Water and lipids, with their ability to bind to other components and act as solvents, are also considered important factors. The functional properties of foods reflect the physico-chemical properties of both the complexed and the individual macromolecules. This is because the complexed molecules obtain new properties as compared to the individual macromolecule. Therefore, the diversity of the food structures is contributed to by the interactions between macromolecules (Tolstoguzov 1996, 2003).

The interactions of macromolecules in food systems are not specific in nature and occur during many stages of food handling and processing. These processes facilitate unfolding of the macromolecules, a prerequisite for the interactions. The chemical and enzymatic reactions also lead to altered intra- and intermolecular interactions, which at the macromolecular level are reflected in conformation changes, complexation, and aggregation. The interactions may also result in the formation of colloids and 3D networks, which are essential in the development of macroscopic structures like suspensions, foams, emulsions, gels, and others (Tolstoguzov 1996).

Consequently, in processing of food and food products comprising either two of the three macromolecules, it is advisable to understand the potential behaviours of interactions of these macromolecules within a real food matrix. Macromolecular interactions highly influence the food quality therefore it is of immense importance to improve our understanding of macromolecular interactions in foods. To maintain the palatable properties and bioactivity in food products, interactions between the bioactive molecules and other food components, including macro- and micro-components, needs to be curtailed. These interactions vary in nature and can be physical, chemical, or a combination of the two. Physical interactions between ingredients could involve changes in adsorption, absorption, drying, evaporation, and particle size properties. Physico-chemical properties and interactions pertain to formation of gels, emulsions, foams, etc. Chemical interactions may be dependent upon the composition of the food including proteins, amino acids constituents, and concentrations of individual components. They may also include interactions between the food and external surfaces such as packaging material.

The intensity of chemical interactions is also dependent upon external physical conditions such as temperature, pH, moisture, and time duration of the process. The interactions and their intensity may either be desirable or undesirable, depending upon the food application and end product required. For example, the Maillard action is a chemical carbonyl–amine reaction leading to browning in egg whites which is undesirable; however, the same reaction is favourable in baking for generating aroma, colour, and visual effects.

4.3 Types of Macromolecular Interactions

Most of the foods and food products contain all the macromolecules, though some macromolecules can be present in large or very little amounts. Complexation may take place within a type of biopolymer (e.g. protein–protein) or between different types of biopolymers (e.g. protein–lipid, protein–carbohydrate). The interactions taking place within the same type of macromolecules are termed intramolecular interactions and those taking place between different types of macromolecules are termed intermolecular interactions. The interactions may also be binary (occurring

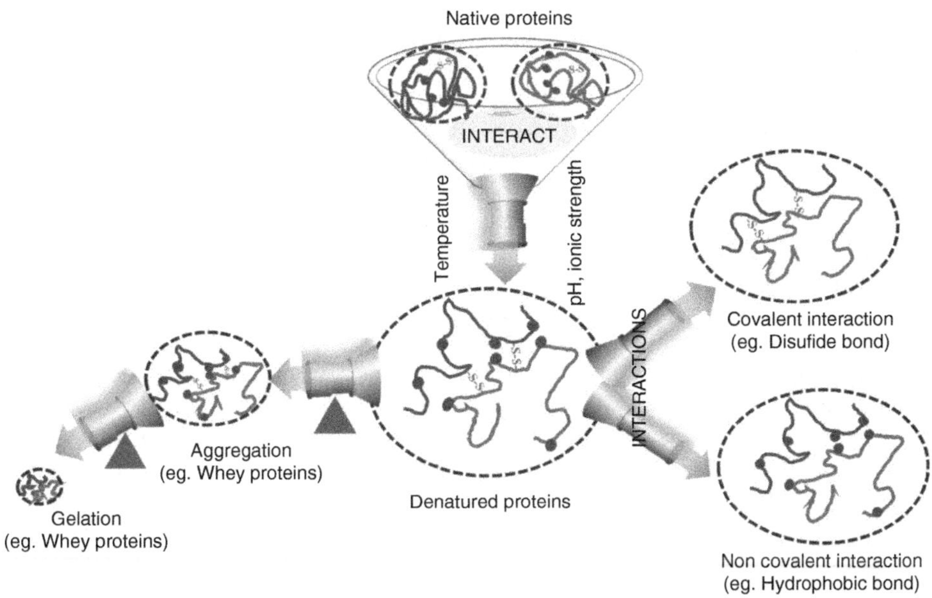

Figure 4.1 Illustration of heat-induced changes in native protein structures. Formation of interactions (covalent and noncovalent in dairy proteins), right side of figure. Severe heat treatment to protein (whey) solutions inducing denaturation, aggregation, and gelation/ excessive thickening, left side of figure.

between two macromolecules) or ternary (occurring between three macromolecules). The classification and various types of macromolecular interactions are displayed in Figure 4.1. The various types of important interactions significant to food processing are described in the following sections.

4.3.1 Protein–Protein Interactions

van der Waals' forces, such as hydrophobic, hydrogen bonds, and covalent interactions, are responsible for the formation of secondary and tertiary structures of protein molecules. The tertiary structure is the 3D arrangement, whereby both inter- and intramolecular interactions such as noncovalent (electrostatic interactions and van der Waals') or covalent bonds (disulphide bonds) of the several arrays of the proteins are balanced in a delicate manner by different interactions, as shown in Figure 4.1. However, formation of the quaternary structure is a result of protein–protein noncovalent interactions that result in the spatial arrangement. Pressure, heat, shear, or any further changes in formulation conditions (i.e. pH or ionic strength) result in loss of proteins identity from their native structure. Denaturation or unfolding of proteins is the alteration of stabilizing forces in their native structure in which the compact protein molecules unfold into a random structure (Figure 4.1) by losing its native state.

The interactions between proteins chiefly occur through disulphide linkages. Such interactions are called protein–protein cross-linking. Protein cross-linking usually refers to the formation of covalent bonds between polypeptide chains within a protein (intramolecular cross-links) or between proteins (intermolecular cross-links) (Feeney and Whitaker 1988). Disulphide cross-links, and cross-links derived from dehydroprotein, tyrosine, the Maillard reaction, and those formed via transglutaminase catalysis, are the major types of cross-links. Besides these cross-links, severe heat treatment can result in the formation of isopeptide cross-links during processing of low carbohydrate foods (Gerrard 2002).

Protein–protein interactions (PPIs) occur in raw as well as cooked food products, including straight after harvesting, transportation, material handling, processing, and storage. These interactions can result in the formation of new food structures, resulting in their use as specific tailor-made foods, which can be enhanced by sensory attributes like body, texture, appearance, mouthfeel, etc. Various types of protein interactions (covalent and non-covalent in dairy proteins) depend on the food system and its environment, which are summarized in Figure 4.1. The PPIs and protein structure knowledge are used in the development of different new ingredients with specific functionalities like dairy ingredients (whey proteins, caseins), in various food products (bakery, etc.) and beverages. The milk proteins are used as functional ingredients after denaturation, as they possess emulsifying, stabilizing, and good water-holding capacites as their new properties. Specific PPIs can be tailored by modifying external factors such as temperature, shearing, solvent, and pH to achieve functionalities such as solubility, emulsification, and nutritional properties.

Thermally-induced conformational changes in proteins lead to covalent and non-covalent interactions between exposed reactive groups and depending upon protein concentration, gelation may also occur. The oxidative coupling of two cysteine residues, which are adjacent within the food protein matrix, form the disulphide bonds in protein cross-linking. The disulphide cross-links are produced by accepting the hydrogen atoms via a suitable oxidant from the thiol groups of the cysteine residues. The intermolecular disulphide cross-link's formation ability of proteins upon heat processing is very important for the gelling of some food proteins, including milk proteins, surimi, soybeans, eggs, meat, and some vegetable proteins (Zayas 1997). These changes in proteins may be scrutinized using a technique known as differential scanning calorimetry (DSC). Gelation of whey protein solutions was observed by heating, concentrating, and reheating the protein solutions (Figure 4.1).

This cross-linking of protein molecules generates a 3D solid network providing the desirable texture to the foods (Dickinson 1997).

The globular dairy protein, β-lactoglobulin, is one of the main constituents stabilized by two disulphide bonds, susceptible to denaturation by heating the milk to around 72 °C under regular conditions. The dimer in β-lactoglobulin may also be destabilized by lowering the pH to 3.5 or by raising the temperature to more 40 °C (Dalgleish and Hunt 1995). Heating the globular proteins prior to other treatments may cause precipitation more readily by changing their protein structure, which includes changes in viscosity, surface hydrophobicity, and absorption of UV radiation. Studies in the literature also reported that soy protein precipitates easily if heated (90 °C) prior to precipitation of the proteins by using chemical treatments like HCl or $CaCl_2$ (Renkema et al. 2000; Ziegler and Foegeding 1990). Diluting the proteins to 20% and reheating up to 95 °C resulted in formation of gels (Aoki 1969). Thus, heating of proteins and the disulphide interchange led to intermolecular cross-linkage, thus stabilizing the soy gels. It has also been observed that intermolecular interactions between protein molecules resulted in a 3D gel network providing structural rigidity (Catsimpoolas and Meyer 1970; Gaonkar and McPherson 2016; Su et al. 2015).

Protein concentration is a critical parameter in the gel formation process. Aggregation and increased viscosity were observed for a protein concentration of more than 8%, while between 8 and 14%, gelation of soy proteins could be attained by heating and cooling of protein dispersions (Catsimpoolas and Meyer 1970). As stated earlier, protein interactions are dependent upon factors like temperature and pH, as does gelation. Highly acidity (pH = 1 or 2) or high alkalinity (pH > 10) can result in poor gelation of the proteins, a functionality desirable for food products such as boiled eggs, yoghourt, frankfurters, flans, crab leg analogues, tofu, pie fillings, etc. (Aguilera and Rademacher 2004).

In summary, PPs in food ingredients result in the formation of new food products with different flavour, structure, and texture, as demanded by today's consumers. Food can be processed to become a more nutritious and better tasting new product and beverage. PPIs can provide a new food ingredient which is unique, so as to provide an even wider field for food processing researchers to help industry to meet and even exceed consumer expectations.

4.3.2 Protein–Polysaccharide Interactions

Protein–polysaccharide interactions (PPIs) have been increasingly studied and appreciated during the last few years. Novel functional food formulations containing protein and starch, such as bakery products, infant foods, desserts, and snack foods, have been studied extensively to understand the interactions affecting functionality, nutritional value, and bioactivity (Colombo et al. 2011; Giuberti et al. 2015; Sarabhai and Prabhasankar 2015; Yang et al. 2013). According to Tolstoguzov (1986) there are three different systems that can result from mixing aqueous solutions of proteins and polysaccharides:

1. A liquid two-phase system (water-in-water emulsion), in which the two macromolecular components occur mainly in different phases. This is because of the poor thermodynamic compatibility of proteins and polysaccharides in aqueous media.
2. A two-phase system in which both the macromolecular components occur in about the same single concentrated phase. This phenomenon, commonly termed

as complex coacervation, is due to the formation of an insoluble electrostatic protein–anionic polysaccharide complex.
3. Homogeneous stable solutions in which the two macromolecular components either exist as soluble complexes or they do not chemically interact with each other.

The protein–polysaccharide complexation is chiefly driven by the various non-covalent interactions like electrostatic, H-bonding, hydrophobic, and steric interactions (Kruif and Tuinier 2001). Electrostatic attraction is the main driving force for protein–polysaccharide complexation; however, hydrogen bonding and hydrophobic interactions also a play secondary role in the stability of protein–polysaccharide aggregates (McClements 2006). This complexation, resulting from various non-covalent bonds, is influenced by various factors, mainly temperature and pH. Pea protein and gum acacia complexes stabilize at low temperatures due to an increase in hydrogen bonding and they destabilize at high temperatures (Liu et al. 2009). Ghosh and Bandyopadhyay (2012) observed that the concentration of polysaccharide and pH have a major influence on the solubility of protein–polysaccharide aggregates. Polysaccharide interaction with protein influences its properties like emulsification, gelation, foaming, surface activity, etc. Their interaction (aggregation and gelation) contributes textural and structural properties to food products (Benichou et al. 2002). PPIs are influenced by processing conditions such as pH, ionic strength and type, biopolymer ratios, etc. (Figure 4.2). These interactions are initiated, if desirable, during

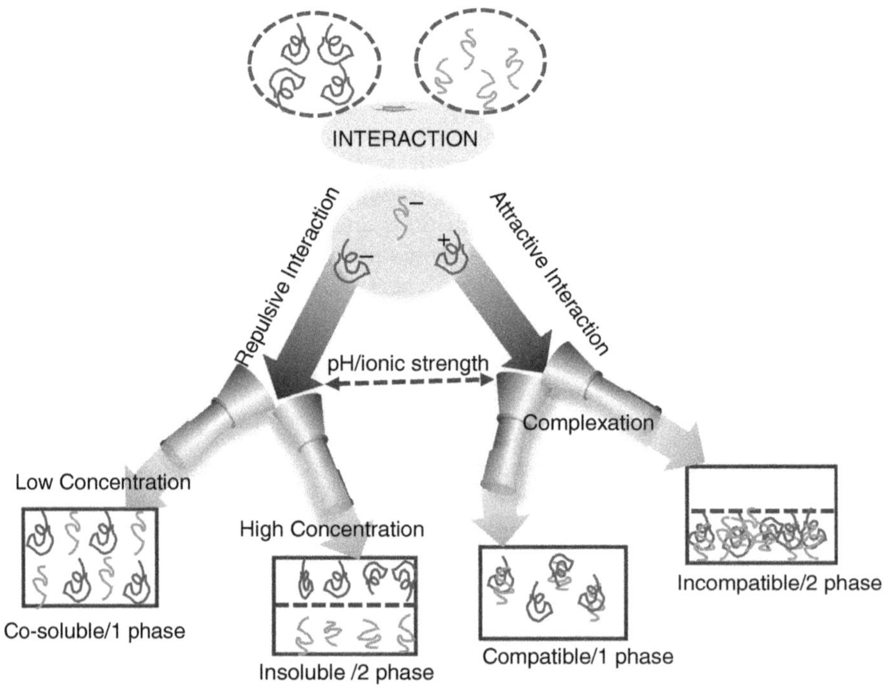

Figure 4.2 Illustration of a protein–polysaccharide interaction through repulsive and attractive interaction for the formation of co-soluble/insoluble phases and compatible/incompatible phases, respectively.

processing for the stabilization of structures formed in different food systems like gelation, thickening, hydration, and surface/interface stabilizing properties (Cgde and Tuinier 2001).

These interactions may be attractive/repulsive phase separation through thermodynamic compatible/incompatible processes (Le et al. 2016). A repulsive interaction (same charged groups) at a low concentration may form a stable single-phase solution (co-soluble) (Khalesi et al. 2016), and at a higher concentration may form an unstable two-phase solution (one rich in polysaccharide and other in protein) (Tolstoguzov 1991). In a gel network formation, gelling molecules may trap the segregation phase to form a two-phase co-gel, having a dispersed and continuous phase. Higher gel strength in a phase-separated network (between biopolymers) was reported by Spotti et al. (2012).

In contrast, attractive interaction (oppositely charged groups) occurs between biopolymers due to their net attraction (pKa polysaccharide $<$pH$<$pI protein) and form an associative phase separation or complex coacervation (Tolstoguzov 1991). Laneuville et al. (2006) concluded from their study that a coupled gel can be formed between native β-lactoglobulin and xanthan gum (oppositely-charged biopolymers) due their interactive behaviour. Similarly, Le et al. (2016) reported that under associative interactions, coupled networks are formed by the electrostatic attraction between proteins and polysaccharides. The biopolymer interactions are also used in the preparation of foams/emulsions (making them self-stable) in liquid foods such as soft drinks and in semi-solid foods such as dressings, sauces, ketchup, ice-cream, etc. These interactions (depending on the concentrations of biopolymers) also contribute to tailoring the composition of foods and can be used for encapsulating bioactive compounds or fat mimetics, etc. (Dickinson 2009; Wagoner et al. 2016).

As compared to any ingredient interaction, the most popular and widely studied interaction between protein–polysaccharide is the Maillard's reaction. Martins et al. (2000) had put together various pathways contributing to Maillard reaction products (MRPs) in food products (Figure 4.3). The aldehyde or ketone group of the reducing sugar and the amino group of the protein react together to form a Schiff's base, leading to the formation of intermediate reaction products, which eventually lead to the formation of covalent aggregates known as MRPs. MRPs are formed through various pathways, of which Hydroxymethylfurfural (HMF) formation is the usual route, leading to the formation of browning compounds desirable in baked products.

Other reaction pathways (pH$>$7) involve sugar dehydration and fragmentation, amino acid degradation, and finally polymerization to form melanoidins. Reductones, fission products (acetol, pyruvaldehyde, and diacetyl), aldehydes, aldols and N-free polymers, and aldimines are the intermediate products in the series of reactions. These intermediate products can react with amino acids to form low molecular weight compounds, leading to the production of high molecular weight MRPs by polymerization (Martins et al. 2000; Pathania et al. 2019).

4.3.3 Protein–Lipid Interactions

Lipids can be present at different levels of interaction with the other components in the complex food matrix. Lipids can be embedded within hydrogel or protein structures in solid matrices, like in cheese, meat, fish, or nuts. However, in milk (liquid matrices), lipids are less bounded to the structure, which makes them easily accessible to enzyme action (Calvo-Lerma et al. 2018). These hydrophobic or amphiphilic

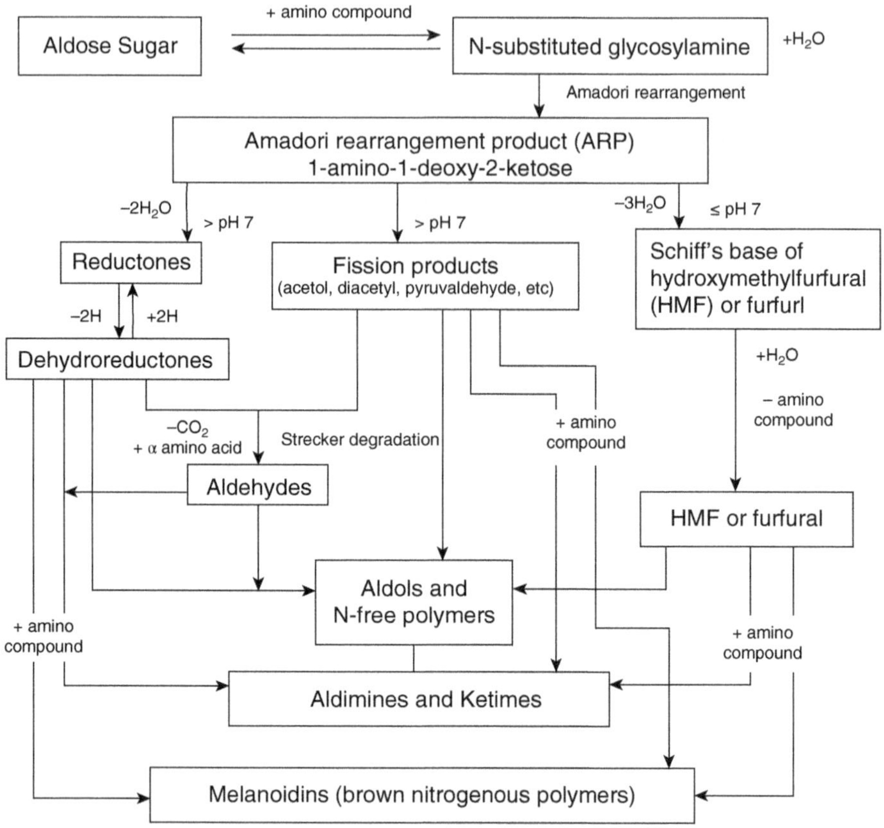

Figure 4.3 Maillard reaction pathways. *Source*: Adapted from Hodge (1953), Martins et al. (2000).

molecules contribute either positively or negatively to the texture and flavour of foods (e.g. bread, biscuits, cakes) and beverages (e.g. beer) (Marion et al. 2003). In a study to establish the influence of lipid organization within the food matrix and interactions with non-lipid components, Calvo-Lerma et al. (2018) classified 52 different types of foods into 3 groups:

1. Foods with complex solid structure:
   a. Lipids inclusion in a protein matrix (e.g., cheese)
   b. Lipids inclusion in carbohydrate and protein matrix (e.g., cookies).
2. Foods with a continuous aqueous phase:
   a. Lipids in foods with a continuous aqueous phase such as oil in water emulsion (e.g., milk)
   b. Intracellular lipid droplets and membrane structures (tissues) (e.g., meat, egg yolk, vegetables).
3. Foods with a continuous lipid phase:
   a. Free fat (e.g., oil, lard), foods with particles dispersed in solid fat (e.g., chocolate)
   b. Water in oil emulsions (e.g., butter or margarine).

The structural linkages between lipids and proteins in naturally-occurring protein–lipid complexes are such that they cannot be readily dissociated by simple manipulations of pH, ionic strength, and ultra-centrifugal force fields (Kamat et al. 1978). The types of bonding which are known to be involved in protein–lipid interactions include hydrogen bonding, electrostatic bonding, covalent bonding, hydrophobic bonding, and van der Waals' bonding; more than one type of bonding may be involved in the interactions in the same protein–lipid complex (Brunner 1962). Electrostatic and hydrophobic bonds are thought to be especially important in protein–lipid interactions (Chapman 1969). van der Waals' forces may occur between non-polar lipid chains and non-polar amino acid side chains in proteins. In the presence of water, such interaction will be strengthened by hydrophobic bonding. Electrostatic binding may occur between positively-charged protein residues and ionized fatty acids or negatively-charged phosphate groups of phospholipids. Hydrogen bonding may occur between hydroxyl groups of fatty acids, di- and mono-glycerides, or the head groups of phosphatidylethanolamine and phosphatidylserine with carbonyl groups of proteins (Izzo and Ho 1989). The most probable mechanism involves the interaction between dipole bonds or electrostatic forces in polar molecules. Protein–lipid interactions can be induced in food systems during processing, for example, high shear mixing to make emulsions (ice cream, cream), heating and external factors such as various salt concentrations (as shown in Figure 4.4), pH of the solution, etc. Induced interactions are the result of covalent and multiple hydrogen bonds which make these complexes more stable than naturally occurring protein–lipid complexes. These interactions are responsible for the functional properties of proteins, which correlate with characteristics of processed foods like water-holding capacity, emulsification, gelation, foaming, and solubility (Kamat et al. 1978).

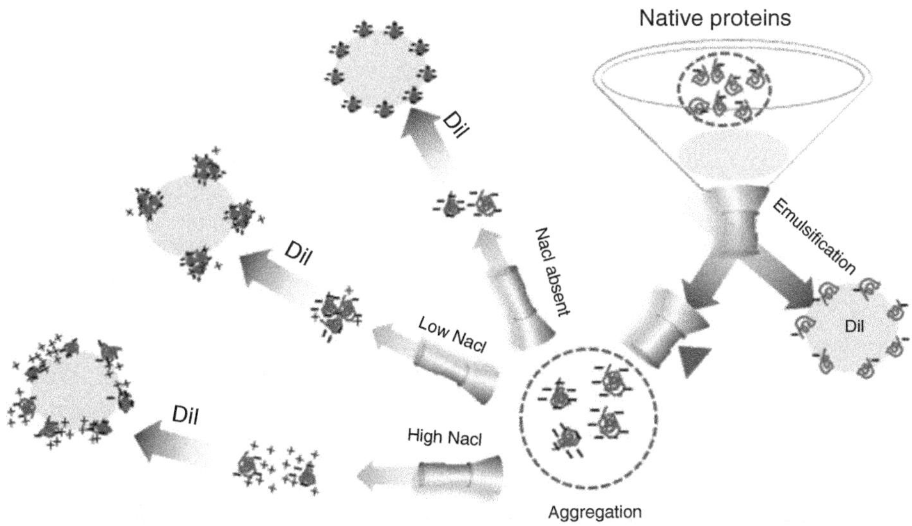

Figure 4.4 Illustration of protein–lipid interaction. Severe heat treatment to protein solutions induce denaturation, aggregation, and gelation/excessive thickening. Interactions of aggregates with lipid under different environment conditions, left side of figure. *Source*: Adapted from Xu et al. (2015).

Figure 4.4 compares the effect of salt concentration on the aggregation behaviour of protein particles. In aqueous suspensions, protein particles exhibit net negative charges on their surface, leading to electrostatic repulsion interactions between the prevailing particles, which can hold the particles apart and make them kinetically stable in aqueous suspensions. At lower salt concentration, the positively-charged sodium ions (Na^+) may interact with the negatively-charged protein particles by means of electrostatic screening, thereby decreasing electrostatic repulsions and resulting in a higher tendency for the formation of particle aggregates (the 'salting-out' effect). At higher concentrations, sodium ions are adsorbed on the protein particles and hence the number of charges on the particle surface may increase again, which results in enhanced electrostatic repulsive forces between particles. The enhancement of electrostatic repulsive interactions thereby increases the apparent dispersibility of the particles (Xu et al. 2015).

The protein–lipid interactions, in milk and milk products, have been classified into two categories. The first category constitutes those interactions in which the lipid molecules are bound to the proteins through distinct reaction sites, while the second category constitutes those interactions which are characterized by the combination of proteins with lipid molecules in combination with other lipid molecules (Gurd 1960). β-lactoglobulin is observed to bind with dipalmitoylphosphatidylcholine through lysyl, thus interacting with the negatively-charged lipid (Brownlow et al. 1997; Brown et al. 1988; Cornell and Patterson 1989; Sawyer 2003). β-lactoglobulin also has the capacity to bind with other lipids such as fatty acids and hydrophobic ligands like retinol, alkanones, and protoporphyrin (Gaonkar and McPherson 2016). Similarly, in infant formula, protein–lipid interactions manifest by addition of κ-casein or β-lactoglobulin to skimmed milk (Gliguem and Birlouez-Aragon 2005; Rowley and Richardson 1985). Upon ultracentrifugation, three different fractions were formed, the first with a lipid phase and some free protein, the second with free casein micelles and the third, a fluid phase. The protein (κ-casein or β-lactoglobulin) was diversely distributed and the content was widely varied.

It has been observed that variations in processing and changes in recipe induce protein–lipid interactions. Addition of emulsifying salts during cheese processing enhanced the protein–lipid interactions.

In the bakery industry, lipid–protein interactions are vital to give texture, and enhance aroma and taste, to the bakery products. There are two kinds of lipids present: lipids already present in the cereal being used (mostly wheat), i.e. endogenous lipids, and external sources of lipids, i.e. exogenous lipids. Lipid content may account for only 1–2% (as in cereals); however, their impact on the processing and quality of cereal-based products is quite significant. Endogenous lipids, especially monoglycerides, are mainly associated with starch granules/amylose and thus may prevent retrogradation and staling by controlling the starch swelling and leaching of amylose into solution (Hoseney 1994). Upon addition of water, polar lipids such as phospholipids form lamellar liquid crystals in the water and are effective in improving the quality of bread (Le Meste et al. 2006). Moreover, lipid content, especially polar lipids present in cereals, tend to act as a stabilizing agent for gas bubbles prior to starch gelatinization during the bread-making process (Gallagher 2009; Gan et al. 1995).

High-temperature short-time extrusion cooking involves high temperatures and screw speeds, which impacts the protein–lipid interactions during the process, resulting in increased proteolysis (Blond et al. 1988; Kilara 2006; Sun-Waterhouse et al. 2014). Similarly, corn snacks prepared with zein and corn oil using extrusion at high temperatures (i.e. 120 and 165 °C), were studied for protein–lipid interactions. Moisture content during the processing affected the extraction rate of lipids and high temperature–low moisture content led to increased lipid–protein interactions, while a high moisture–high temperature combination during extrusion resulted in lower lipid–protein interactions.

4.3.4 Protein–Water Interactions

Protein–water interactions have a direct impact on the protein's functionality and its interaction with other components in the food matrices (Kinsella 1984). As water is one of the most abundant constituents present in many foods such as fruits, vegetables, fish, eggs, dairy foods, and meat among others, its presence affects the molecular and conformational organization of proteins (Le Meste et al. 2006). Many functional properties of macro components, especially proteins in foods, are associated with their interaction with water such as solubility, water absorption, viscosity, gelation, foaming, emulsification, etc. Solubility is an important property of proteins in beverages. The presence of ions, type of ions, and their concentration are important factors affecting hydrophilicity mechanisms. Thus, protein interaction with water results in functional properties like gelation, emulsification, and foaming, which can be used as functional food additives in food processing (Sikorski et al. 1976).

The amino acid profile of proteins is also important to determine the level and scale of interaction between protein and water molecules and is also. critical for other functional properties like foaming, viscosity, and gelation (Gaonkar and McPherson 2016). pH and temperature are the key parameters affecting protein solubility in water. The isoelectric point of proteins is relative to the minimum solubility, while pH variation above or below the isoelectric point enables protein–water interactions. Similarly, increasing temperature distorts the structural properties of proteins, thereby causing a loss in solubility (Kilara 2006; Nakai 2018; Tang et al. 2009). Solubility is also a precursor to other functional protein properties, i.e. emulsification and foaming. Foaming occurs at the interface of air–water interactions, while emulsions are formed at the oil–water interface. Protein solubility allows it to rearrange its structure at the interface to reduce surface tension and aids in formation of emulsions/foams. Another important characteristic of protein–water interactions is the viscosity of the system which greatly affects the sensory properties of foods. Macromolecules are known to increase viscosity at lower concentrations, depending upon the shape, size, flexibility, and hydration properties of the proteins (Damodaran 1997). Increase in hydrodynamic radius of the protein due to heating and increased concentration in protein solutions tends to increase viscosity exponentially. External processing parameters such as pH, temperature, ionic strength, etc. greatly impact the viscosity of protein solutions. Past literature on whey protein concentrate showed that at similar protein concentrations, the thermal treatment causing denaturation combined with acidic pH increased the viscosity of the whey protein solution by 1000% (Kessler and Beyer 1991; Modler and Emmons 1977; Wang and Guo 2019). Gelation of proteins can be attained by addition of salts, enzymes, or altering pH of the protein solutions. The above agents lead to alteration in conformation and structure of the proteins. allowing protein-–protein and protein–water interactions, thereby leading to gelation.

4.3.5 Polysaccharide–Lipid Interactions

Polysaccharide–lipid interactions result in the formation of particular structures showing specific properties. As starch is the main polysaccharide present in food products, extensive work has been done in studying the complexation of starch with other macromolecules which could impact the processability, texture, structure, and digestibility of the final product. Within the starch molecules, amylose fraction can interact with lipids and form a V-type inclusion complex under suitable conditions. Waxy wheat starch (0–5% amylose) lipid complexes showed no diffraction or dissociation peaks of the inclusion complex in the diffractograms or thermograms, representing that the number of inclusion complexes is too few to be detected by X-ray diffraction and differential scanning calorimeter (Wang et al. 2016). Generally, enzymatic debranching methods are used to improve the starch complexation ability. The Pullulanase debranching method produces more debranched starch chains and favours complex formation (Liu et al. 2019). Combinations of isoamylase debranching and β-amylase pre-treatment methods are also used to facilitate the V-type inclusion complex formation for both potato and maize starches (Arijaje and Wang 2016).

Tang and Copeland (2007) reported that the amylose–lipid complex forms two types of structures, structure I and structure II. The structure type-I is reported to melt at temperatures 10–30 °C lower than the melting temperature of structure type-II. Furthermore, the structure type-II may be detected by X-ray crystallography, which is unable to detect the structure type-I. This suggests that the structure type-I is probably in an amorphous state. It is found that the amylose has an ability to form structures of both type I and II, while amylopectin may bond to a lipid only through an individual chain (Parada and Santos 2016).

Many factors are reported to affect the interaction and formation of the starch–lipid complex. The kind of starch–lipid structure formed depends on the type of lipid and the conditions under which it is formed. Lipids possessing the charged head groups have a lower tendency to form type-II structures. With other lipids, both structure types have been reported, based on formation of the complexes. It is also observed that if the complex formation occurs rapidly and at low temperatures (between 70 and 80 °C), the complexes formed of type-I are dominant, while if the complex formation occurs at higher temperatures, then the most ordered complexes of type-II are favoured (Eliasson and Wahlgren 2004).

Interaction of amylose and lipids and their complexes have a profound effect on many properties and functionalities (e.g. reduction of starch solubility, swelling capacity, increased gelatinization temperature, etc.) of different food systems. Presence of lipid which lead to interaction of starch and lipid may also prevent gelatinization by inhibiting hydration of amylopectin chains, and retard retrogradation, which affecting digestibility (Henry 2009).

4.3.6 Polysaccharide–Protein–Lipid Interactions

Complexation between starch amylose and lipids has been documented in the literature and results in V-type crystallites with relatively high melting temperatures (Mikus et al. 1946, Zhang and Hamaker, 2004). The interactions between starch, proteins, and lipids during food processing and their effects on the quality and nutritional value of final food products are of great importance. Zhang and Hamaker (2004) studied the changes in heat flow and pasting properties of the sorghum starch-free fatty acid (FFA) complexation in the presence of whey protein. They observed that the addition of whey protein decreased the endothermic enthalpy of the starch-FFA complexation, confirming its interference in the complex formation process. This was only possible if some FFA molecules were binding to the protein component, leading to formation of the three-way complex during the Rapid Visco Analyser (RVA) pasting process. The complex formation is believed to be initiated by whey protein because the negatively-charged FFAs and the positively-charged basic groups of the denatured protein interact through the electrostatic force of attraction and ultimately would bring the components together (Zhang et al. 2003, 2010; Zhang and Hamaker 2004).

The study of Zhang et al. (2010) concluded that the FFAs may act as a bond between amylose and thermodynamically incompatible protein molecules, whose functional carboxyl group plays a vital role in the formation of a stable ternary complex. Their study proposes that the electrostatic interactions amongst negatively-charged carboxyl groups of FFAs and the poly-ionic protein are the fundamental basis for the self-assembly of the complex.

In a model system, the presence of serum protein diminishes starch-FFA complex formation. The formation of a ternary complex between starch, FFAs, and proteins is related to these changes (Zhang and Hamaker 2004). The presence of salt has a significant effect on the properties of the complex formed by the interaction of starch, protein, and FFAs (Zhang et al. 2015). The type of fatty acids and the sequence of addition of the components also affect the formation of the complex by interaction of starch, protein, and lipids. The starch–protein–lipid complex has a greater degree of ordered structures than a starch–lipid complex. The starch–protein–lipid complex, however, has lower enzymatic digestibility. The starch–protein–lipid complex can be utilized in food processing as a carrier material for the encapsulation of hydrophobic bioactive substances, including food functional ingredients (Wang et al. 2019).

4.4 Role of Macromolecular Interactions in Various Food Products

4.4.1 Cereal Products

The understanding of starch–protein interactions is important in producing quality cereal-based food products. It is also vital in gaining knowledge of the behaviour of cereal proteins in the presence of starch and lipids in products like bread and cakes. Starch granules become dispersed in a continuous gluten matrix in a wheat flour dough system. There is great opportunity for starch–gluten interactions that may be of importance for baking performance.

Protein–lipid interactions take place during the bread-making process, and gluten–lipid interactions are important for achieving the appropriate and desirable texture in baked products (Chung 1986). It was only with the understanding of the macromolecular interactions that a high-quality, nutritional gluten-free bread with acceptable overall acceptability was formulated by food processors.

In dough and bread, the gluten network depends on the existence of protein–lipid complexes, which is essential for the development of gas impermeability for good gas retention and adequate loaf volume, and is also necessary for satisfactory bread structure

(Chobert and Haertle 1997; Finney 1971). The interaction of glycolipids mainly occurs with gluten rather than with the soluble wheat flour proteins during dough preparation. This phenomenon is important, as gluten is the skeleton or framework of wheat flour dough; it is also responsible for the retention of gas which is required in the preparation of low density, yeast-leavened products. It is observed that much of the interaction of macromolecules occurs with the starch in the baked bread. This observation is also of significance because the retention of freshness is largely governed by the starch (Pomeranz 1973).

It was suggested that the glutenin fraction has lipid binding capacity (Lasztity et al. 1996). The developed gluten formed by wetting of flour and by mechanical input during dough-making contains the glutenin-polar–lipid-gliadin complex (Pomeranz 1980; Pomeranz and Chung 1983). Lasztity et al. (1996) devised two hypothetical possibilities indicating the position of lipid molecules in the gluten complex. For both variations, aggregation of protein molecules and gluten complex network formation is influenced by the lipid molecules. Furthermore, it was also observed that the free polar lipids, particularly glycolipids, are bound to the gliadin and glutenin proteins by hydrophilic and hydrophobic bonds, respectively (Hoseney et al. 1970). The lipid is apparently bound to both the protein groups at the same time in non-fractionated gluten. The interaction and associative binding of polar lipids to gliadin, and glutenin at the same time, may result in structural contribution towards gas-retaining complexes in gluten. Wehrli and Pomeranz (1970) supported the model proposed by Hoseney et al. (1970), by studying the interaction of glycolipids with the macromolecules of wheat flour. The study investigated the complexes between galactolipids and raw starch, gelatinized starch, gliadin, and glutenin with the help of infrared and nuclear magnetic resonance (NMR) spectroscopy. Infrared spectroscopy revealed the presence of hydrogen bonds between glycolipids and gelatinized starch or gluten components, and van der Waals' bonds between glycolipids and gluten components. Wehrli (1969) explained the mechanism of interaction between lipids and proteins and starch, and methods to study these interactions. It is well known that the interaction between starch and gluten has a major role to play in governing the staling process. It is observed that the hydrogen bonding between the hydroxyl group of gelatinized starch granules and the amide groups of glutenin, gliadin, and possibly, albumin in the gluten network controls the staling process (Erlander and Erlander 1969). The freshness of bread restores upon reheating as the cross-links (hydrogen bonds) between starch polymer and gluten molecules are easily broken (Soltanizadeh et al. 2014). It is also found that the reduced glycemic response and decreased rate of digestion is attributed to the presence of the starch–protein interaction in white wheat flour. The interaction of starch with proteins in foods is important in the context of the effects of foods on the gastrointestinal tract and specifically where absorptive efficiency is compromised as in case of celiac disease (Jenkins et al. 1987).

Stickiness in cooked rice was found to be the result of starch–protein interactions. Amylopectin and/or amylose are leached out of the grain due to the disintegration of the starch molecules during the cooking process and they tend to bind with the storage protein, oryzenin. Chrastil (1990) established that this binding is directly related to the stickiness of cooked rice. Storage temperature of the rice grains was observed to have a negative effect on this interaction, as rice grains stored at higher temperatures exhibited decreased stickiness in the cooked rice (Chrastil 1990).

Madeka and Kokini (1992) observed that mixtures of waxy maize-zein or waxy maize-gliadin depict a synergistic increase in viscosity at 20–40% moisture content when heated above a threshold temperature of 100–150 °C. The protein–starch interactions were held responsible for this increase in viscosity. These interactions, however, could be balanced by the disruption of hydrogen and hydrophobic bonds and the breakdown of starch, protein denaturation, and loss of functionality when the temperature is further increased.

Extrusion of cereal flours is an established food processing technique to develop a range of food structures and textures. Extensive research has been done to solve the riddle of transformational changes taking place during extrusion process. The protein–lipid interactions occurring during this process was documented by Izzo and Ho (1989). The results of their study on zein and corn oil indicated that the diglycerides were formed during the extrusion process and these diglycerides then interacted with the proteins present in the feed. They suggested that the mechanism of these protein–lipid interactions did not involve degradation and covalent incorporation of lipids into proteins and that the protein–lipid interaction did not increase as the barrel temperature increased. Liu and Hsieh (2008) studied the PPIs occurring during the high moisture extrusion for fibrous meat analogues. Results of this study indicated that in comparison to non-covalent bonds, the disulphide bonding is chiefly responsible for not only holding the rigid structure of extrudates, but also in the formation of fibrous texture.

4.4.2 Meat Products

In comminuted meat systems, the biophysical properties of the proteinaceous membrane surrounding the lipid particles determine the role of protein–lipid interactions (Jones 1984). This proteinaceous membrane at the oil–water interface has a major role to play in deciding the stability during thermal processing (Jones and Mandigo 1982; Li-Chan et al. 1984). However, it is not comprehensively understood at what level this interfacial protein membrane contributes to the fat binding or holding ability of the comminuted meat system and its subsequent fatting-out inhibition effect. The monolayer of the most polar proteins occurs first, followed by random PPIs. These interactions are usually influenced by various factors including hydrophobic forces, covalent bonding energies, hydrogen bonding, and weak attractive forces of the solubilized proteins (Jones 1984). The structural stability of the interaction between proteins and lipids in comminuted meat systems is believed to be affected by three different phenomena: (i) the occurrence of an interfacial protein membrane or film which surrounds lipid particles; (ii) the gel strength of the batter matrix; and (iii) the physical characteristics of the fat (Jones 1984).

The gelation properties of mix obtained by addition of potato starch to high water content (84 and 86%) fish protein were examined by Hui-Huang and Yu (2008). The study reported that the storage modulus increased on addition of an incremental level of starch. The increase in storage modulus is attributed to the absorption of water by the granules from added starch in the fish protein, which also leads to making the fish proteins more rigid. The storage modulus for the starch added fish proteins did not decrease during the gel resolution stage upon thermal scanning. It was also observed that the breaking force, deformation, and gel strength of fish protein increased on addition of starch. The interaction between starch and proteins in the food system is believed to be responsible for the increased gel strength (Jamilah et al. 2009). The study found that at least 10% starch addition is required for the gel strength of 86% water containing fish protein to become detectable. The granules from added starch were found to act as a filler reinforcement agent and are also thought to be water absorbent in fish protein (Hui-Huang and Yu 2008). This increased gel strength is attributed to an interaction between starch and protein in food systems.

During storage of meat products, myoglobin can interact with oxidized lipids and results in the formation of protein–lipid complexes. In meat batter systems, the myosin has been identified as the most significant protein for encapsulation of fat particles (Jones 1984). Egg yolk contains Low density lipoproteins, i.e. LDL (11% proteins, 61.5% fats, and 27.5% phospholipids); and High density lipoproteins, i.e. HDL (78% proteins, 10% fats, and 12% phospholipids).

4.4.3 Dairy Products

The lipoprotein complex of the milk fat globule membrane is the classic representation of principal protein–lipid interaction in fresh milk. Protein–lipid interactions are also formed during food processing operations such as homogenization and mixing of milk and milk products. In ice cream, lipids interact with caseins to form protein–lipid complexes, while in butter-making, fat globules concentrate at an air–water interface and interact with proteins to form complexes at the interface (Alzagtat 2002; Karel et al. 1975).

Acidified milk drinks are prepared by adding fermented milk and direct acidification with fruit juices and/or acids. The pH of these milk drinks ranges from 3.4 to 4.6 and at such lower pH values, casein micelles associate and precipitate. However, casein micelles adsorb carrageenan or pectin (Maroziene and de Kruif 2000; Syrbe et al. 1998) and prevent the association of casein micelles. The addition of charged polysaccharides to the protein solutions results in formation of complexes, depending on the pH and the ionic strength. Acidified milk drinks are stabilized by the addition of pectin. Casein micelles adsorbs polysaccharide chains of high methoxyl pectin on its surface via the charged blocks of the pectin chains, while the uncharged blocks of high methoxyl pectin extend into solution, contributing to stabilization via steric repulsion (Pereyra et al. 1997; Tuinier et al. 2002).

The effect of the exopolysaccharide–protein interaction on quality of yoghurt is studied by Van de Velde et al. (2015). The protein–polysaccharide interactions arise from two distinct origins in yoghurt: (i) the interactions resulting from the polysaccharides added before the fermentation process; and (ii) the interactions resulting from polysaccharides produced by lactic acid bacteria during the fermentation process. In the first type, mainly starches and gums are used to improve the texture and mouthfeel of low-fat yoghurt or for reducing the protein content in yoghurt. In the second case, the exopolysaccharide–protein interactions may be of an aggregative or segregative type, depending on the charge density of the polymer. The neutral exopolysaccharides tend to phase out separately from the protein phase, whereas the charged exopolysaccharides tend to associate or aggregate with the protein phase in yoghurt.

Proteins and free amino acids with riboflavin in milk powder are good examples in which the macromolecular interactions occur and result in both protein aggregation and significant loss of riboflavin. These interactions result in a product with degraded quality and lower nutritional value (Dalsgaard et al. 2007).

4.5 Conclusions

The macromolecules, carbohydrate, protein, and lipids are usually found in almost all foods. These macromolecules interact within themselves (intramolecular interactions) or with other macromolecules (intermolecular interactions). These interactions result in formation of complexes with varying physicochemical and functional properties. The macromolecular interactions are attributed to electrostatic, covalent bonding, van der Waals' force of attraction, adsorption, and others. These highly-structured composite complexes have the ability to exhibit versatile and better functional properties like hydration, solubility, rheological, interfacial, and adsorption properties, than proteins lipids and polysaccharides alone. These complexes can be used practically in foods, especially in cereal, meat, and dairy products, and as potential food ingredients such as fat replacers, texturising agents, and stabilizers of dispersed systems. The specific properties of these complexes resulting from macromolecular interactions can be employed for various purposes including the stabilization of oil-in-water emulsion, the formation of microcapsules, purification, and concentration of protein fraction. However, a significant amount of research work is required to understand and optimize the functionality of these complexes as ingredients for food applications in a wide range of processing conditions.

References

Aguilera, J.M. and Rademacher, B. (2004). Protein gels. In: *Proteins in Food Processing* (ed. R.Y. Yada), 468–482. Woodhead Publishing Series in Food Science, Technology and Nutrition.

Alzagtat, A.A. (2002). Effects of protein-lipid interactions on physiochemical and functional properties of food proteins. PhD thesis. McGill University, Montreal, Quebec

Aoki, H. (1969). Studies on the gelation of soybean protein. On the effects of heat denaturation. *Nippon Nogei Kagaku Kaishi* 43: 448–456.

Appelqvist, I.A.M. and Debet, M.R.M. (1997). Starch-biopolymer interactions – a review. *Food Reviews International* 13: 163–224.

Arijaje, E.O. and Wang, Y.J. (2016). Effects of enzymatic modifications and botanical source on starch–stearic acid complex formation. *Starch-Stärke* 68 (7-8): 700–708.

Benichou, A., Aserin, A., and Garti, N. (2002). Protein-polysaccharide interactions for stabilization of food emulsions. *Journal of Dispersion Science and Technology* 23 (1–3): 93–123.

Blond, G., Haury, E., and Lorient, D. (1988). Effects of batch- and extrusion-cooking on lipids-proteins interactions of processed cheese. *Sciences des Aliments (France)* 8 (3): 325–340.

Brown, E.M., Pfeffer, P.E., Kumosinski, T.F., and Greenberg, R. (1988). Accessibility and mobility of lysine residues in beta-lactoglobulin. *Biochemistry* 27 (15): 5601–5610.

Brownlow, S., Cabral, J.H.M., Cooper, R. et al. (1997). Bovine ⊠-lactoglobulin at 1.8 Å resolution – still an enigmatic lipocalin. *Structure* 5 (4): 481–495.

Brunner, J.R. (1962). Protein-lipid interactions and their relation to the physical chemical stability of concentrated milk: a review. *Journal of Dairy Science* 45: 943–951.

Calvo-Lerma, J., Fornés-Ferrer, V., Heredia, A., and Andrés, A. (2018). *In vitro* digestion of lipids in real foods: influence of lipid organization within the food matrix and interactions with nonlipid components. *Journal of Food Science* 83 (10): 2629–2637.

Catsimpoolas, N. and Meyer, E. (1970). Gelation phenomena of soybean globulins. I: Protein-protein interactions. *Cereal Chemistry* 47: 559–570.

Cgde, K. and Tuinier, R. (2001). Polysaccharide protein interactions. *Food Hydrocolloids* 15 (4–6): 555–563.

Chapman, D. (1969). Physical studies of lipid-lipid and lipid–protein interactions. *Lipids* 4: 251–260.

Chobert, J.M. and Haertle, T. (1997). Protein–lipid and protein–flavor interactions. In: *Food Proteins and Their Applications* (eds. S. Damodaran and A. Paraf), 143–170. New York: Marcel Dekker.

Chrastil, J. (1990). Protein-starch interactions in rice grains. Influence of storage on oryzenin and starch. *Journal of Agricultural and Food Chemistry* 38: 1804–1809.

Chung, O.K. (1986). Protein-lipid interactions in wheat flour, dough, gluten, and protein fraction. *Cereal Foods World* 31: 242–256.

Colombo, A., León, A.E., and Ribotta, P.D. (2011). Rheological and calorimetric properties of corn-, wheat-, and cassava-starches and soybean protein concentrate composites. *Starch-Stärke* 63: 83–95.

Cornell, D.G. and Patterson, D.L. (1989). Interaction of phospholipids in monolayers with beta-lactoglobulin adsorbed from solution. *Journal of Agricultural and Food Chemistry* 37 (6): 1455–1459.

Dalgleish, D. and Hunt, J. (1995). Protein-protein interactions in food materials. In: *Ingredients Interactions: Effects on Food Quality* (ed. A.G. Gaonkar), 199–233. New York: Marcel Dekker.

Dalsgaard, T.K., Otzan, D., Nielsen, J.H., and Larsen, L.B. (2007). Changes in structures of milk proteins upon photo-oxidation. *Journal of Agricultural and Food Chemistry* 55: 10968–10976.

Damodaran, S. (1997). Protein-stabilized foams and emulsions. In: *Food Proteins and Their Applications* (eds. S. Damodaran and A. Paraf), 57–110. New York: Marcel Dekker.

Dickinson, E. (1997). Enzymic crosslinking as a tool for food colloid rheology control and interfacial stabilization. *Trends in Food Science and Technology* 8: 334–339.

Dickinson, E. (2009). Hydrocolloids as emulsifiers and emulsion stabilizers. *Food Hydrocolloids* 23 (6): 1473–1482.

Eliasson, A.C. and Wahlgren, M. (2004). Starch–lipid interactions and their relevance in food products. In: *Starch in Food-Structure, Function and Application* (ed. W. Bergthaller), 441–460. Cambridge, UK: Woodhead.

Erlander, S.R. and Erlander, L.G. (1969). Explanation of ionic sequences in various phenomena X. protein-carbohydrate interactions and the mechanism for the staling of bread. *Starch-Starke* 21: 305–315.

Feeney, R.E. and Whitaker, J.R. (1988). Importance of cross-linking reactions in proteins. *Advances in Cereal Science and Technology* 9: 21–43.

Finney, K.F. (1971). Fractionating· and reconstituting techniques to relate functional (bread makîng) to biochemical properties of wheat flour components. *Cereal Science Today* 16: 342–356.

Gallagher, E. (2009). *Gluten-Free Food Science and Technology*. Wiley.

Gan, Z., Ellis, P., and Schofield, J. (1995). Gas cell stabilisation and gas retention in wheat bread dough. *Journal of Cereal Science* 21 (3): 215–230.

Gaonkar, A.G. and McPherson, A. (2016). *Ingredient Interactions: Effects on Food Quality*. CRC press.

Gerrard, J.A. (2002). Protein-protein crosslinking in food: methods, consequences, applications. *Trends in Food Science and Technology* 13: 91–399.

Ghosh, A.K. and Bandyopadhyay, P. (2012). Polysaccharide–protein interactions and their relevance in food colloids. In: *The Complex World of Polysaccharides* (ed. D.N. Karunaratne), 395–408. Rijeka: InTech.

Giuberti, G., Gallo, A., Cerioli, C. et al. (2015). Cooking quality and starch digestibility of gluten free pasta using new bean flour. *Food Chemistry* 175: 43–49.

Gliguem, H. and Birlouez-Aragon, I. (2005). Effects of sterilization, packaging, and storage on vitamin C degradation, protein denaturation, and glycation in fortified milks. *Journal of Dairy Science* 88 (3): 891–899.

Gurd, F.R.N. (1960). Association of lipids with proteins. In: *Lipid Chemistry* (ed. D.J. Hanahan), 208–259. New York-London: Wiley.

Henry, J. (2009). Processing, manufacturing, uses and labeling of fats in the food supply. *Annals of Nutrition and Metabolism* 55: 273–300.

Hodge, J.E. (1953). Dehydrated foods, chemistry of browning reactions in model systems. *Journal of Agricultural and Food Chemistry* 1 (15): 928–943.

Hoseney, R.C. (1994). *Principles of Cereal Science and Technology*. American Association of Cereal Chemists (AACC).

Hoseney, R.C., Finney, K.F., and Pomeranz, Y. (1970). Functional (bread making) and biochemical properties of wheat flour components VI. Gliadin-lipid-glutenin interaction in wheat gluten. *Cereal Chemistry* 47: 135–140.

Hui-Huang, C. and Yu, C.H. (2008). Rheological properties of HPMC enhanced surimi analyzed by small- and large-strain tests. II: effect of water content and ingredients. *Journal of Food Hydrocolloids* 22: 313–322.

Izzo, M.T. and Ho, C. (1989). Protein-lipid interaction during single screw extrusion of zein and corn oil. *Cereal Chemistry* 66: 47–51.

Jamilah, B., Mohamed, A., Abbas, K.A. et al. (2009). Protein-starch interaction and their effect on thermal and rheological characteristics of a food system: a review. *Journal of Food, Agriculture and Environment* 7: 169–174.

Jenkins, D., Thorne, M.J., Wolever, T. et al. (1987). The effect of starch-protein interaction in wheat on the glycemic response and rate of in vitro digestion. *The American Journal of Clinical Nutrition* 45: 946–951.

Jones, K.W. (1984). Protein-lipid interactions in processed meats. *Reciprocal Meat Conference Proceedings* 37: 52–57.

Jones, K.W. and Mandigo, R.W. (1982). Effects of chopping temperature on the microstructure of meat emulsions. *Journal of Food Science* 47: 1930–1935.

Kamat, V.B., Graham, G.E., and Davis, M.A.F. (1978). Vegetable protein:lipid interaction. *Cereal Chemistry* 55: 295–307.

Karel, M., Schaich, K., and Roy, R.H. (1975). Interaction of peroxidizing methyllinoleate with sorne proteins and aminoacids. *Journal of Agricultural and Food Chemistry* 23: 159–163.

Kessler, H.-G. and Beyer, H.-J. (1991). Thermal denaturation of whey proteins and its effect in dairy technology. *International Journal of Biological Macromolecules* 13 (3): 165–173.

Khalesi, H., Emadzadeh, B., Kadkhodaee, R., and Fang, Y. (2016). Whey protein isolate-Persian gum interaction at neutral pH. *Food Hydrocolloids* 59: 45–49.

Kilara, A. (2006). Interactions of ingredients in food systems: an introduction. *Food Science and Technology-New York-Marcel Dekker* 154: 1.

Kinsella, J.E. (1984). Milk proteins: physiochemical and functional properties. *Critical Reviews in Food Science and Nutrition* 21: 197–262.

Kruif, d.C.G. and Tuinier, R. (2001). Polysaccharide-protein interactions. *Food Hydrocolloids* 15: 555–563.

Laneuville, S.I., Turgeon, S.L., Sanchez, C., and Paquin, P. (2006). Gelation of native beta-lactoglobulin induced by electrostatic attractive interaction with xanthan gum. *Langmuir American Chemical Society Journal of Surface Colloids* 22 (17): 7351.

Lasztity, R., Bekes, F., Orsi, F. et al. (1996). Protein-lipid and protein-carbohydrate interactions in the gluten complex. *Periodica Polytechnica, Chemical Engineering* 40: 29–40.

Le Meste, M., Roudaut, G., Champion, D. et al. (2006). Interaction of water with food components. *Food Science and Technology – New York-Marcel Dekker* 154: 87.

Le, X.T., Rioux, L.E., and Turgeon, S.L. (2016). Formation and functional properties of protein-polysaccharide electrostatic hydrogels in comparison to protein or polysaccharide hydrogels. *Advances in Colloid and Interface Science* 239: 127–135.

Li-Chan, E., Nakai, S., and Wood, D.F. (1984). Hydrophobicity and solubility of meat proteins and their relationship to emulsifying properties. *Journal of Food Science* 49: 345–350.

Liu, K. and Hsieh, F.G. (2008). Protein-protein interactions during high-moisture extrusion for fibrous meat analogues and comparison of protein solubility methods using different solvent systems. *Journal of Agricultural and Food Chemistry* 56: 2681–2687.

Liu, S., Low, N.H., and Nickerson, M.T. (2009). Effect of pH, salt and biopolymer ratio on the formation of pea protein isolate-gum Arabic complexes. *Journal of Agricultural and Food Chemistry* 57: 1521–1526.

Liu, P., Kang, X., Cui, B. et al. (2019). Effects of amylose content and enzymatic debranching on the properties of maize starch-glycerol monolaurate complexes. *Carbohydrate Polymers* 222: 115000.

Lopes da Silva, J.A. and Rao, M.A. (2007). Rheological behaviour of food gels. In: *Rheology of Fluid and Semisolid Foods Principles and Applications* (ed. M.A. Rao), 339–401. New York: Springer.

Madeka, H. and Kokini, J.L. (1992). Effect of addition of zein and gliadin on the theological properties of amylopectin starch with low to intermediate moisture. *Cereal Chemistry* 69: 489–494.

Marion, D., Dubreil, L., and Douliez, J.P. (2003). Functionality of lipids and lipid-protein interactions in cereal-derived food products. *Oléagineux, Corps Gras, Lipides* 10: 47–56.

Maroziene, A. and de Kruif, C.G. (2000). Interaction of pectin and casein micelles. *Food Hydrocolloids* 14: 391–394.

Martins, S.I., Jongen, W.M., and Van Boekel, M.A. (2000). A review of Maillard reaction in food and implications to kinetic modelling. *Trends in Food Science & Technology* 11 (9-10): 364–373.

McClements, D.J. (2006). Non-covalent interactions between proteins and polysaccharides. *Biotechnology Advances* 24: 621–625.

Mikus, F.F., Hixon, R.M., and Rundle, R.E. (1946). The complex of fatty acids with amylose. *Journal of the American Chemical Society* 68: 1115–1123.

Modler, H. and Emmons, D. (1977). Properties of whey protein concentrate prepared by heating under acidic conditions. *Journal of Dairy Science* 60 (2): 177–184.

Nakai, S. (2018). *Hydrophobic Interactions in Food Systems*. CRC Press.

Parada, J. and Santos, J.L. (2016). Interactions between starch, lipids, and proteins in foods: microstructure control for glycemic response modulation. *Critical Reviews in Food Science and Nutrition* 56: 2362–2369.

Pathania, S., Parmar, P., and Tiwari, B.K. (2019). Stability of proteins during processing and storage. In: *Proteins: Sustainable Source, Processing and Applications* (ed. C.M. Galanakis), 295–330. Academic Press.

Pereyra, R., Schmidt, K.A., and Wicker, L. (1997). Interaction and stabilization of acidified casein dispersions with low and high methoxyl pectins. *Journal of Agricultural and Food Chemistry* 45: 3448–3451.

Pomeranz, Y. (1973). Interaction between glycolipid and wheat flour macromolecules in breadmaking. *Advances in Food Research* 20: 153–188.

Pomeranz, Y. (1980). Interaction between lipids and gluten proteins. *Annales Technolgie Agricole* 29: 385.

Pomeranz, Y. and Chung, O.K. (1983). Lipid-proteins interactions: new findings and interpretations. In: *Developments in Food Science, Vol 5.B. Progress in Cereal Chemistry and Technology* (eds. J. Holas and J. Kratochwil), 825–830. Amsterdam: Elsevier.

Renkema, J.M., Lakemond, C.M., De Jongh, H.H. et al. (2000). The effect of pH on heat denaturation and gel forming properties of soy proteins. *Journal of Biotechnology* 79 (3): 223–230.

Rowley, B. and Richardson, T. (1985). Protein-lipid interactions in concentrated infant formula. *Journal of Dairy Science* 68 (12): 3180–3188.

Sarabhai, S. and Prabhasankar, P. (2015). Influence of whey protein concentrate and potato starch on rheological properties and baking performance of Indian water chestnut flour based gluten free cookie dough. *LWT - Food Science and Technology* 63: 1301–1308.

Sawyer, L. (2003). *β-Lactoglobulin Advanced Dairy Chemistry—1 Proteins*, 319–386. Springer.

Sikorski, Z.E., Olley, J., and Kostuch, S. (1976). Protein changes in frozen fish. *CRC Critrical Reviews in Food Science and Nutrition* 8: 97–129.

Soltanizadeh, N., Mirmoghtadaie, L., Nejati, F. et al. (2014). Solid-state protein-carbohydrate interactions and their applications in the food industry. *Comprehensive Reviews in Food Science and Food Safety* 13: 860–870.

Spotti, M.J., Santiago, L.G., Rubiolo, A.C., and Carrara, C.R. (2012). Mechanical and microstructural properties of milk whey protein/espina corona gum mixed gels. *LWT Food Science and Technology* 48 (1): 69–74.

Su, Y., Dong, Y., Niu, F. et al. (2015). Study on the gel properties and secondary structure of soybean protein isolate/egg white composite gels. *European Food Research and Technology* 240 (2): 367–378.

Sun-Waterhouse, D., Zhao, M., and Waterhouse, G.I. (2014). Protein modification during ingredient preparation and food processing: approaches to improve food processability and nutrition. *Food and Bioprocess Technology* 7 (7): 1853–1893.

Syrbe, A., Bauer, W.J., and Klostermeyer, H. (1998). Polymer science concepts in dairy systems-an overview of milk protein and food hydrocolloid interaction. *International Dairy Journal* 8: 179–193.

Tang, M.C. and Copeland, L. (2007). Analysis of complexes between lipids and wheat starch. *Carbohydrate Polymers* 67: 80–85.

Tang, C.-H., Sun, X., and Yin, S.-W. (2009). Physicochemical, functional and structural properties of vicilin-rich protein isolates from three Phaseolus legumes: effect of heat treatment. *Food Hydrocolloids* 23 (7): 1771–1778.

Tolstoguzov, V.B. (1986). Functional properties of food macromolecules. *Applied Sciences*: 385–416.

Tolstoguzov, V.B. (1991). Functional properties of food proteins and role of protein-polysaccharide interaction. *Food Hydrocolloids* 4: 429–468.

Tolstoguzov, V. (1996). Structure-property relationships in foods. In: *Macromolecular Interactions in Food Technology* (eds. N. Parris, A. Kato, L.K. Creamer and J. Pearce), 2–14. Washington DC: American Chemical Society.

Tolstoguzov, V. (2003). Some thermodynamic considerations in food formulation. *Food Hydrocolloids* 17: 1–23.

Tuinier, R., Rolin, C., and Kruif, d.C.G. (2002). Electrosorption of pectin onto casein micelles. *Biomacromolecules* 3: 632–638.

Van de Velde, F., de Hoog, E.H.A., Oosterveld, A., and Tromp, R.H. (2015). Protein-polysaccharide interactions to alter texture. *Annual Review of Food Science and Technology* 6: 371–388.

Wagoner, T., Vardhanabhuti, B., and Foegeding, E.A. (2016). Designing whey protein–polysaccharide particles for colloidal stability. *Annual Review of Food Science and Technology* 7 (1): 93–116.

Wang, G. and Guo, M. (2019). Manufacturing Technologies of Whey Protein Products. In: *Whey Protein Production, Chemistry, Functionality, and Applications* (ed. M. Guo), 13–37. Hoboken, NJ: Wiley.

Wang, S., Wang, J., Yu, J., and Wang, S. (2016). Effect of fatty acids on functional properties of normal wheat and waxy wheat starches: a structural basis. *Food Chemistry* 190: 285–292.

Wang, S., Zheng, M., and Chao, C. (2019). Interactions between starch, proteins and lipids and the formation of ternary complexes with distinct properties. In: *Reference module in food science, Encyclopedia of Food Chemistry* (eds. L. Melton, F. Shahidi and P. Varelis), 487–493. Amsterdam: Elsevier.

Wehrli, H.P. (1969). The synthesis of glycolipids and their role in break making. Ph.D. dissertation. Kansas State University.

Wehrli, H.P. and Pomeranz, Y. (1970). A note on the interaction between glycolipid and wheat flour macromolecules. *Cereal Chemistry* 47: 160–166.

Xu, H.N., Liu, Y., and Zhang, L. (2015). Salting-out and salting-in: competitive effects of salt on the aggregation behavior of soy protein particles and their emulsifying properties. *Soft Matter* 11 (29): 5926–5932.

Yang, N., Liu, Y., Ashton, J. et al. (2013). Phase behaviour and in vitro hydrolysis of wheat starch in mixture with whey protein. *Food Chemistry* 137: 76–82.

Zayas, J.F. (1997). *Functionality of Proteins in Food*. Berlin: Springer Verlag.

Zhang, G. and Hamaker, B.R. (2004). Starch-free fatty acid complexation in the presence of whey protein. *Carbohydrate Polymers* 55: 419–424.

Zhang, G., Maladen, M.D., and Hamaker, B.R. (2003). Detection of a novel three component complex consisting of starch, protein, and free fatty acids. *Journal of Agricultural and Food Chemistry* 51: 2801–2805.

Zhang, G., Maladen, M., Campanella, O.H., and Hamaker, B.R. (2010). Free fatty acids electronically bridge the self-assembly of a three-component nanocomplex consisting of amylose, protein, and free fatty acids. *Journal of Agricultural and Food Chemistry* 58: 9164–9170.

Zhang, G., Bhopatkar, D., Hamaker, B.R., and Campanella, O.H. (2015). Self-assembly of amylose, protein, and lipid as a nanoparticle carrier of hydrophobic small molecules. In: *Nanotechnology and Functional Foods: Effective Delivery of Bioactive Ingredients* (eds. C. Sabliov, H. Chen and R. Yada), 263–271. Hoboken, NJ: Wiley.

Ziegler, G.R. and Foegeding, E.A. (1990). The gelation of proteins. *Advances in Food and Nutrition Research* 34: 203–298.

5

Effect of Ingredient Interactions on Techno-Functional Properties

Hanuman Bobade, Savita Sharma, and Baljit Singh

Department of Food Science and Technology, Punjab Agricultural University, Ludhiana, Punjab, India

Summary

Our food is composed of many ingredients with varying technical and functional properties. These ingredients interact with each other during various stages of food handling, from harvesting to consumption. The complexes formed from the interaction of these ingredients display entirely different properties than the inherent properties of separate ingredients in the food system. The macromolecules, carbohydrates, proteins, and fat, as well as micromolecules, interact with each other resulting in the modification of many properties including water holding, gelation, viscosity, emulsification, and others. Many properties of foods and food products, including microstructure, particle size distribution (PSD), and texture, are affected during food formulation with varying ingredients and their interactions. The interaction of ingredients used in food formulation as well as naturally present in the food significantly influence the emulsification behaviour, water absorption characteristics, solubility, water holding capacity (WHC), viscosity, gelation, foaming and foam stability, and flavour characteristics of the food system.

5.1 Introduction

There has been an ever-increasing demand for nutritive and high-quality food that has an optimum level of calorific value to cater to consumer needs. The spotlight on calorific and nutritional aspects of food has only recently taken shape and traditionally producers were mainly concerned with only qualitative aspects. To address this recent demand, producers and processors alike are focusing on reformulating products with reduced salt, fat and sugar content, as well as tailoring new formulations for the younger, active, and ageing populace. As discussed in the previous chapter, the interactions between

Food Formulation: Novel Ingredients and Processing Techniques, First Edition.
Edited by Shivani Pathania and Brijesh K. Tiwari.

macromolecules affect the functional characteristics of food products like texture, solubility, water holding, and others. Additionally, properties like rheology, microstructure, and other traits are playing critical roles in reformulation of processed food products (Feinle et al. 2002; Lundin et al. 2008; Tieken et al. 2007).

Extensive research has been carried out to understand the interactions between food components to develop low salt and sugar variations, while retaining the sensory attributes of the products. It has been realized and largely agreed by the scientific and processing sector that reducing the amount of key functional ingredients leads to new interaction mechanisms, therefore influencing food textures and shelf life. As food is a complex system, component reduction of one ingredient cannot be replaced by other components in the recipe; for example, in bakery products, sugar reduction levels cannot be adjusted by replacing the sugar with flour or water. This chapter is a collective review of the impact of ingredient interactions on most of the techno-functional properties of food.

5.2 Effect of Food Formulation on Structural Properties

5.2.1 Effect on Microstructure

Food microstructure is the governing property behind the interplay between formulation, processing, and functional attributes like taste, nutrition, texture, etc. (McClements et al. 2009). Structural components such as strands, networks, crystals, droplets, and air cells are spatially arranged to give structural coherence to the food microstructure. These structural components interact and form major structural constituents, i.e. fat, proteins, polysaccharides, etc. of varying size, structural arrangement, and length. Different techniques, such as scanning electron microscopy and confocal laser scanning microscopy, are used to visualize strands of constituent polymers, polymer networks, and fat globules.

The literature reveals that the food microstructure is affected by the quantity and type of fat or protein added to the product. For example, development of a meat emulsion by substituting beef fat with rendered beef fat, canola oil, and palm oil at varying protein concentrations was studied in emulsified beef batters (Youssef and Barbut 2010). They observed that the changes in emulsion stability were reflected by the variations in fat dispersion and fat globule size. Canola oil replacing the beef fat reduced in size more easily compared to other fats studied in their experiment. Increased protein content also resulted in the reduction of fat globule size and formed a denser protein matrix (attributed to more interactions between extracted proteins). This led to protein aggregation, therefore affecting the cook yield and texture, which was also observed by Schmidt et al. (1985) in comminuted meat products.

More protein concentration in dairy cheese production leads to increased casein–casein interaction, thereby providing stability to the cheese matrix (Lucey et al. 2003), while more moisture content reduces the firmness of processed cheese (Gupta and Reuter 1993). The effects of basil seed gum (BSG) on the microstructural properties of a model processed cheese was studied and it was noted that increasing the BSG from 1 to 8% protein concentration in the model cheese increased the protein–protein interactions and thus the viscosity significantly. The increased viscosity prohibited comprehensive emulsification leading to a more compact microstructure. The viscosity was observed to be five times (6880 ± 70 Pa s v/s 1260 ± 5 Pa s) higher with 1% BSG added to the model cheese compared to the no-BSG cheese (Hosseini-Parvar et al. 2015). A higher sol–gel transition temperature of BSG-containing processed cheese leads to strengthening effects, thus allowing more protein–protein interactions in a dense protein matrix. Azzollini et al. (2018) studied the effects of formulation on microstructural and textural properties of extruded insect-enriched snacks. They identified the effect of adding mealworms on the size of extruded snack, size of pores, and porosity of the snack. They highlighted the negative linear relationship between level of mealworm enrichment and the expansion ratio of the extruded product. Porosity dropped by 12% with the use of different formulated mixtures, while an increase in mealworm addition from 10 to 20% increased the average pore size and pore wall thickness. While keeping the concentration of mealworm powder fixed, an increase in processing parameters, i.e. barrel temperature and screw speed, induced a higher pore size and expansion ratio. The results observed in their study strengthened expectations upon supplementation of products using fat- or protein-rich ingredients (Foschia et al. 2013; Robin et al. 2011).

Microstructural properties are dependent upon the viscoelastic nature of the mixture, which ultimately hinges upon the properties of the ingredients (Moraru and Kokini 2003). During extrusion, lower concentrations of lipids were found to stabilize the process, with extrudates exhibiting a higher expansion ratio as compared to higher fat concentrations, which lead to loss in viscosity of the melt and mechanical energy of the process (Ilo et al. 2000). Scanning electron micrographs of the pseudo-cereal flours revealed differences in particle size and shape (Figure 5.1). Potato starch and wheat flour have the smallest particle size, whereas the size of a starch granule is smallest for amaranth and quinoa flour (< 2 μm). With regards to shape, amaranth, quinoa, and buckwheat granules are polygonal, while rice starch granules are irregular in shape, and wheat and potato granules show an oval shape.

5.2.2 Effect on Particle Size Distribution

With the increased competitiveness in food manufacturing, there is an increased need for tightly monitored and controlled processing, to increase operating and utilization efficiencies, and also the importance of controlling factors affecting the fluidic performance during high solid content processing. Sensitivity of palate, perception, and subsequent acceptance are highly dependent upon the particle size distribution (PSD) in food processing. For example, the optimum particle size of chocolate is limited at 35 μm, with anything over this limit being considered as coarse or gritty and so lead to the product's sensory rejection (Servais et al. 2002). Despite this upper constraint on PSD, it still plays a critical role in fluidity/viscosity of many food products (Chevalley 1994) and optimization of PSD is often the best course of action. PSD optimizations lead to better process efficiency and enhanced yields and are classified as monomodal, bimodal, or trimodal PSD. Bimodal PSD has been observed to have a favourable impact on viscosity reduction and better mixing, thus leading to improved shear, time, and temperature stability of the final products (Afoakwa et al. 2007; Aguilar et al. 1991; Do et al. 2007). PSD is thus essential to rheological characteristics and has a direct impact on sensory appeal.

Many studies in the past have looked at the effect of particle size on functional properties of different kinds of food. Different techniques are being employed to improve PSD for applications in food systems right from production, processing, and storage to development of innovative materials, processes, products, and applications.

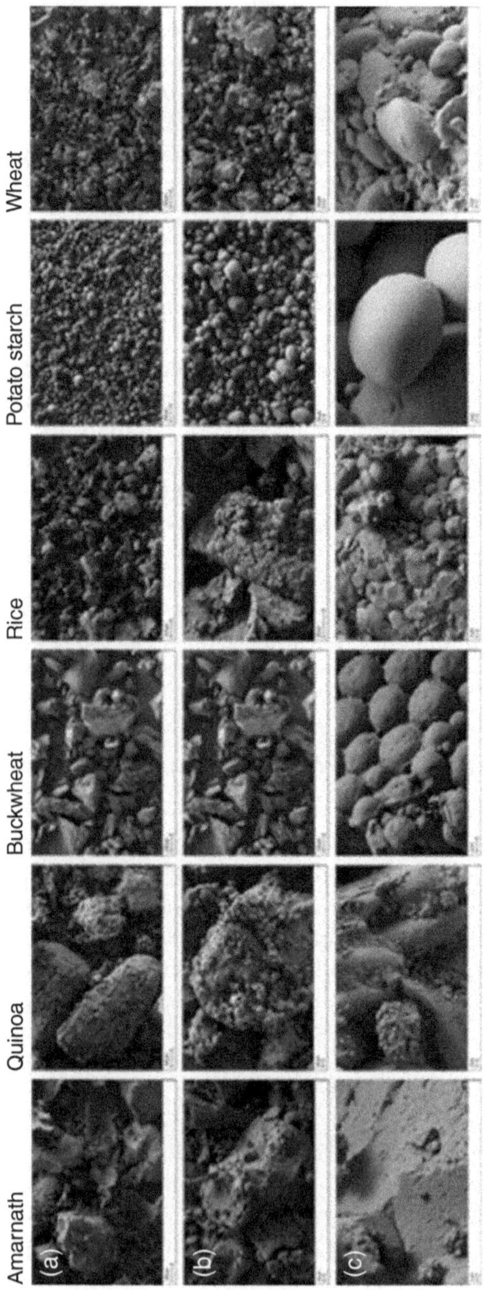

Figure 5.1 Scanning electron micrographs of amaranth, quinoa, buckwheat and rice flour, potato starch, and wheat flour. Scale bars row (a) 100 μm; (b) 20 μm; (c) 2 μm. *Source:* Alvarez-Jubete et al. (2010), Figure 5.1, p.440. Reproduced with permission of Springer Nature.

Micro- and nano-encapsulation are being touted as the game-changer, especially in agriculture and food industries. This process can greatly benefit the food sector by generating innovation in food properties such as taste, texture, colouring strength, processability, and stability in storage. It can also enable enhanced properties like water solubility, thermal stability, and bioavailability of bioactive compounds (Ezhilarasi et al. 2013; Huang et al. 2010; Silva et al. 2012). Encapsulation is a process of protecting bioactive compounds like polyphenols, enzymes, and micronutrients from degradation by packing them within a wall material. Microcapsules include particles having a diameter in the range of 3–800 µm, while nanocapsules particle size ranges between 10 and 1000 nm (Konan et al. 2002; Meena et al. 2011). The reduction in particle size significantly increases the surface-to-volume ratio and thus enhances reactivity of compounds by inducing changes in mechanical and electrical properties (Neethirajan and Jayas 2011). It has been established that particle size greatly affects the delivery of bioactive compounds (Hughes 2017) inside the body, therefore formulations using nano-encapsulation can significantly improve the bioavailability, controlled release of bioactive compounds, and target specific streams/areas within a system (Mozafari et al. 2006). Certain examples of nano-encapsulation highlight its significant benefits to food processing and food systems in general. Capsaicin particles were reduced to approximately 100 nm size by using a simple coacervation process and encapsulation in gelatin. The encapsulation of cross-linked gelatin particles over capsaicin significantly improved the thermal pyrolysis temperature and melting point (Wang et al. 2008). Another study by Hădărugă et al. (2006) encapsulated linoleic acid in cyclodextrin using a technique called inclusion complexation to improve thermal stability of linoleic acid.

Osborn and Akoh (2004) studied structured lipids, such as customized fats or lipids having value-added nutritional and physical properties brought about by addition of new fatty acids or rearrangement of the present fatty acid structure. The effect of droplet size on lipid oxidation was evaluated in oil-in-water emulsions prepared from canola oil or 98% pure caprylic acid. Lipid oxidation rates were significantly improved as the particle size reduction achieved by an increase in the surface area exposed to the reactive zone in aqueous solution. Particle size also greatly influences the powder, powder-blends, and powder-based products in terms of their flowability and moisture-acquiring nature leading to caking. Production efficiency and movement of product are important during any production process or manufacturing and caking or agglomeration of particles to form larger particles, which contribute significantly to lost functionality and poor quality (Aguilera et al. 1995). Stoklosa et al. (2012) analysed the impact of formulation and particle size on flowability and caking characteristics of deliquescent ingredients in single and binary mixtures. They observed that smaller-sized particles absorbed more moisture in higher relative humidity as compared to larger particle sizes. Smaller particle size and thus smaller pores posed significant issues in removal of water from the powder blends (Tuller et al. 1999). Surface moisture adsorption and capillary action occurred to different degrees based on the PSD. Moisture adsorption on the surface of particles may lead to adhesion of the particles to each other (Cleaver and Looi 2007). Smaller particle size (~50 µm) had a higher degree of adsorption, leading to lesser flowability and more caking compared to larger particles. PSD has also been widely studied for its application in the baking industry, wherein larger particle size could be associated with rough or coarse texture and the optimal size of particles was greatly influential on the dough characteristics like volume, expansion ratio, and firmness of bread (Sangnark and Noomhorm 2003).

Sangnark and Noomhorm (2003) evaluated the PSD effect on functional properties of dietary fibre (DF) and its impact on baking properties. They found that the larger particle size (>0.3 mm) limited the expansion rate by 15% compared to only 2% if the particle size was less than 0.075 mm. This was attributed to higher rate of moisture absorption by larger particles restraining the yeast action on the gluten structure. Substituting wheat flour with alkaline hydrogen peroxide treated sugarcane bagasse enhanced the firmness of bread by four times, while reducing loaf volume by 10% due to dilution of gluten content and crumb structure. Reducing the particle size significantly improved the quality of bread, making it more soft and elastic. Smaller particle sizes also impacted the sensory evaluation with texture and overall acceptability moderately liked by the sensory panel with particle size of less than 0.075 mm.

Variations in particle size are also known to affect the distribution of nutrients and functional properties in food systems. The particle size obtained after ultrafine grinding and their impact on hydration and anti-oxidant properties of DF were assessed (Zhu et al. 2010). Ultrafine grinding led to reduction in particle size (between 300 and 620 nm, with an average size of 343.5 nm). Total DF content was redistributed with the insoluble DF content reducing from 81.1 to 68.6%, while the soluble DF content increased from 2.9 to 11.5%. It has been noted in past literature that soluble DF content plays a more important role in human health than insoluble components (Galisteo et al. 2008). Therefore, grinding and reduction in particle size is important for redistribution of DFs. Particle size also played an important role in determining related characteristics such as water retention capacity (WRC), water holding capacity (WHC), and swelling capacity. This is similar to other studies conducted on coconut DF (Raghavendra et al. 2006), sugarcane bagasse DF (Sangnark and Noomhorm 2003), and wheat bran, sugar-beet, and citrus fibre (Auffret et al. 1994). Anti-oxidative properties like total phenolic content were also studied, indicating that DF after ultrafine grinding might be a primary reactant with free radicals, limiting the occurrence of free radical damage in the human body. This could be attributed to fibre matrix damage causing linked phenolic compounds to be released. Therefore, it can be easily deduced that powder with reduced particle size has higher dispersibility and solubility in food systems. Powders with such PSD are also easier to adsorb in the body, thus improving the quality and safety aspects of food products.

5.2.3 Effect on Textural Attributes

Torbica et al. (2010) measured the textural properties of bread formulations based on buckwheat and rice flour in the production of gluten-free breads. They found that an increase in buckwheat flour content did not significantly affect the textural properties of the final bread. This aligned with the assumption that preparation of bread with gluten-free buckwheat flour did not affect the textural characteristics but enhanced the nutritional value of the bread (Lin et al. 2009). However, the increased buckwheat content in formulations resulted in the appearance of a cracked surface of the upper crust due to reduced quality of the protein structure. Softness of the bread from husked buckwheat flour also scored maximum points compared to the bread prepared from unhusked buckwheat flour. This is in line with the results of Villanueva et al. (2019), who studied the effect of replacing traditional ingredients (native rice flour) in dough and bread manufacture by rice flour with reduced moisture content on textural and other physical properties. The use of reduced moisture rice flour led to increased softness and reduced crumb firmness and springiness. Partial substitution of native rice flour with other ingredients also decreased firmness significantly. One of the substitutes used in bread preparation (30% moisture reduced rice flour) showed higher L*, h, and C* parameters, which meant breads were lighter and more yellowish in colour. The authors concluded that formulated ingredients in manufacture of rice flour resulted in significant variations in viscoelastic and pasting properties of dough, thus producing breads with lower firmness and slower staling.

The effect of rice flour and potato starch along with pseudo-cereals like amaranth, quinoa, and buckwheat as replacers for wheat in gluten-free bread formulations was studied by Alvarez-Jubete et al. (2010). Viscosity for the wheat replacers were observed to follow the trend as rice > buckwheat > quinoa > amaranth. Pseudo-cereals were observed to have higher heat resistance and shear strength detected through lesser breakdown. Similarly, loaf volume, i.e. expansion of the product, was noted to be higher for pseudo-cereals (buckwheat and quinoa) as compared to potato starch (Alvarez-Jubete et al. 2010). Lipid content, especially polar lipids present in cereals, tend to act as a stabilizing agent for gas bubbles prior to starch gelatinization during the bread-making process (Gallagher 2009; Gan et al. 1995). A softer crumb structure in gluten-free breads was observed as a higher content of monoglycerides in pseudo-cereals led to fewer attachments between starch and amylose controlling the starch swelling and leaching of amylose into solution (Hoseney 1994).

During extrusion, higher protein content in the formulation affects the expansion ratio negatively, as proteins interact with water molecules affecting water distribution and forming covalent and nonbonding interactions, thus limiting the expansion potential (Moraru and Kokini 2003). Yu et al. (2013) assessed the effect of protein enrichment in extruded snacks using soy protein and corn flour blends. They showed significant correlations between total protein content in the snack and extrudate properties, i.e. water solubility index, expansion ratio, and lightness. Interactive effects of protein and moisture also affected rehydration ratio, breaking stress, bulk density, and lightness.

Baixauli et al. (2003) assessed the effect of formulation (addition of dextrin or dried egg) on the rheological and textural properties of batters for fried foods. It was noted that hardness of the batter was reduced when dextrin or dried egg was added to the product. Generally, fried foods with a higher crispiness are preferred due to more satisfactory sensory appeal. The crispiness of battered products using dextrin was measured by penetrometry curves indicating higher, numerous peaks corresponding with a crispy texture, while batters with dried eggs showed lesser peaks, and lower crispiness due to retention of water.

Cross-linking and cell wall degrading may also be applied to formulate products that may alter textural properties of the final products (Jeon et al. 2013). Enzymes such as transglutaminase enable cross-linking between glutamine and lysine residues in proteins, thus enabling enhanced control over the textural properties and functionality of different foods (Motoki and Seguro 1998). Use of enzymes and their effect on fat absorption and textural properties of French fries was studied, which identified destruction of cell structure by enzymes, thus leading to lesser oil/fat absorption into the internal section of French fries. Blanching reduced the hardness of potato strips and further reduction in hardness was also observed when the strips were kept soaked in water containing enzymes. Pectolytic and hemicellulytic enzymes destroyed the cell structure, thus allowing ease of cutting of the potato strips (Lisińska et al. 2007). Texture profile analysis includes the assessment of parameters such as hardness, cohesiveness, chewiness, and springiness. These profile parameters were analysed for effects of replacing wheat flour with amaranth flour as chicken nuggets batter. All these parameters increased significantly by increasing the content of amaranth flour. It was noted that the addition of amaranth in chicken paste greatly amplified the hardness of the chicken nugget. This increase in textural profile attributes, i.e. hardness, cohesiveness, chewiness, and springiness, was attained due to non-covalent linkage between the amaranth's amino acids, i.e. arginine, glutamic acid, and aspartic acid with lysine, glutamic acid, and aspartic acid present in the chicken's myofibrillar proteins (Tamsen et al. 2018).

In the dairy industry, formulation may also be applied in production of cheese by altering the textural characteristics of the end product. Cheeses with a higher salt content (1.5%) were perceived as least firm and springy, and were also the stickiest. These properties were only observed in cheeses with a low-fat content. Low-fat cheese (20% fat) was considered to be more springy and firm compared to high-fat cheese (40% fat) (Saint-Eve et al. 2009). Protein bars and healthy snacks are fast becoming a popular trend in the food industry, with a lot of emphasis on various functional properties of the end product. Several studies in the past have focused on the effects of different dairy ingredients and formulations to assess the textural properties of protein bars. Whey protein hydrolysate (WPH) and its effect on bar hardening was studied, comparing reduced hardness in bars made with WPH against hardness of bars made from whey protein isolate (McMahon et al. 2009). Similarly, protein bar hardening in a 20% protein bar made with milk protein concentrate was studied (Loveday et al. 2009) and it was found that physico-chemical changes in polyhydroxyl carbohydrate/water were the reason for bar hardening.

A more common food product, i.e. mayonnaise, has been widely studied, most commonly to reduce the fat content to develop a healthier version of the product. Presence of harmful preservatives and agents such as ethylenediaminetetraacetic acid (EDTA) to prevent deterioration has been a source of criticism and public concern. Plant materials rich in phenolic compounds and thus offering antimicrobial, antioxidative, and anti-inflammatory properties, are being tested to provide a healthier option. Processed beetroot in mayonnaise as a formulation was studied to identify the effects of formulation on the sensory and texture profile of the end product. As a result of the addition of beetroot, the viscosity of the mayonnaise was increased and affected all the textural characteristics. Hardness, adhesiveness, and force required to shear the mayonnaise increased with addition of beetroot, while spreadability (less spreadable) and graininess (more grainy), was also impacted (Raikos et al. 2016). The formulation of different ingredients and/or changes introduced into recipes of different products affect the physical, chemical, and mechanical properties of the end product, leading to variations in textural characteristics such as firmness, springiness, chewiness, and friability along with other functional properties.

5.3 Effect of Food Formulation on Functional Properties

5.3.1 Emulsification Behaviour

Emulsions form a significant part of many processed food formulations (Singh et al. 2009) and the occurrences of other compounds in multi-phase systems makes emulsions relatively complex (Phillips et al. 1994). It is necessary to supply energy for disrupting and intermingling the oil and water phase to form an emulsion. This is generally accomplished by either mechanical agitation or a homogenization process, if the emulsifier is not used. In the absence of an emulsifier, the kind of emulsion formed depends largely on the concentration in which the two phases of the emulsion system occur at the initial level during its preparation. The mere dispersions of two or more immiscible liquid phases are not true emulsions until such dispersions have adequate stability to stay dispersed for a considerable period of time.

Emulsion stabilization is typically attained by use of little surfactant molecules like phospholipids, polysorbates, and/or proteins, such as milk proteins and egg proteins. These surfactant molecules are called emulsifiers or emulsifying agents, which facilitate the creation of emulsions by reducing the tension at the oil–water interface and thus giving stability through formation of a protective film surrounding the dispersed droplets (Rousseau 2000). Surface active molecules like proteins are routinely used to stabilize food emulsions like mayonnaise, yoghurt, etc. Protein molecules tend to rapidly adsorb on the surface of droplets, reducing interfacial tensions and provide a protective cover on the molecules (Guo and Mu 2011).

Various parameters such as droplet size, interfacial tensions, and microstructures impact the emulsifying properties and stability. Physico-chemical properties like pH, protein concentration, and presence of ions along with different processing conditions also bring about different changes in the functional properties of the foods and its constituents. The amount of protein adsorbed at the interface is largely dependent on the pH of the system. Hence, the stability of emulsion formed is affected considerably by the pH and the amount of protein adsorbed at interface (Pearce and Kinsella 1978).

The stability of the emulsion is also governed by a few other factors, like the rate of adsorption of the emulsifier, concentration of emulsifier, and others. The emulsion system fails to hold its original droplet size at low concentration of emulsifiers. Many mechanisms are responsible for this type of destabilization. In instances of limited coverage at the interface by liquid droplets, they can merge in one another resulting in the formation of a larger droplet. The stable oil droplets are formed when the interfaces at liquid droplets become properly covered, which takes place in emulsions prepared with a high concentration of emulsifiers (Liu et al. 2011). Soy protein isolates show more emulsification capability than whey protein concentrates (WPC) (Utsumi et al. 1997). Lipoproteins have a good ability to form strong films surrounding the oil droplets, and thereby many lipoproteins, like those of egg origin, act as good emulsifying agents (Karel 1973). Low density lipoprotein (LDL) also possess excellent emulsification capability and the emulsions prepared by use of LDL are fairly stable (Mizutani and Nakamura 1987); the lipid part present surrounding the apoprotein of LDL is believed to offer the hydrophobic environment to enhance the binding of the apoprotein to the oil droplet surface during the process of emulsification.

The stability of the emulsions prepared with the complexes of protein–polysaccharide interactions is likely to be more in comparison to the emulsions stabilized by either protein or surfactant molecules alone, because the relative thickness of the liquid film between two adjacently placed droplets is highest for the protein–polysaccharide complex (5–10 nm) followed by proteins (1–5 nm) and surfactants (0.5–1 nm), as depicted in Figure 5.2. The rate of desorption of emulsifier from the interface is also a critical factor in governing the stability of the emulsion (Tcholakova et al. 2008). Adsorption of emulsifier molecules (i.e. surfactants, proteins, etc.) at the liquid interface is highly reversible. Desorption of the emulsifiers from the liquid interface

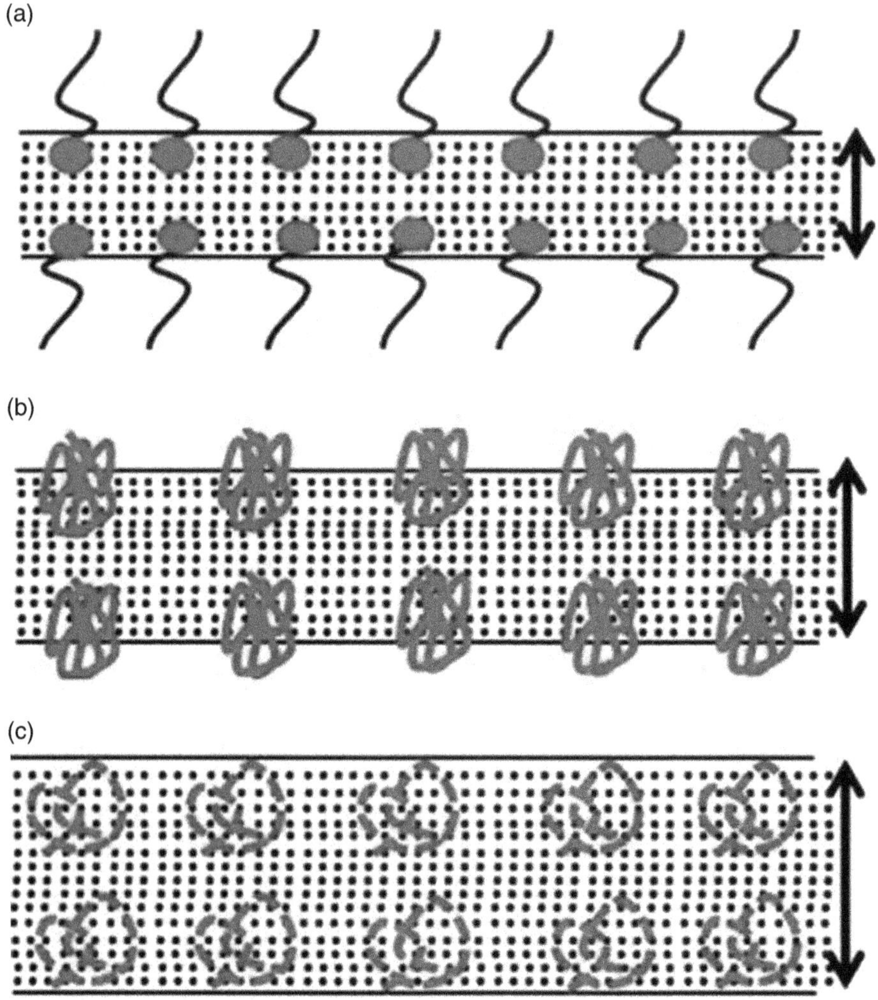

Figure 5.2 Relative thickness of a thin liquid film between adjacently located droplets stabilized by (a) surfactant, (b) protein, and (c) protein-polysaccharide complex. *Source*: Ghosh and Bandyopadhyay (2012), Figure 5, p. 403. CC BY 3.0.

governs the instability of the system. Based on this theory, as the liquid film in between the two adjacently placed droplets has maximum thickness, the emulsions stabilized by particles which range in size from about 10 nm to several micrometres are expected to have indefinite stability. This stability of emulsions is also attributed to the highest desorption energy of the particles from the liquid interface. However, the results of a study carried out by Tcholakova et al. (2008) are not in agreement with the hypothesis that the emulsions stabilized by surfactant, protein, or hydrocolloids are less stable than those of particle stabilized emulsions.

The emulsions stabilized by the complex formed by the interaction of proteins and polysaccharides are likely to be prepared with two alternative approaches: (i) addition of an aqueous solution consisting of the complex formed as a result of protein and polysaccharide interaction as an emulsifying agent following the homogenization

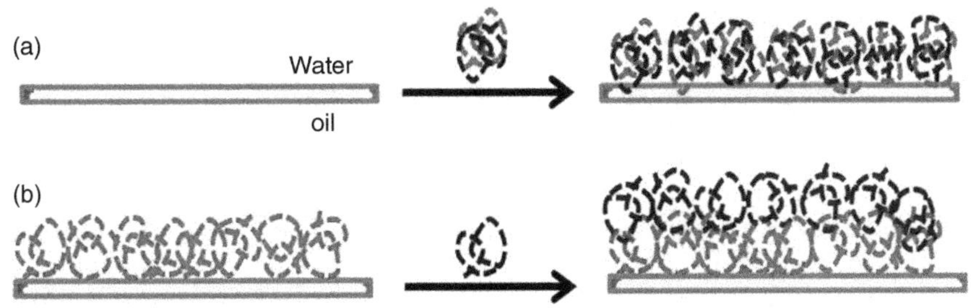

Figure 5.3 Schematic illustration of two alternative approaches of preparing oil–water emulsion using protein–polysaccharide complexes as an emulsifier. (a) Mixed emulsions, both protein and polysaccharide present together during emulsification. (b) Bilayer emulsion, polysaccharide added to emulsion prior stabilized by protein. *Source*: Ghosh and Bandyopadhyay (2012), Figure 6, p. 404. CC BY 3.0.

as depicted in Figure 5.3a; and (ii) addition of solution with charged polysaccharide into a protein-stabilized primary emulsion, in order to form emulsion droplets which will have a bilayer surface coating of protein-polysaccharide, as shown in Figure 5.3b. The first method of emulsion formation is called the 'mixed emulsion', while the second method is called the 'bilayer emulsion'. The bilayer method is also popularly known as the 'layer-by-layer' method. Lately, as this bilayer method of emulsion formation is being increasingly used in the process of nano-encapsulation and for protection of many emulsions against several environmental stresses, this method has gained noticeable significance. The chief concern in stability of emulsions occurs in the layer-by-layer approach, wherein the droplets of emulsion are likely to flocculate. The two major mechanisms are responsible for this flocculation of emulsion droplets in the layer-by-layer method of emulsion preparation. These two mechanisms are: (i) bridging flocculation, which occurs when polysaccharide concentration is low and when droplet collisions take place quicker than the rate of polysaccharide saturation of the droplet surfaces coated with protein; and (ii) depletion flocculation, which happens at relatively greater polysaccharide concentration and when non-adsorbed polysaccharide surpasses the critical value. For this reason, it is suitable to prepare emulsions with protein and polysaccharide added collectively before the process of homogenization than to prepare the emulsions by the layer-by-layer method (Camino et al. 2011).

High-pressure application on proteins causes development of new bonds leading to a modified structure with different functional properties (Molina et al. 2001). Similarly, addition of different functional ingredients in food formulation enables improvement in the functional properties of the foods and enhances the nutritive value. Use of soybean protein isolates has been widely studied as a functional ingredient due to their high functionality and nutritive value. Soybean protein impact on emulsifying properties has been widely evaluated (Mitidieri and Wagner 2002; Molina et al. 2001; Puppo and Añón 1999). The study by Puppo et al. (2005) demonstrated the improvement in emulsification properties of soybean emulsions upon treatment with high pressure. The emulsification properties of the soybean emulsion varied significantly with changes in pH. It was shown that there was a decrease in droplet size, increased depletion flocculation, and increase in adsorption of proteins at pH 8 while dropping to a pH to 3 led to an initial denaturing effect inducing increase in droplet size, a slight decrease in bridging flocculation, and increase in depletion flocculation (Puppo et al. 2005).

Another study assessed the effect of adding epigallocatechin-3-gallate (EGCG) and sodium chloride on the emulsifying properties and oxidation of myofibrillar protein emulsion (pork) (Feng et al. 2017). As lysine, arginine, proline, and threonine residues became readily oxidized into aldehydes and ketones, the addition of EGCG lowered the oxidative stress on proteins while addition of NaCl significantly affected ($p<0.001$) the free amine and carbonyl content. Surface hydrophobicity is a marker for protein unfolding and this was notably higher for oxidized myofibrillar proteins in the absence of EGCG. Large doses of EGCG reduced the surface hydrophobicity, which is attributed to protein aggregation leading to loss of emulsion gel structure, induced instability to the emulsifying properties, and higher cooking losses and texture deterioration. Guo and Mu (2011) assessed the impact of protein concentration and oil volume fraction on emulsions made with sweet potatoes and also analysed the stability and properties of the emulsion at neutral pH. They noted that the emulsion activity index (EAI) was significantly reduced (from 174.6 m^2/g at 0.1% to 14.9 m^2/g at 2.0%) with increase in protein concentration, while the EAI increased significantly with increased oil volume fraction. The emulsion stability index increased with protein concentration ($p<0.05$), while the ESI decreased significantly (dropped from 221 to 92.6 minutes) with increase in oil volume fraction from 5 to 35%. Emulsion microstructure was also impacted with lowered protein concentrations leading to pronounced coalescence and flocculation.

Cofrades et al. (2013) developed multiple emulsions and analysed their utility as a replacer for pork backfat in various meat emulsion systems. They prepared a water-in-oil-in-water emulsion using lipophilic and hydrophilic emulsifiers. The lipophilic emulsifier used was olive oil (lipid) and polyglycerol ester of polyricinoleic acid (PGPR), whereas WPC and sodium caseinate (SC) were used as hydrophilic emulsifiers. Emulsions made with SC showed initial stability after heating and creaming, while under similar conditions, emulsion made with WP showed lesser stability and no creaming. Emulsion stability reduced with storage time and was approximately 85% lower for both emulsions after seven days of storage, while creaming was approximately 90 and 85% and higher in emulsions with SC than WP ($p<0.05$). Biopolymers like polysaccharides have been evaluated for their broad range of applications in food emulsions. Chitosan is one such biopolymer or polysaccharide, a copolymer of glucosamine and N-acetyl glucosamine which finds use in the food industry due to its functional, nutritional, and physiological properties (Muzzarelli 1996; Shahidi et al. 1999). Speiciene et al. (2007) evaluated the effect of chitosan concentration on the properties of emulsions containing whey protein isolate. Addition of chitosan caused a gradual reduction in the particle size of the emulsions. The average particle size reduced from $10.95\pm0.62\,\mu m$ (emulsion with 4% WPI alone) to $3.54\pm0.19\,\mu m$ (in emulsion with 4% WPI and 0.5% chitosan). Droplets also flocculated due to the presence of chitosan, with the intensity of flocculation increasing with an increase in the chitosan content. This could be attributed to protein–polysaccharide interactions important for flocculation mechanisms. Creaming measurements noted for emulsion with WPI alone implied that the emulsion was not stable. Addition of chitosan could improve creaming stability by reducing the rate of diffusion of droplets and flocs (Speiciene et al. 2007).

5.3.2 Water Absorption Behaviour

Mohamed et al. (2006) prepared a low carbohydrate bread supplemented with soy protein and vital gluten and assessed the various rheological related characteristics of the dough such as farinograph profile and water absorption. Four different dough were prepared, namely PB1 containing gluten and soy in proportions of 70:30 and substituted with 30% of wheat flour; PB2 which contained gluten and soy in the same proportions as in PB1 but substituted with 50% of wheat flour; PB3 containing equal proportions of gluten and soy (50:50) and replaced with 30% of wheat flour; PB4 which also contained equal proportions of gluten and soy (50:50) but was replaced with 50% of wheat flour; and the control being 100% wheat flour bread. The effect of extra protein on water absorption of flour was monitored and shown to be higher by up to 40% compared to 100% in wheat flour bread. The highest water absorption was noted in the PB2 mixture with more protein and gluten content, indicating the impact of hydration on wheat gluten functional properties. Samples with higher protein content – PB2 and PB4 – needed more water, i.e. 123% to prepare a consistent dough compared to PB1 (99%) and PB3 (103%). Functional pasta was prepared from a blend of 20% finer millet flour, 12% pearl millet flour, 4% carboxy methyl cellulose, and the rest of the proportion (64%) contained durum semolina and carrot pomace (Gull et al. 2018). The pasta depicted better functional and other quality characteristics like swelling index, cooked weight, and water absorption properties. It was noted that the cooked weight of the functional pasta (35.10 ± 1.17) increased significantly ($p < 0.05$) in comparison to the control (33.93 ± 1.28). This may have been attributed to the presence of carboxy-methyl cellulose and carrot pomace, which have higher water binding and water holding (WHC) capactities. The swelling index of the pasta is a marker for water absorption by starch and protein utilized for gelatinization and hydration. The swelling index increased from 1.92 ± 0.04 for the control to 2.58 ± 0.15 for the functional pasta, while water absorption (%) increased from 181 ± 3.44 to 202 ± 2.40. Replacement of semolina flour with protein powder obtained from fish also impacted the swelling index and water absorption capacity of pasta (Desai et al. 2018). The cooking losses increased as the content of protein powder increased from 3.99 ± 0.27 g/100 g for control pasta to 5.85 ± 0.40 g/100 g for pasta enriched with 20 g fish powder. This may be attributed to a disruption in the protein gluten network. Cooking time for enriched pasta was also reduced from 6.30 to 5.00 minutes for 20 g fish powder enriched pasta due to reduced water absorption (98.9 ± 6.2 g/100 g for control; 87.8–74.1% for enriched pasta with 5–20 g of fish powder). Similarly, the swelling index for enriched pasta was considerably lower than the control pasta reported by Desai et al. (2018).

The study of Marti et al. (2013) assessed the effect of different cooking (extrusion cooking and conventional extrusion) methods on various premixes used for development of gluten-free pasta and analysed the water absorption behaviour of different rice flour premixes. Four different types of rice flours, native rice flour (NF), pregelatinized rice flour (PGF), mild gelatinized flour (MGF), and severe parboiled flour (SPF). Pasta samples prepared from different flours showed varying water absorption characteristics. Parboiled flours (PGF) promoted the formation of less-hydrophilic starchy structure, resulting in lower water absorption compared to native rice flour (91 g/100 g) or commercially available semolina pasta (99/100 g). Substitution of 50% SPF with PGF also improved the cooking loss and water absorption of the pasta (Marti et al. 2013). Whey protein, a milk processing by-product, is an important nutritional and functional ingredient gaining acceptance in food formulations. To consider this importance, the effect of enzymatic hydrolysis on functional properties was studied and also to develop a beverage formulation incorporating WPC (Sinha et al. 2007). Fungal protease and papain were used for enzymatic hydrolysis and it was noted that the water absorption capacity of WPC increased with an increase in hydrolysis time (max. for 60 minutes; Control = 10 ± 0.1 ml/100 g, fungal protease at

60 minutes = 32.5 ± 0.1 ml/100 g, Papain treated at 60 minutes = 34.0 ± 0.1 ml/100 g). This could be attributed to severance of proteins to smaller units, thus enabling more water binding sites (Catsimpoolas et al. 1970). Fortification of hard red spring (HRS) wheat flour with freeze dried okra extract was evaluated for development of high-soluble fibre bread. Okra extract powder at varying concentrations (4, 7, 10, and 13%) was used to replace HRS wheat flour for different formulations. The water absorption capacity increased by up to 4.4% as a result of the addition of okra extract flour (OE). This can be attributed to WHC and solubility of hydrophilic polysaccharides. The addition of gums and emulsifiers gives better moisture retention (Seyhun et al. 2003). Water absorption was noted at 60.1 ± 0.91% for 4% okra extract to 62.8 ± 1.73% at 13% OE. Addition of various DFs to bread formulations enhances the dough water absorption characteristic and also influences the loaf volume (Gómez et al. 2003). Addition of banana powder to HRS wheat flour was analysed to study the effect on yeast leavened bread formulation (Mohamed et al. 2010). In general, it has been observed that the presence of high water binding capacity components negatively affect the flour water absorption property, dough mixing, and bread quality like loaf volume, texture, and colour (McCleary 1986; Michniewicz et al. 1990; Shelton and D'Appolonia 1985). The addition of banana powder to HRS flour increased the water absorption of the mix, with water absorption increasing by 5.2, 7.5, 10.3, 12.5, and 15.2% for 10, 15, 20, 25, and 30% banana powder concentration, respectively.

5.3.3 Solubility

Solubility of food, particularly dry and powdered foods, is one of the important functional properties. Some of the functional properties like gelation, emulsification, and foaming are also affected by the solubility, which may also be necessary for effective utilization of different proteins as a functional food additive in many food processing operations. Solubility depends on many parameters including concentration of solute, pH, temperature, characteristics of solvent in the system, and surface hydrophobicity (Sikorski 1997).

The complex formed from the interaction of two macromolecules can be considered as a new type of biopolymer whose characteristics, in general, strongly differ from the characteristics of the original macromolecules involved in the complexation process (Tolstoguzov 1999). The complex formed from two different macromolecules (intermolecular complex) involving electrostatic force of attraction can be soluble and insoluble. For example, the interaction between proteins and polyphenols results in the formation of reversible or irreversible complex of either soluble or insoluble type (De Freitas and Mateus 2001; Luck et al. 1994). This solubility and insolubility of complexes depends on the mutual neutralization of opposite charges of the macromolecules taking part in the complexation process. The maximum yield of insoluble complexes is normally attributed to the ratio in which the oppositely-charged macromolecules take part in the complexation reaction. It means that when there is complete mutual neutralization of opposite charges of the macromolecules, the newly-formed complex will be of the insoluble type. Such insoluble complexes formed represent a separate concentrated liquid or gel-like phase, in which both the interacting macromolecules are present. This type of phase is called a complex coacervate. The soluble complexes, resulting from the interaction of the different types of macromolecules, are usually non-stoichiometric and are charged (Tolstoguzov 1996).

The interactions of protein with water or that of protein with a solvent are critically significant in food systems, particularly in cases where the protein performance is noticeably influenced by its interactions with some of the other components present in the food, and by the processing conditions (Kinsella 1984). The solubility of proteins is determined by average hydrophobicity of amino acid residues and charge frequency. The solubility is high if the average hydrophobicity is low and the charge frequency is high (Damodaran 1997a).

Proteins are larger in comparison to the lipid molecules such as phospholipids. The unfolding of protein at the water–lipid interface results in insertion of the hydrophobic groups in the lipid phase. Therefore, in certain cases, the surface of soluble proteins is characterized by charged groups, which continue to be in contact with the water molecules (Mangino 1994). Proteins orient and generally modify their conformation at an interface for reducing the interfacial energy. Proteins can interact with both types of solvents, polar as well as non-polar, due to their amphiphathic nature. This also encourages adsorption of proteins at the interaction of lipid–water systems (Hung and Zayas 1991). The solubility effect is accomplished by diffusion of protein molecules from the solution to interfacial tension (Tornberg et al. 1990). During milk processing, such as operations like homogenization, the oil surface exposed has the ability to adsorb the extra molecules like phospholipid and proteins in order to contribute a stable suspension system, since there are adequate phospholipids and proteins in milk to establish a suspension on a significantly extended lipid–water interface (Cornell 1991).

Takahashi et al. (1983) found that solubility as well as the swelling power of potato and sago starch decreases with the addition of soybean protein. The complexation between lipid and amylose results in decreased solubility of starch in the water (Copeland et al. 2009). Raphaelides and Karkalas (1988) observed the formation of an insoluble amylose–lipid complex on the interaction of amylose and lipid. According to Shah et al. (2011), the poorly soluble long-chain fatty acids can be brought to the soluble form by hydrophobic interactions between the denatured protein and the lipophilic molecules. The study further hypothesized that the protein molecule with a charged group is essential to facilitate the solubilization of insoluble starch lipid complexes, which results in the formation of soluble ordered complexes of high molecular weight from the three components.

5.3.4 Water Holding Capacity

WHC is usually identified as the capacity of a molecule or substance to hold its original water and may be the water which is added during processing operations, if any. The WHC is generally considered as a functional property of protein-based foods. The various food characteristics like texture, juiciness, and taste are affected by WHC (Xiong 1997); for instance, the capacity of meat and many meat products to hold its water before, after, and during several kinds of processing affects the juiciness and texture of such products and is critical in determining the palatability and consumer acceptance. Physico-chemically, the water in a food product is found in either the free or bound state. The bound form of water is strongly held by the proteins through their charged groups and dipolar sites present on their surface. Several intrinsic factors like protein type, structure of protein, and composition of its amino acids, as well as extrinsic factors like temperature, pH, and ionic strength, influence the WHC of foods.

The occurrence of lipids in protein gels increases the WHC; for example, in the presence of lipids, the gels of egg white protein exhibit a highly dense type of primary structure along with many open pockets binding a huge amount of water, which leads to elevating the WHC of the gels (Montejano et al. 1984). The organization of protein molecules around the lipid molecules in a way that entraps the water within the gel matrix, resulting in elevated WHC, could be responsible for this phenomenon (Catsimpoolas and Meyer 1971). Yuno-Ohta et al. (1996) found that the presence of fatty acid salts like sodium caprate and sodium laurate increase the WHC of ovalbumin gels in comparison to the WHC of ovalbumin alone. The uniformity of the protein gel matrix is enhanced in the presence of lipids in protein gels. Similarly, the uniformity of the gel structure is enhanced by protein–protein and protein–water interactions, which result in increasing the WHC of the gels (Nayak et al. 1998). The starch properties and its functionality are modified by complexation of amylose component of starch and lipids, including monoglycerides, lysophospholipids, and fatty acids; for example, solubility of starch in water is decreased, rheological characteristics of pasta are changed, swelling ability is lowered, temperature required to bring about gelatinization is increased, retrogradation is inhibited, and susceptibility to enzymatic hydrolysis is also decreased (Copeland et al. 2009).

Citrates and carbonates are widely used to increase the ionic strength and muscle fibre structure in poultry products by improving the water binding capacity (Feiner 2006; Petracci et al. 2013). Sodium bicarbonate is now widely used as a water binding agent in poultry products by reducing shear forces and improving yield (Petracci et al. 2012).

Certain properties of the meat such as colour, yield, and WHC treated with sodium bicarbonate were similar to the conventional method of phosphate treatment. But this bicarbonate treatment was found to impact the structure leading to increased coarseness attributed to CO_2 generation during cooking (Sheard and Tali 2004). The increased swelling of poultry meat could be attributed to increased electrostatic repulsion (a net increase in negative charge) and movement of pH away from isoelectric point of the myofibrillar proteins (Petracci et al. 2012).

5.3.5 Viscosity

The molecular properties of biopolymers, such as molecular weight, shape of the biopolymer, and flexibility of the chain, mainly govern the rheological properties such as viscosity and gelling. Beside these molecular properties, several other factors like biopolymer concentration, interaction between the biopolymers and that of water, as well as solution parameters such as ionic strength, pH, and presence of other components/ligands, also influence the viscosity of foods. The functional properties of each individual biopolymer are widened by the interactions between polysaccharide and proteins. The rheological properties of polysaccharide–protein complexes are altogether completely different from the properties of the individual biopolymer.

The biopolymeric interactions result in the formation of large-sized entities, which increases the viscosity of the food system. The rheological behaviours of most of the protein–polysaccharide complexes vary from viscous to viscoelastic. The increased viscosity of hydrated polysaccharide–protein complexes is largely dependent on the nature and structure of the polysaccharide. For instance, the globular conformation of the polysaccharide is supposed to be responsible for the viscous property of gum acacia-protein coacervates, while the interaction of the same protein with pectin results in a gel-like system. Apart from the nature of the polymer, the rheology of the system is also influenced by the solution properties and concentration of bio-polymers interacting with each other (Dickinson 2011). For example, it was observed that in deciding the viscosity of the coacervate phase, the pH played a major role. In this case, it was found that the maximum viscosity of the coacervates is possible when the pH is 4 at maximum amount of whey protein and gum arabic and highest possible degree of electrostatic attraction. This indicates that the highly viscous behaviour of the system is attributed to the electrostatic interactions of the biopolymers. If the pH of the system is increased above the isoelectric pH of the protein, the mixture of the same concentrations of these two biopolymers results in the formation of a system which is more elastic than viscous. The rheology of protein–polysaccharide systems is also known to be influenced significantly by the ionic strength and the ratio in which these two biopolymers are present. The complexation of β-lactoglobulin with a greater amount of pectin is favoured at optimal salt concentrations (0.21 M NaCl), which subsequently results in formation of a gel of stronger strength. It is essential to manage the parameters influencing the complex formation in order to obtain the gel with better gelling characteristics, because decrease in the solubility of complexes results from the strong interaction of biopolymers. This phenomenon is responsible for lowering the hydration capacity of the complex and eventually leading to a decrease in the viscosity (de Kruif et al. 2004; Schmitt and Turgeon 2011).

The interaction of sorghum starch, serum protein, and free fatty acids mixed in the ratio of 20:2:1 was studied with a rapid visco-analyzer (RVA) by Zhang and Hamaker (2003). It was observed that when the three components, different from binary systems constituted only by starch and protein or free fatty acids, were present in the system, a significant rise in the viscosity during the cooling phase occurs. However, the same viscosity rise has not been observed or was very little in the presence of only a binary system, besides, the order of addition of components to the ternary system as well as the type of fatty acids used affected peaks. The formation of the complex through interaction of proteins, starch, and fatty acids by self-assembly was reported by Zhang et al. (2015) while assessing the pasting characteristics of sorghum flour. This study found that the increased viscosity of cooked flour is attributed to the development of three component complexes, which is correlated to the continuous increase in the level of fatty acids over the storage period of three months. A three-component interaction is also shown by a pure system containing starch, whey protein, and free fatty acid mixed in a ratio of 20:1:2. This system explains the occurrence of viscosity rise or peak during the cooling stage (Zhang and Hamaker 2003).

5.3.6 Gelation

Several proteins exhibit the property of producing gels, which are composed of a 3D network of bulky molecules or aggregates. The network formed retains the water and is also accountable for the elasticity as well as the textural strength of the gel (Phillips et al. 1994). The protein gel development is a two-step process; the first step consists of dispersing the molecules of proteins in the aqueous medium. The protein molecules then undergo change in the conformation or become partially denatured. This step is generally heat induced. In the second step, the denaturation proceeds further. This step is also characterized by continuous unfolding of the protein molecules and increased changes in its 3D structure, which is responsible for the increased viscosity of the dispersion. The individual proteins then slowly associate or form aggregates (Sikorski 1997). The capacity of protein to produce a gel presents a structural matrix for retaining and holding water and various other food ingredients in its structure. The gel properties are principally influenced by the protein characteristics, cross-linking mechanisms, either inter cross-linking or intra cross-linking, as well as the nature of the protein chains and protein flexibility. Moreover, the process of denaturation and thus the cross-linking which occur during gel formation is also affected by the concentration of proteins, temperature during heating and cooling, time of heating, and some environmental factors like ionic strength, pH, and others (Boye et al. 1995; Wang and Damodaran 1990).

The gelation properties of proteins are modified or interfered with by the occurrence of lipids in protein dispersions (Kinsella 1984; Ziegler and Foegeding 1990). The presence of lipid droplets, which are coated by proteins resulting in improvement in characteristics of gel obtained from protein, also affect the protein gel networks (Dickinson and Yamamoto 1996; King et al. 1984). The proteins and lipids interacting with each other can enhance gelling properties in some cases, whereas the viscosity-related characteristics of proteins may be adversely affected by the protein–lipid interaction under other conditions. For example, the addition of soy bean proteins can enhance the viscosity of lipid-containing soups and sauces (Seal 1980). The presence of lipids with small-sized fat globules and having narrow PSD produce the protein gels with smooth texture and increased gel strength (Sikorski 1997). Heat set protein gels also have higher gel strength in the presence of lipids compared to protein gel produced without lipids; this behaviour is consistent with a strengthening of the gel network formed from protein by 'filler' lipid particles in the gel matrix (Van-Vliet 1988). The gels prepared with complex were formed from the interaction of whey proteins and lecithin depicts increased gel strength. The aqueous protein layer which surrounds the lecithin droplets acts as an impermeable layer and modifies the gel characteristics (Dickinson and Yamamoto 1996). The properties of protein gel may be affected by the interaction reaction of soybean protein isolate and soybean oil due to synthesis of the protein matrix or structure network, which entraps fat droplets (Farnum et al. 1976). It is also found that the gel strength of proteins increases with the addition of starch. This may be due to an increase in the density of the protein matrix and formation of elastic starch globules.

5.3.7 Foaming and Foam Stability

The foam, in case of an air–water system, can be defined as the entrapment or occlusion of air by the thin film of liquid; this thin liquid film is stabilized by surface active molecules. Foam is a colloidal system which contains the bubbles of air dispersed in a continuous aqueous phase. Several processed food products, including cakes, marshmallows, whipped cream, and ice cream, are kinds of foam products. Proteins play the major role of surface-active agents in such products, which assist in forming and contributing the stability to the dispersed gas phase (Damodaran 1997b). The proteins are required to be water-soluble and flexible for the formation of a cohesive film at the interface of air and water. Accordingly, a balance of flexibility and intermolecular cohesiveness is necessary to form a foam with good stability. The structure and stability of foam contributes and affects many quality characteristics like mouthfeel, texture, lightness, appearance, consistency, and other quality attributes of food products are affected by the structure of foams (Hettiarachchy and Kalapathy 1997; Kinsella 1984). The globulin proteins have excellent surface activity and thus easily envelope and incorporate air bubbles on whipping. The mixture of proteins, in general, produces excellent food foams (Johnson and Zabik 1981).

The foaming capacity of proteins is usually hampered when the lipids are present with the proteins. This is evidenced by the observation that the lipid-free soy protein, lipid-free egg white, defatted whey proteins, or whey protein isolate, exhibit excellent foaming characteristics when compared with the foaming properties displayed by these proteins when lipid is present (Zayas 1997). It seems that the lipids easily locate the proteins at the air–water interface owing to their surface activity and polarity and then interfere with the most desirable conformation of adsorbed protein films. Moreover, the different types of lipids affect the foaming characteristics of proteins differently. In comparison to some lipids like corn oil and butter fat, lecithin shows higher levels of impairment (Zayas 1997). The WPC produces poorly stabilized foam in the presence of lipids, particularly monoglycerides and polar lipids. The destabilization of foam prepared from protein in the presence of lipids may be due to the competition of lipids, particularly phospholipids and other surface active molecules, with the proteins becoming adsorbed onto the surface leading to the partial displacement of the proteins at the air–water interface (Kinsella 1984).

The stability of foam in many types of food products is governed by the viscoelastic properties of protein–polysaccharide complexes. It is observed that the increase in stability of the interfacial liquid film results in increased foam stability because, if the stability of interfacial liquid film is low, it will result in diffusion of entrapped air in the foam itself. The rate of diffusion of air entrapped by interfacial liquid film in the foam depends on the viscosity of the liquid film. Hence, increased stability of foam resulting from the lower diffusion rate of air entrapped by the interfacial liquid film is possible when the viscosity and stability of this liquid film is high. The air–water interfacial properties of the complex formed by interaction of β-lactoglobulin-acacia gum at pH 4.2 were studied by Schmitt et al. (2005). The findings of this study indicate that this complex forms a strong viscoelastic interfacial film with a thickness of about 250 Å, even though surface activity of this complex is comparable with that of the pure protein. This results in significantly reduced gas permeability of the thin liquid film stabilized by the complexes (0.021 cm/s) against the gas permeability of pure β-lactoglobulin (0.521 cm/s). This phenomenon reveals that the stability of protein–polysaccharide complex stabilized foam is more in comparison to the foam stabilized by only proteins. The probable elucidation of the increased foam stability and varying interfacial properties of the protein–polysaccharide complex over the individual polymer is that the former is able to reorganize at the interface by coalescence resulting in the formation of interfacial microgel. The results of these findings were applied for the ice cream formulation to improve the stability of air bubbles (Schmitt and Kolodziejczyk 2010).

Likewise, for improving the stability of air bubbles in chilled dairy products, gelatin has been substituted by the complex formed from the interaction of whey protein isolate and gum acacia. The size of the complexes formed and thus their surface activity can be managed by varying the ratio of biopolymers in the interaction reactions, wherever these complexes are used for stabilization purposes. However, it is possible to adjust the viscoelastic properties of the air–water interfacial film by either adsorbing two biopolymers at once or by the sequential adsorption of protein and then polysaccharide. For example, the air–water interfaces are being popularly stabilized by complexes formed from β-lactoglobulin and pectin. In this instance, the film thickness resulting from the sequential adsorption of protein followed by polysaccharide is greater (450 Å) than the film thickness obtained from the adsorption of complexes (250 Å) (Ganzelves et al. 2008). The uses of complexes resulting from the interaction of protein and polysaccharide for the stabilization of oil–water emulsion (Martinez et al. 2007) have received much more attention in comparison to their use for the stabilization of air–water foam.

5.4 Effect of Food Formulation on Flavour

Flavour modification in foods has long been practised worldwide, with salt as a flavour enhancer and the use of heat to cook food. Such has been the growth in cooking of food over the past decades that has revolutionized the pattern of food consumption and production. Nowadays, much of the human diet is made up of convenience foods or pre-packed food that can be prepared by just heating in a microwave to be ready to eat. Development of new foods on a small scale is quite different from mass-scale production, thus making consistent flavours difficult to achieve. One way to adapt to this dilemma is the addition of flavouring agents which are not affected by shelf life, cooking conditions, or reheating cycles. Addition of flavouring agents finds extensive application in confectionary, cakes, biscuits, beverages, and snack foods. The base flavour of the product is modified to provide the desired characteristics to the food. With the advent and mass acceptance of flavourings and flavouring agents, a wave of criticism and concern has arisen attributing the 'unhealthiness' of foods to these flavouring agents. This has moved food processing organizations to modify the food formulations to describe their products as natural or free from artificial/chemical flavouring agents (Taylor and Hort 2007).

One such variation was the preparation of peach ice cream and its use as a symbiotic food. Skimmed milk powder, inulin, peach flavouring/aromatizing agents, without added fat and probiotic culture (0.1 g/l), was added to test the effect of formulation on sensory properties like taste, flavour, and aroma. Flavour as a characteristic was found to be 75% adequate; inulin added in the formulation had a synergistic effect with sucralose and thus positively influenced the sensory characteristics of the formulation (Villalva et al. 2017). Common salt is a widely used flavouring ingredient in many food products including meat products. Salt reduction to cater to health-conscious consumers not only reduces the perceived saltiness but also weakens the flavour profile of the product. Fat content and saltiness increased in a linear relationship in low-sodium ground meat patties without phosphate. Patties made with phosphate had slightly reduced flavour intensity with increase in fat content (Ruusunen et al. 2005). This is in line with a previous study which shows that perceived saltiness of the fatty meat products is more when compared to lean products (Matulis et al. 1994). In baking and cereals, gluten is the main structural protein in flour and is vital for viscoelastic properties of dough and crumb texture and appearance of the final product. Removal of gluten to serve coeliacs leads to a loss in mouthfeel and flavour of the product (Arendt et al. 2002). Replacing gluten-rich wheat with rice flours (80%) or potato starch (20%) to produce gluten-free breads was tested. It was found that gluten-free bread loaves from medium-grain rice flours displayed a higher standard in terms of flavour and colour amongst other properties compared to long-grain rice flour (Ylimaki et al. 1988). Preparation of pre-cooked pasta-like products using different extrusion-cooking parameters at different moisture content was carried out to analyse certain quality characteristics. The highest flavour scores (4.8 ± 0.49) were obtained for 30% moisture content at 100 rpm screw speed and best taste score (4.8 ± 0.32) (Wójtowicz and Mościcki 2009). Requirements of low- or zero-added sugar in food products combined with an increasing number of cases in diabetes around the world has focused research towards development of products with a low glycemic index (GI). The GI is defined as the measure of the ability of a food to raise sugar/glucose levels in blood after consumption compared to an equivalent dose of glucose. Foods containing low GI are vital in dietary and food intake management for diabetic patients, sports performers, maintaining and/or reducing weight, along with reduction in cardiac and hypertension cases (Jenkins et al. 1981). Whelan et al. (2008) assessed the use of alternative sweeteners for production of low-GI ice cream to enhance the sensory traits of the formulation. They identified that ice cream formulations with 14% polydextrose were very bland in taste and had an alcohol-like aftertaste. It was concluded that all the experimental ice cream formulations were unacceptable and a reduction in polydextrose with an increase in dairy flavour was recommended to improve acceptability. Another study on preparation of weaning foods for babies using wheat and mungbean flour evaluated the functional properties of weaning foods. Formulation F3 (36% mungbean flour, 44% wheat flour, 10% full fat milk powder, and 10% sucrose) was found to have the highest scores on flavour (7.40 ± 0.10), mouthfeel (7.40 ± 0.10), and overall acceptability (7.45 ± 0.10) (Imtiaz et al. 2011).

It is worth mentioning that our increased understanding of flavour receptors in mouth and nose has greatly added to flavour research. Research on taste receptors and transduction mechanisms have greatly increased our understanding of the way tastants act on the system (Taylor and Hort 2007). Research coupling the knowledge of taste receptors and molecular biology has enabled companies to develop compounds screened for tastes, such as to block bitterness or enhance flavour.

5.5 Conclusions

Techno functional properties of the food products hinges upon the type, amount, and properties of the ingredients used in its formulation. The processing of ingredients such as mixing, shearing, cooking, drying, packaging, and storage can lead to ingredient interactions which influence the quality and acceptability of the final product. Ingredient interaction modifies the foods' microstructure, PSD which may further contribute to functional properties like emulsion formation, viscosity, texture, stability, and others. The formulation of different ingredients and/or changes introduced in recipes of different products affect the physical, chemical, and mechanical properties of the end product. Interaction of ingredients used in food formulation greatly enhances many functional properties of the food system. The desired functional properties of food can be manoeuvred by appropriate selection and precise control of ingredient interaction. Moreover, homogeneity of the process and uniform product properties are important to processors for saleability of their product, which is why an understanding of the functionality of each ingredient and its interaction with other ingredients is all the more important. This can be extremely helpful in formulation of new foods, as well as reformulation of the existing food products to cater for evolving consumer demands.

References

Afoakwa, E.O., Paterson, A., and Fowler, M. (2007). Factors influencing rheological and textural qualities in chocolate–a review. *Trends in Food Science & Technology* 18 (6): 290–298.

Aguilar, C., Rizvi, S.S.H., Ramirez, J.F., and Inda, A. (1991). Rheological behavior of processed mustard II: storage effects. *Journal of Texture Studies* 22 (1): 85–103.

Aguilera, J., del Valle, J., and Karel, M. (1995). Caking phenomena in amorphous food powders. *Trends in Food Science & Technology* 6 (5): 149–155.

Alvarez-Jubete, L., Auty, M., Arendt, E.K., and Gallagher, E. (2010). Baking properties and microstructure of pseudocereal flours in gluten-free bread formulations. *European Food Research and Technology* 230 (3): 437–445. https://doi.org/10.1007/s00217-009-1184-z.

Arendt, E.K., O'Brien, C.M., Schober, T.J., and Gormley, T.R. (2002). Development of Guten-free cereal products. *Farm Food* 12: 21–27.

Auffret, A., Ralet, M.C., Guillon, F. et al. (1994). Effect of grinding and experimental conditions on the measurement of hydration properties of dietary fibres. *LWT - Food Science and Technology* 27 (2): 166–172. https://doi.org/10.1006/fstl.1994.1033.

Azzollini, D., Derossi, A., Fogliano, V. et al. (2018). Effects of formulation and process conditions on microstructure, texture and digestibility of extruded insect-riched snacks. *Innovative Food Science & Emerging Technologies* 45: 344–353.

Baixauli, R., Sanz, T., Salvador, A., and Fiszman, S.M. (2003). Effect of the addition of dextrin or dried egg on the rheological and textural properties of batters for fried foods. *Food Hydrocolloids* 17 (3): 305–310. https://doi.org/10.1016/S0268-005X(02)00091-7.

Boye, J.L., Alli, L., Ismail, A.A. et al. (1995). Factors affecting molecular characteristics of whey protein gelation. *International Dairy Journal* 5: 337–353.

Camino, N.A., Sánchez, C.C., Rodríguez Patino, J.M., and Pilosof, A.M.R. (2011). Hydroxypropylmethylcellulose at the oil–water interface. Part I. Bulk behaviour and dynamic adsorption as affected by pH. *Food Hydrocolloids* 25: 1–11.

Catsimpoolas, N. and Meyer, E.W. (1971). Gelation phenomenon of soybean globulins III. Protein-lipid interactions. *Cereal Chemistry* 48: 158–167.

Catsimpoolas, N., Funk, S.K., and Meyer, E.W. (1970). Thermal aggregation of glycinin subunits. *Cereal Chemistry* 47: 331.

Chevalley, J. (1994). Chocolate flow properties. In: *Industrial Chocolate Manufacture and Use* (ed. S.T. Beckett), 139–155. Boston, MA: Springer.

Cleaver, J. and Looi, L. (2007). AFM study of adhesion between polystyrene particles-the influence of relative humidity and applied load. *Powder Technology* 174 (1–2): 34–37.

Cofrades, S., Antoniou, I., Solas, M.T. et al. (2013). Preparation and impact of multiple (water-in-oil-in-water) emulsions in meat systems. *Food Chemistry* 141 (1): 338–346.

Copeland, L., Blazek, J., Salman, H., and Tang, M.C. (2009). Form and functionality of starch. *Food Hydrocolloids* 23: 1527–1534.

Cornell, D.G. (1991). Surface activity of bovine whey proteins at the phospholipid-water interface. In: *Interactions of Food Proteins* (eds. N. Parris and R. Barford), 123–136. Washington DC: American Chemical Society.

Damodaran, S. (1997a). Food proteins: An overview. In: *Food Proteins and Their Applications* (eds. S. Damodaran and A. Paraf), 1–56. New York;: Marcel Dekker Inc.

Damodaran, S. (1997b). Protein-stabilized foams and emulsions. In: *Food Proteins and Their Applications* (eds. S. Damodaran and A. Paraf), 57–110. New York;: Marcel Dekker Inc.

De Freitas, V. and Mateus, N. (2001). Structural features of procyanidin interactions with salivary proteins. *Journal of Agricultural and Food Chemistry* 49: 940–945.

Desai, A., Brennan, M.A., and Brennan, C.S. (2018). The effect of semolina replacement with protein powder from fish (Pseudophycis bachus) on the physicochemical characteristics of pasta. *LWT - Food Science and Technology* 89: 52–57.

Dickinson, E. (2011). Mixed biopolymers at interfaces: competitive adsorption and multilayer structures. *Food Hydrocolloids* 25: 1966–1983.

Dickinson, E. and Yamamoto, Y. (1996). Viscoelastic properties of heat-set why protein stabilized emulsion gels with added lecithin. *Journal of Food Science* 61: 811–816.

Do, T.A., Hargreaves, J., Wolf, B. et al. (2007). Impact of particle size distribution on rheological and textural properties of chocolate models with reduced fat content. *Journal of Food Science* 72 (9): E541–E552.

Ezhilarasi, P., Karthik, P., Chhanwal, N., and Anandharamakrishnan, C. (2013). Nanoencapsulation techniques for food bioactive components: a review. *Food and Bioprocess Technology* 6 (3): 628–647.

Farnum, C., Stanly, D.W., and Gray, L.L. (1976). Protein-lipid interactions in soy films. *Canadian Institute of Food Science and Technology Journal* 9: 201–206.

Feiner, G. (2006). *Meat Products Handbook: Practical Science and Technology*. Elsevier.

Feinle, C., Christen, M., Grundy, D. et al. (2002). Effects of duodenal fat, protein or mixed-nutrient infusions on epigastric sensations during sustained gastric distension in healthy humans. *Neurogastroenterology and Motility* 14 (2): 205–213.

Feng, X., Chen, L., Lei, N. et al. (2017). Emulsifying properties of oxidatively stressed myofibrillar protein emulsion gels prepared with (−)-epigallocatechin-3-gallate and NaCl. *Journal of Agricultural and Food Chemistry* 65 (13): 2816–2826.

Foschia, M., Peressini, D., Sensidoni, A., and Brennan, C.S. (2013). The effects of dietary fibre addition on the quality of common cereal products. *Journal of Cereal Science* 58 (2): 216–227.

Galisteo, M., Duarte, J., and Zarzuelo, A. (2008). Effects of dietary fibers on disturbances clustered in the metabolic syndrome. *The Journal of Nutritional Biochemistry* 19 (2): 71–84.

Gallagher, E. (2009). *Gluten-Free Food Science and Technology*. Wiley.

Gan, Z., Ellis, P., and Schofield, J. (1995). Gas cell stabilisation and gas retention in wheat bread dough. *Journal of Cereal Science* 21 (3): 215–230.

Ganzelves, R.A., Fokkink, R., van Villet, T. et al. (2008). Structures of mixed β-lactoglobulin/pectin adsorbed layers at air/water interfaces: a spectroscopy study. *Journal of Colloid and Interface Science* 317: 137–147.

Ghosh, A.K. and Bandyopadhyay, P. (2012). Polysaccharide-protein interactions and their relevance in food colloids. In: *The Complex World of Polysaccharides* (ed. D.N. Karunaratne), 395–408. Rijeka: InTech.

Gómez, M., Ronda, F., Blanco, C.A. et al. (2003). Effect of dietary fibre on dough rheology and bread quality. *European Food Research and Technology* 216 (1): 51–56.

Gull, A., Prasad, K., and Kumar, P. (2018). Nutritional, antioxidant, microstructural and pasting properties of functional pasta. *Journal of the Saudi Society of Agricultural Sciences* 17 (2): 147–153.

Guo, Q. and Mu, T. (2011). Emulsifying properties of sweet potato protein: effect of protein concentration and oil volume fraction. *Food Hydrocolloids* 25 (1): 98–106.

Gupta, V. and Reuter, H. (1993). Firmness and melting quality of processed cheese foods with added whey protein concentrates. *Le Lait* 73 (4): 381–388.

Hădărugă, N.G., Hădărugă, D.I., Păunescu, V. et al. (2006). Thermal stability of the linoleic acid/α-and β-cyclodextrin complexes. *Food Chemistry* 99 (3): 500–508.

Hettiarachchy, N.S. and Kalapathy, U. (1997). Soybean protein products. In: *Soybeans: Chemistry, Technology and Utilization* (ed. K. Liu), 379–411. New York;: Chapman & Hall.

Hoseney, R.C. (1994). *Principles of Cereal Science and Technology*. American Association of Cereal Chemists (AACC).

Hosseini-Parvar, S.H., Matia-Merino, L., and Golding, M. (2015). Effect of basil seed gum (BSG) on textural, rheological and microstructural properties of model processed cheese. *Food Hydrocolloids* 43: 557–567.

Huang, Q., Yu, H., and Ru, Q. (2010). Bioavailability and delivery of nutraceuticals using nanotechnology. *Journal of Food Science* 75 (1): R50–R57.

Hughes, G.A. (2017). Nanostructure-mediated drug delivery. In: *Nanomedicine* (ed. L.P. Balogh), 47–72. Singapore: Stanford.

Hung, S.C. and Zayas, J.F. (1991). Emulsifying capacity and emulsion stability of milk proteins and corn germ protein flour. *Food Science* 56: 1216–1218.

Ilo, S., Schoenlechner, R., and Berghofe, E. (2000). Role of lipids in the extrusion cooking processes. *Grasas y Aceites* 51 (1–2): 97–110.

Imtiaz, H., BurhanUddin, M., and Gulzar, M. (2011). Evaluation of weaning foods formulated from germinated wheat and mungbean from Bangladesh. *African Journal of Food Science* 5 (17): 897–903.

Jenkins, D.J.A., Wolever, T.M.S., Taylor, R.H. et al. (1981). Glycemic index of foods: a physiological basis for carbohydrate exchange. *The American Journal of Clinical Nutrition* 34 (3): 362–366.

Jeon, S., Lim, J., Inglett, G.E., and Lee, S. (2013). Effect of enzymatic treatments on the rheological and oil-resisting properties of wheat flour-based frying batters. *Journal of Food Engineering* 115 (2): 215–219. https://doi.org/10.1016/j.jfoodeng.2012.10.015.

Johnson, T. and Zabik, M.E. (1981). Ultrastructural examination of egg albumin protein foams. *Journal of Food Science* 46: 1237–1240.

Karel, M. (1973). Protein-lipid interactions. *Journal of Food Science* 38: 756–763.

King, A.J., Ball, H.R., Catignani, G.L., and Swaisgood, H.E. (1984). Modification of egg white proteins with oleic acid. *Food Science* 49: 1240–1244.

Kinsella, J.E. (1984). Milk proteins: physiochemical and functional properties. *Critical Reviews in Food Science and Nutrition* 21: 197–262.

Konan, Y.N., Gurny, R., and Allémann, E. (2002). Preparation and characterization of sterile and freeze-dried sub-200 nm nanoparticles. *International Journal of Pharmaceutics* 233 (1–2): 239–252.

de Kruif, C.G., Weinbreck, F., and de Vries, R. (2004). Complex coacervation of proteins and anionic polysaccharides. *Current Opinion in Colloid and Interface Science* 9: 340–349.

Lin, L.Y., Liu, H.M., Yu, Y.W. et al. (2009). Quality and antioxidant property of buckwheat enhanced wheat bread. *Food Chemistry* 112 (4): 987–991.

Lisińska, G., Tajner-Czopek, A., and Kalum, L. (2007). The effects of enzymes on fat content and texture of French fries. *Food Chemistry* 102 (4): 1055–1060.

Liu, L., Zhao, Q., Liu, T., and Zhao, M. (2011). Dynamic surface pressure and dilatation viscoelasticity of sodium caseinate/xanthan gum mixtures at the oil-water interface. *Food Hydrocolloids* 25: 921–927.

Loveday, S.M., Hindmarsh, J.P., Creamer, L.K., and Singh, H. (2009). Physicochemical changes in a model protein bar during storage. *Food Research International* 42 (7): 798–806.

Lucey, J., Johnson, M., and Horne, D. (2003). Invited review: perspectives on the basis of the rheology and texture properties of cheese. *Journal of Dairy Science* 86 (9): 2725–2743.

Luck, G., Liao, H., Murray, N.J. et al. (1994). Polyphenols astringency and proline-rich proteins. *Phytochemistry* 37: 357–371.

Lundin, L., Golding, M., and Wooster, T.J. (2008). Understanding food structure and function in developing food for appetite control. *Nutrition and Dietetics* 65: S79–S85.

Mangino, E.M. (1994). Protein interactions in emulsions: protein-lipid interactions. In: *Protein Functionality in Food Systems* (eds. S.N. Hettiarachchy and R.G. Ziegler), 147–178. New York;: Marcel Dekker Inc.

Marti, A., Caramanico, R., Bottega, G., and Pagani, M.A. (2013). Cooking behavior of rice pasta: effect of thermal treatments and extrusion conditions. *LWT - Food Science and Technology* 54 (1): 229–235.

Martinez, K.D., Sanchez, C.C., Ruiz-Henestrosa, V.P. et al. (2007). Effect of limited hydrolysis of soy protein on the interactions with polysaccharides at the air-water interface. *Food Hydrocolloids* 21: 813–822.

Matulis, R.J., Mckeith, F.K., and Brewer, M.S. (1994). Physical and sensory characteristics of commercially available frankfurters. *Journal of Food Quality* 17 (4): 263–271.

McCleary, B.V. (1986). Enzymatic modification of plant polysaccharides. *International Journal of Biological Macromolecules* 8 (6): 349–354.

McClements, D.J., Decker, E.A., Park, Y., and Weiss, J. (2009). Structural design principles for delivery of bioactive components in nutraceuticals and functional foods. *Critical Reviews in Food Science and Nutrition* 49 (6): 577–606.

McMahon, D.J., Adams, S., and McManus, W. (2009). Hardening of high-protein nutrition bars and sugar/polyol-protein phase separation. *Journal of Food Science* 74 (6): E312–E321.

Meena, K., Bairwa, N., and Parashar, B. (2011). Formulation and *in-vitro* evaluation of verapamil hydrochloride loaded microcapsule using different polymer. *Asian Journal of Biochemical and Pharmaceutical Research* 1 (3): 528–538.

Michniewicz, J., Biliaderis, C., and Bushuk, W. (1990). Water-insoluble pentosans of wheat: composition and some physical properties. *Cereal Chemistry* 67 (5): 434–439.

Mitidieri, F.E. and Wagner, J.R. (2002). Coalescence of o/w emulsions stabilized by whey and isolate soybean proteins. Influence of thermal denaturation, salt addition and competitive interfacial adsorption. *Food Research International* 35 (6): 547–557.

Mizutani, R. and Nakamura, R. (1987). Emulsifying properties of a complex between apoprotein from hen's egg yolk lecithin. *Agricultural and Biological Chemistry* 51: 1115–1119.

Mohamed, A.A., Rayas-Duarte, P., Shogren, R., and Sessa, D. (2006). Low carbohydrates bread: formulation, processing and sensory quality. *Food Chemistry* 99 (4): 686–692.

Mohamed, A., Xu, J., and Singh, M. (2010). Yeast leavened banana-bread: formulation, processing, colour and texture analysis. *Food Chemistry* 118 (3): 620–626.

Molina, E., Papadopoulou, A., and Ledward, D. (2001). Emulsifying properties of high pressure treated soy protein isolate and 7S and 11S globulins. *Food Hydrocolloids* 15 (3): 263–269.

Montejano, J.G., Hamann, D.D., Bau, H.R., and Lanier, T.C. (1984). Thermally induced gelation of native and modified egg white rheological changes during processing final strength and microstructures. *Food Science* 49: 1249–1256.

Moraru, C. and Kokini, J. (2003). Nucleation and expansion during extrusion and microwave heating of cereal foods. *Comprehensive Reviews in Food Science and Food Safety* 2 (4): 147–165.

Motoki, M. and Seguro, K. (1998). Transglutaminase and its use for food processing. *Trends in Food Science & Technology* 9 (5): 204–210.

Mozafari, M.R., Flanagan, J., Matia-Merino, L. et al. (2006). Recent trends in the lipid-based nanoencapsulation of antioxidants and their role in foods. *Journal of the Science of Food and Agriculture* 86 (13): 2038–2045.

Muzzarelli, R.A. (1996). Chitosan-based dietary foods. *Carbohydrate Polymers* 29 (4): 309–316.

Nayak, R., Kenney, P., Slider, S. et al. (1998). Cook yield, texture and gel ultrastructure of model beef batters as affected by low levels of calcium, magnesium and zinc chloride. *Journal of Food Science* 63: 945–950.

Neethirajan, S. and Jayas, D.S. (2011). Nanotechnology for the food and bioprocessing industries. *Food and Bioprocess Technology* 4 (1): 39–47.

Osborn, H.T. and Akoh, C.C. (2004). Effect of emulsifier type, droplet size, and oil concentration on lipid oxidation in structured lipid-based oil-in-water emulsions. *Food Chemistry* 84 (3): 451–456.

Pearce, K. and Kinsella, J. (1978). Emulsifying properties of protein: evaluation of a turbidimetric technique. *Journal of Agricultural and Food Chemistry* 26: 716–723.

Petracci, M., Laghi, L., Rocculi, P. et al. (2012). The use of sodium bicarbonate for marination of broiler breast meat. *Poultry Science* 91 (2): 526–534.

Petracci, M., Bianchi, M., Mudalal, S., and Cavani, C. (2013). Functional ingredients for poultry meat products. *Trends in Food Science & Technology* 33 (1): 27–39.

Phillips, L., Whitehead, D.M., and Kinsella, J. (1994). *Structure-Function Properties of Food Proteins*. San Diego: Academic Press, Inc.

Puppo, M. and Afñón, M. (1999). Soybean protein dispersions at acid pH: thermal and rheological properties. *Journal of Food Science* 64 (1): 50–56.

Puppo, M., Speroni, F., Chapleau, N. et al. (2005). Effect of high-pressure treatment on emulsifying properties of soybean proteins. *Food Hydrocolloids* 19 (2): 289–296.

Raghavendra, S., Swamy, S.R., Rastogi, N. et al. (2006). Grinding characteristics and hydration properties of coconut residue: a source of dietary fiber. *Journal of Food Engineering* 72 (3): 281–286.

Raikos, V., McDonagh, A., Ranawana, V., and Duthie, G. (2016). Processed beetroot (*Beta vulgaris L.*) as a natural antioxidant in mayonnaise: effects on physical stability, texture and sensory attributes. *Food Science and Human Wellness* 5 (4): 191–198. https://doi.org/10.1016/j.fshw.2016.10.002.

Raphaelides, S. and Karkalas, J. (1988). Thermal dissociation of amylose–fatty acid complexes. *Carbohydrate Research* 172: 65–82.

Robin, F., Dubois, C., Pineau, N. et al. (2011). Expansion mechanism of extruded foams supplemented with wheat bran. *Journal of Food Engineering* 107 (1): 80–89.

Rousseau, D. (2000). Fat crystals emulsion stability: a review. *Food Research International* 33: 3–14.

Ruusunen, M., Vainionpää, J., Lyly, M. et al. (2005). Reducing the sodium content in meat products: the effect of the formulation in low-sodium ground meat patties. *Meat Science* 69 (1): 53–60. https://doi.org/10.1016/j.meatsci.2004.06.005.

Saint-Eve, A., Lauverjat, C., Magnan, C. et al. (2009). Reducing salt and fat content: impact of composition, texture and cognitive interactions on the perception of flavoured model cheeses. *Food Chemistry* 116 (1): 167–175. https://doi.org/10.1016/j.foodchem.2009.02.027.

Sangnark, A. and Noomhorm, A. (2003). Effect of particle sizes on functional properties of dietary fibre prepared from sugarcane bagasse. *Food Chemistry* 80 (2): 221–229. https://doi.org/10.1016/S0308-8146(02)00257-1.

Schmidt, G., Acton, J., and Ray, F. (1985). Histological techniques to study emulsion microstructure. Proceedings of the 38th Reciprocal Meat Conference, 44–47. Chicago, IL: American Meat Science Association.

Schmitt, C. and Kolodziejczyk, E. (2010). Protein-polysaccharide complexes: from basics to food applications. In: *Gums and Stabilizers for the Food Industry*, vol. 15 (eds. P.A. Williams and G.O. Phillips), 211–222. Cambridge, UK: Royal Society of Chemistry.

Schmitt, C. and Turgeon, S.L. (2011). Protein/polysaccharide complexes and coacervates in food systems. *Advances in Colloid and Interface Science* 167: 63–70.

Schmitt, C., Kolodziejczyk, E., and Leser, M.E. (2005). Interfacial and foam stabilization properties of β-lactoglobulin-acacia gum electrostatic complexes. In: *Food Colloids: Interactions*, Microstructure and Processing (ed. E. Dickinson), 284–300. Cambridge, UK: Royal Society of Chemistry.

Seal, R. (1980). Industrial soy protein technology. In: *Applied Protein Chemistry* (ed. A. Grant), 87–111. London: Applied Science Publishers Ltd.

Servais, C., Jones, R., and Roberts, I. (2002). The influence of particle size distribution on the processing of food. *Journal of Food Engineering* 51 (3): 201–208.

Seyhun, N., Sumnu, G., and Sahin, S. (2003). Effects of different emulsifier types, fat contents, and gum types on retardation of staling of microwave-baked cakes. *Food/Nahrung* 47 (4): 248–251.

Shah, A., Zhang, G., Hamaker, B.R., and Campanella, O.H. (2011). Rheological properties of a soluble self assembled complex from starch, protein and free fatty acids. *Journal of Food Engineering* 105: 444–452.

Shahidi, F., Arachchi, J.K.V., and Jeon, Y.J. (1999). Food applications of chitin and chitosans. *Trends in Food Science & Technology* 10 (2): 37–51.

Sheard, P. and Tali, A. (2004). Injection of salt, tripolyphosphate and bicarbonate marinade solutions to improve the yield and tenderness of cooked pork loin. *Meat Science* 68 (2): 305–311.

Shelton, D.R. and D'Appolonia, B.L. (1985). Carbohydrate functionality in the baking process. *Cereal Foods World* 30: 437–442.

Sikorski, Z.E. (1997). *Chemical and Functional Properties of Food Components*. Lancaster, PA: Technomic Publishing Co. Inc.

Silva, H.D., Cerqueira, M.Â., and Vicente, A.A. (2012). Nanoemulsions for food applications: development and characterization. *Food and Bioprocess Technology* 5 (3): 854–867.

Singh, H., Ye, A., and Horne, D. (2009). Structuring food emulsions in the gastrointestinal tract to modify lipid digestion. *Progress in Lipid Research* 48 (2): 92–100.

Sinha, R., Radha, C., Prakash, J., and Kaul, P. (2007). Whey protein hydrolysate: functional properties, nutritional quality and utilization in beverage formulation. *Food Chemistry* 101 (4): 1484–1491.

Speiciene, V., Guilmineau, F., Kulozik, U., and Leskauskaite, D. (2007). The effect of chitosan on the properties of emulsions stabilized by whey proteins. *Food Chemistry* 102 (4): 1048–1054.

Stoklosa, A.M., Lipasek, R.A., Taylor, L.S., and Mauer, L.J. (2012). Effects of storage conditions, formulation, and particle size on moisture sorption and flowability of powders: a study of deliquescent ingredient blends. *Food Research International* 49 (2): 783–791.

Takahashi, S., Kobayashi, R., Watanabe, T., and Kainuma, K. (1983). Effects of addition of soybean protein on gelatinization and retrogradation of starch. *Nippon Shokuhin Kogyo Gakkaishi* 30: 276–282.

Tamsen, M., Shekarchizadeh, H., and Soltanizadeh, N. (2018). Evaluation of wheat flour substitution with amaranth flour on chicken nugget properties. *LWT - Food Science and Technology* 91: 580–587. https://doi.org/10.1016/j.lwt.2018.02.001.

Taylor, A.J. and Hort, J. (2007). *Modifying Flavour in Food*. Elsevier.

Tcholakova, S., Denkov, N.D., and Lips, A. (2008). Comparison of solid particles, globular proteins and surfactants as emulsifiers. *Physical Chemistry Chemical Physics* 10: 1608–1627.

Tieken, S., Leidy, H., Stull, A. et al. (2007). Effects of solid versus liquid meal-replacement products of similar energy content on hunger, satiety, and appetite-regulating hormones in older adults. *Hormone and Metabolic Research* 39 (5): 389–394.

Tolstoguzov, V.B. (1996). Structure-property relationships in foods. In: *Macromolecular Interactions in Food Technology*, ACS symposium series (eds. N. Parris, A. Kato, L.K. Creamer and J. Pearce), 1–14. Washington, DC: American Chemical Society.

Tolstoguzov, V.B. (1999). Interactions of gelatin with polysaccharides. In: *Gums and Stabilisers for the Food Industry*, vol. 5 (eds. G.O. Phillips, P.A. Williams and D.J. Wedlock), 157–175. Oxford: Oxford University Press.

Torbica, A., Hadnađev, M., and Dapčević, T. (2010). Rheological, textural and sensory properties of gluten-free bread formulations based on rice and buckwheat flour. *Food Hydrocolloids* 24 (6): 626–632. https://doi.org/10.1016/j.foodhyd.2010.03.004.

Tornberg, E., Olsson, A., and Person, K. (1990). The structural and interfacial properties of food in relation to their functions in emulsions. In: *Food Emulsions* (eds. L. Kare and E.F. Stig), 247–326. New York;: Marcel Dekker Inc.

Tuller, M., Or, D., and Dudley, L.M. (1999). Adsorption and capillary condensation in porous media: liquid retention and interfacial configurations in angular pores. *Water Resources Research* 35 (7): 1949–1964.

Utsumi, S., Matsumura, Y., and Mori, T. (1997). Structure-function relationships of soy proteins. In: *Food Proteins and Their Applications* (eds. S. Damodaran and A. Paraf), 257–291. New York;: Marcel Dekker Inc.

Van-Vliet, T. (1988). Rheological properties of filled gels: influence of filler matrix interaction. *Colloid & Polymer Science* 266: 518–524.

Villalva, F.J., Cravero Bruneri, A.P., Vinderola, G. et al. (2017). Formulation of a peach ice cream as potential symbiotic food. *Food Science and Technology* 37 (3): 456–461.

Villanueva, M., Harasym, J., Muñoz, J.M., and Ronda, F. (2019). Rice flour physically modified by microwave radiation improves viscoelastic behavior of doughs and its bread-making performance. *Food Hydrocolloids* 90: 472–481. https://doi.org/10.1016/j.foodhyd.2018.12.048.

Wang, C.H. and Damodaran, S. (1990). Thermal gelation of globular proteins: weight average molecular weight dependence of gel strength. *Journal of Agricultural and Food Chemistry* 38: 1157–1164.

Wang, J.C., Chen, S.H., and Xu, Z.C. (2008). Synthesis and properties research on the nano-capsulated capsaicin by simple coacervation method. *Journal of Dispersion Science and Technology* 29 (5): 687–695.

Whelan, A.P., Vega, C., Kerry, J.P., and Goff, H.D. (2008). Physicochemical and sensory optimisation of a low glycemic index ice cream formulation. *International Journal of Food Science & Technology* 43 (9): 1520–1527.

Wójtowicz, A. and Mościcki, L. (2009). Influence of extrusion-cooking parameters on some quality aspects of precooked pasta-like products. *Journal of Food Science* 74 (5): E226–E233.

Xiong, Y.L. (1997). Structure-function relationships of muscle proteins. In: *Food Proteins and Their Applications* (eds. S. Damodaran and A. Paraf), 341–392. New York; and Hong Kong: Marcel Dekker Inc.

Ylimaki, G., Hawrysh, Z., Hardin, R., and Thomson, A. (1988). Application of response surface methodology for development of rice flour yeast breads-sensory measurements. *Canadian Institute of Food Science and Technology Journal* 21 (4): 361.

Youssef, M. and Barbut, S. (2010). Physicochemical effects of the lipid phase and protein level on meat emulsion stability, texture, and microstructure. *Journal of Food Science* 75 (2): S108–S114.

Yu, L., Ramaswamy, H.S., and Boye, J. (2013). Protein rich extruded products prepared from soy protein isolate-corn flour blends. *LWT - Food Science and Technology* 50 (1): 279–289.

Yuno-Ohta, N., Toryu, H., Higasa, T. et al. (1996). Gelation properties of ovalbumin as affected by fatty acid salts. *Journal of Food Science* 61: 906–910.

Zayas, J.F. (1997). Foaming properties of proteins. In: *Functionality of Proteins in Food* (ed. J.F. Zayas), 260–309. New York;: Springer-Verlag.

Zhang, G. and Hamaker, B.R. (2003). A three-component interaction among starch, protein, and free fatty acids revealed by pasting profiles. *Journal of Agricultural and Food Chemistry* 51: 2797–2800.

Zhang, G., Bhopatkar, D., Hamaker, B.R., and Campanella, O.H. (2015). Self assembly of amylose, protein and lipid as a nanoparticle carrier of hydrophobic small molecules. In: *Nanotechnology and Functional Foods: Effective Delivery of Bioactive Ingredients* (eds. C. Sabliov, H. Chen and R.Y. Yada), 263–271. Wiley.

Zhu, K., Huang, S., Peng, W. et al. (2010). Effect of ultrafine grinding on hydration and antioxidant properties of wheat bran dietary fiber. *Food Research International* 43 (4): 943–948. https://doi.org/10.1016/j.foodres.2010.01.005.

Ziegler, G.R. and Foegeding, E.A. (1990). The gelation of proteins. *Advances in Food and Nutrition Research* 34: 203–298.

6
3D Printing

Arianna Dick, Sangeeta Prakash, and Bhesh Bhandari
School of Agriculture and Food Sciences, The University of Queensland, St Lucia, Queensland, Australia

6.1 Introduction

Three-dimensional (3D) printing is an innovative alternative to common food processing technology that brings into the play the freedom of customization and the interaction of the user to scheme on-demand custom-made goods. Although it is unlikely to displace the conventional practices for food manufacturing, 3D printing is increasingly catching attention at the academic, commercial, and industrial levels, due to the advantages offered over traditional manufacturing methods. Through creative-driven innovation, it offers the possibility of constructing complex geometries in a single manufacturing step, with little or no workforce interaction during printing, while contributing sustainably by employing only the required amount of material to build a construct and allowing the conversion of alternative ingredients into appealing products.

6.2 A Brief History of 3D Food Printing

The beginning of 3D-printed food materials was marked in 2006 by the introduction of the Fab@Home Project, conducted at Cornell University. Researchers from the Department of Mechanical and Aerospace Engineering designed an open-source multi-material 3D printer consisting of a syringe-based extrusion system which, besides chocolate, cheese, and cookie dough, was able to 3D print objects out of plastics, plasticine, hydrogels, sintered stainless steel, amongst other materials (Malone and Lipson 2007). Before its release, 3D printers were proprietary and costly. However, the Fab@Home Project launching expanded the advancement of freeform fabrication systems technology, making them suitable for a wider range of materials and applications; and henceforth, the creation of 3D printers has increased, aiming to develop further 3D printing mechanisms and printing conditions.

Food Formulation: Novel Ingredients and Processing Techniques, First Edition.
Edited by Shivani Pathania and Brijesh K. Tiwari.

6.3 Principle and Application

By means of digitally controlled layering and deposition of the feeding material or 'food ink', 3D printers build up freeform geometries based on computer-aided designs (CAD) with customized complex outer and inner structures. The mechanisms by which the food inks are layered depend on the printing methodology attached to the 3D printer and the state of the feeding material (Sun et al. 2015a). For instance, Selective Sintering technologies (laser and hot air) and Binder Jetting (BJ) are used for fusing and binding thermoplastic powder beds, such as sugar-based beds. Hot-Melt extrusion (HME), also known as Fused Deposition Modelling (FDM), melts thermoplastic materials, such as cheese and chocolate, to form a continuous filament that hardens upon deposition, while Room Temperature Extrusion (RTE) is employed for most soft foods, such as doughs and spreads. However, other soft foods, such as meat paste or surimi, require a cooling temperature throughout the printing process, and currently only a small number of 3D printers meet this trait. Detailed information about the 3D printing mechanisms available for food materials has been described by Godoi et al. (2016) and Sun et al. (2015b).

The growing interest in 3D printing processing amongst researchers in the food field has thrived and their assessment of this technology has been applied to a variety of food groups.

6.4 3D Printed Food Products

Food products can be classified into two groups, based on their printability: natively printable materials and non-printable traditional food materials (Sun et al. 2015a). The former accounts for those processed foods that can be easily extruded from a syringe, such as chocolate and spreads. The latter refers to traditional foods like fruits, vegetables, meat, and rice, which require the addition of viscosity modifiers to support their extrusion ability and/or post-processing suitability.

What follows is a review of the current application of 3D printing for different types of food materials, classifying them into food categories, and further assessing the traits of the methodology required for each category, including printing mechanism, formulation and printability, post-processing suitability, and future improvements for major applications, where applicable.

6.4.1 Fruits, Vegetables, and Legumes

Fruits, vegetables, and legumes probably comprise one of the most researched food categories in 3D printing by extrusion. Several authors have investigated the printability of fruits, vegetables, and legumes, and their blends, despite being non-printable by nature and requiring flowing aids for printability enhancement (Sun et al. 2015b). Both fresh and powdered materials have been used to create extrudable, printable pastes. In both cases, the moisture can be controlled for better printability of the formula by separating out or adding liquid components, respectively, and the particle size can be modified to better adjust the printable characteristics of the ink.

Recently, a powder-dispersed ink system, using spinach powder at varying particle sizes (307, 259, 172, and 50 μm) within a xanthan gum matrix, was developed by Lee et al. (2019) to evaluate the printability and printing performance of the ink by extrusion. The authors reported that by increasing the particle size, the higher hydrodynamic stability observed positively affected the printability of the paste, whereas the increased porosity of the spinach powder was reflected in the microstructure of 3D printed samples, decreasing the bulk density and improving the printing performance (Figure 6.1). Similarly, the effect of particle content (10 and 30%) of powdered ingredients (broccoli, carrot, and spinach) and varying hydrocolloids (xanthan gum, guar gum, hydroxypropyl methylcellulose, and locust bean gum) on the rheology and printability (Figure 6.2) of the inks was studied (Kim et al. 2018). The findings showed the suitability of xanthan gum for printing purposes of powders by inhibiting the swelling of particles, thus presenting no significant differences in shear rheology and printability at higher particle contents, as compared to other hydrocolloids.

Similarly, a blend of fruits and vegetables made from fresh ingredients (carrots, pears, kiwi fruit, broccoli rabe leaves, and avocado) was recently printed by extrusion into a pyramid shape (Severini et al. 2018b). In this study, the sensory evaluation proved enhanced appearance for the printed products, as compared to the food formula, whereas no differences were detected for other attributes (colour, odour, taste, off-flavour, and off-odour). These results encourage the applicability of 3D printing technology to increase the consumption of healthier products, reshaped into more appetizing prototypes.

Other foods in this category include citric fruits derivatives, such as orange concentrate and lemon juice gels with high moisture and free-flowing characteristics, which can be added with different starches (wheat, potato, sweet potato, and corn starch) at varying concentrations to enhance the viscosity, and provide ease of extrusion and shape stability (Azam et al. 2018; Yang et al. 2018a, 2019). The printability of food inks comprises both the extrusion and shape stability to ensure the required printing performance. In this way, the use of flow enhancers is recognized for 3D printing by extrusion, and not only the type but also the concentration of hydrocolloids can contribute significantly (Liu et al. 2019c).

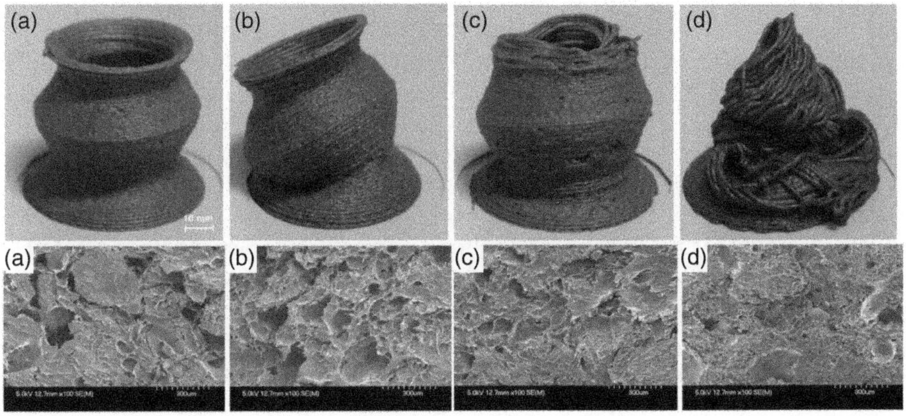

Figure 6.1 Photos (top) and SEM micrographs (bottom) of 3D printed spinach powder dispersions with varying particle sizes. (a) 307 μm; (b) 259 μm; (c) 172 μm; and (d) 50 μm. *Source:* Adapted from Lee et al. (2019), Figures 3, 4, pp. 5–6. Reproduced with permission of Elsevier.

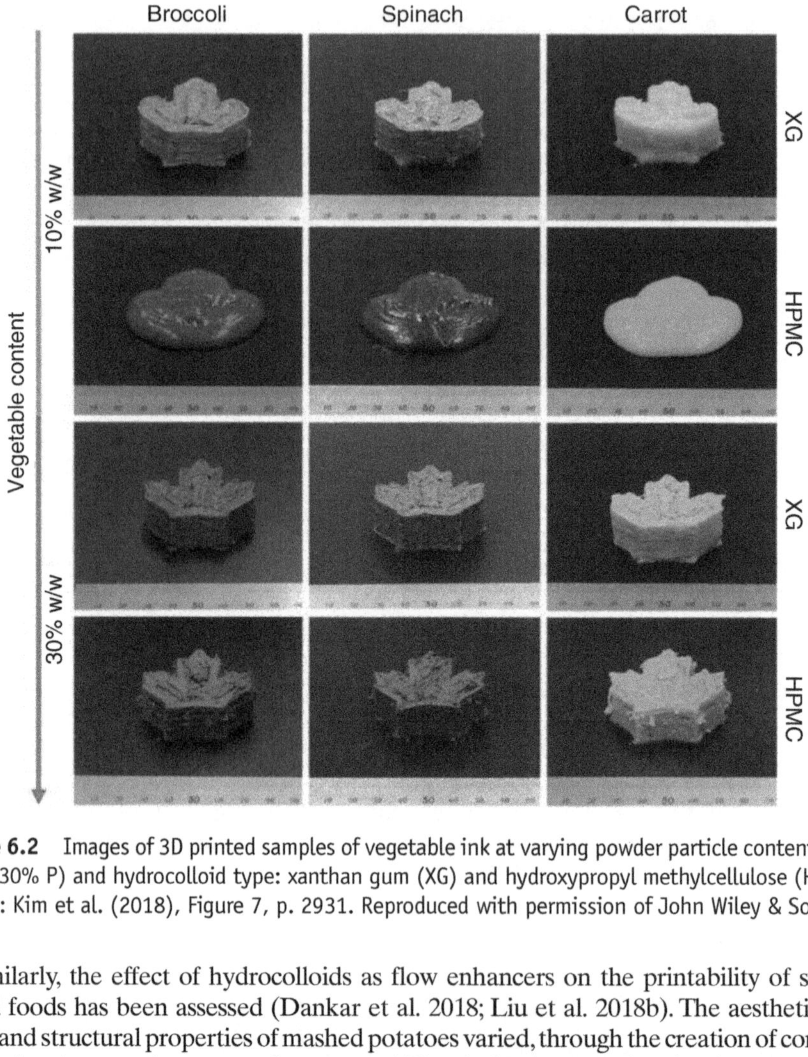

Figure 6.2 Images of 3D printed samples of vegetable ink at varying powder particle content (10% P and 30% P) and hydrocolloid type: xanthan gum (XG) and hydroxypropyl methylcellulose (HPMC). *Source*: Kim et al. (2018), Figure 7, p. 2931. Reproduced with permission of John Wiley & Sons.

Similarly, the effect of hydrocolloids as flow enhancers on the printability of starch-based foods has been assessed (Dankar et al. 2018; Liu et al. 2018b). The aesthetic, textural, and structural properties of mashed potatoes varied, through the creation of complex internal and external structures by means of 3D printing by extrusion (c; Liu et al. 2018a).

However, powdered fruits and vegetables can also be designed into a 3D model by Selective Laser Sintering (SLS). For instance, Diaz et al. (2014) at TNO proposed a method (US 2016/0100621 A1) comprising two formulations of fruit (strawberry) and legume (chickpea) powders, mixed with other structural and binding materials to build up instant 3D fruit beverage and soups. The freeform structures were created by fusing continuous cross-sectional layers of the powder bed with a carbon dioxide laser to generate the desired 3D profile.

6.4.2 Cereals and Grains

Contrary to the category of fruits and vegetables, food products made with cereals and grains usually require post-processing (e.g. baking) to allow or increase its digestibility. Therefore, when 3D printing by extrusion foods in this category, the formulating and processing steps, and the post-processing suitability, are of utmost importance to ensure the shape stability from deposition to cooking, and to attain the expected texture of the final product.

As an example regarding the formulating step, variations in the composition of wheat-based cookie doughs, including butter, sugar, and yolk concentrations, lead to relevant effects on extrusion and deposition, as well as on the shape stability after baking (Lipton et al. 2015; Yang et al. 2018b). However, the amount of leavening agent and processing conditions of the dough, such as mixing and the kneading degree, can modify the microporosity (1–100 μm) of the structures (Figure 6.4) (Noort et al. 2017). Also, the incorporation of food hydrocolloids (methylcellulose and xanthan gum) on the printability and shape stability of 3D printed cookies gives further insight into processing and post-processing in 3D printing (Kim et al. 2019). While methylcellulose (0.5–3%) exposes printing accuracy, the 3D printed shape is not retained after baking at 170 °C. Addition of 0.5% xanthan gum presents both suitable printability and post-processing feasibility (Figure 6.3), while displaying similar textural properties to that of a traditional cookie, as reported by the authors, and thus endorses the suitability of xanthan gum for 3D printing applications, particularly

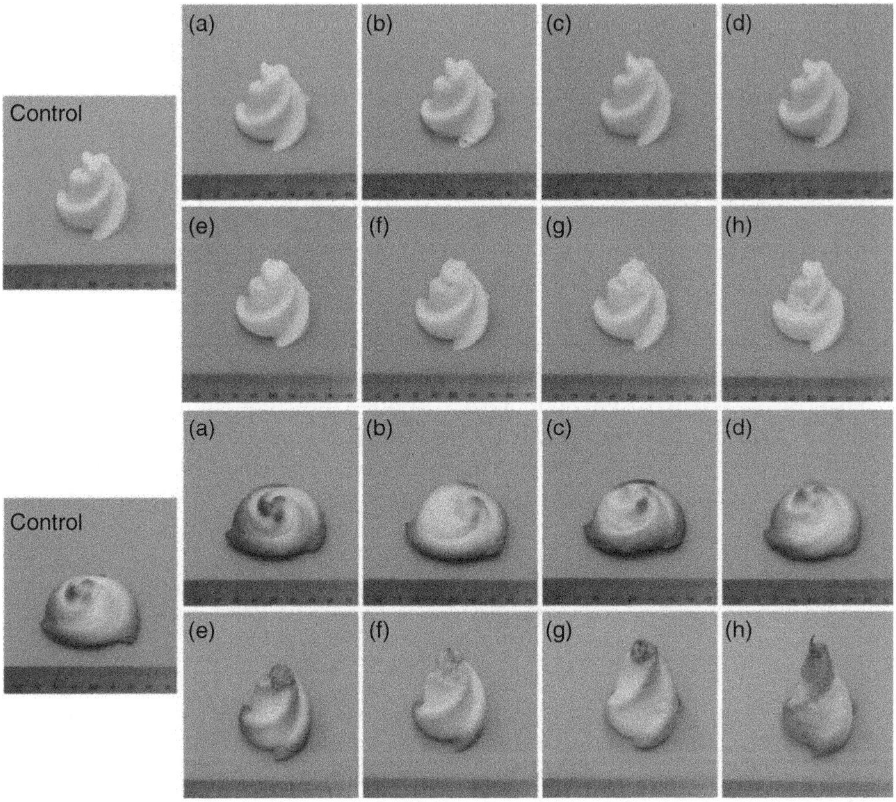

Figure 6.3 3D printed cookie dough. Top: raw, bottom: baked (170 °C). Control: no hydrocolloids added. (a–d) 0.5, 1, 2, and 3% methylcellulose. (e–h) 0.5, 1, 2, and 3% xanthan gum. *Source*: Adapted from Kim et al. (2019), Figure 3, 4, p. 73. Reproduced with permission of Elsevier.

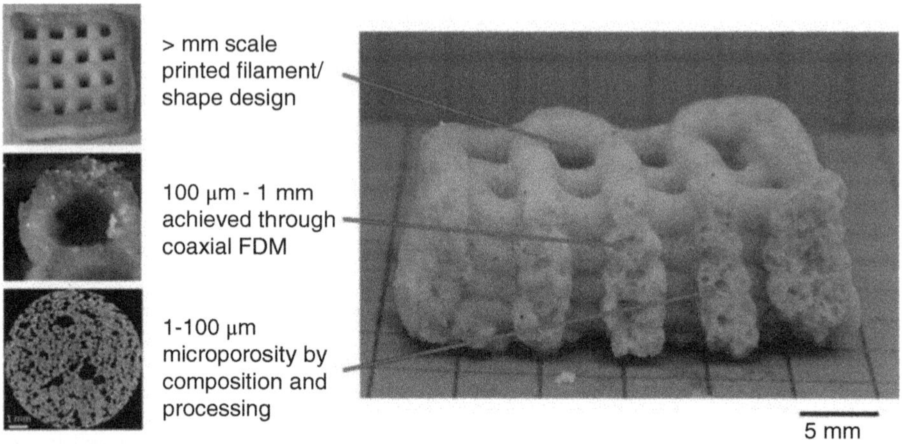

Figure 6.4 3D printed cookies with varying porosity levels printed as independent structural elements (TNO). *Source*: Noort et al. (2017), Figure 6, p. 275. Public domain.

when the formulation is cold- or room temperature-mixed and printed, but requires heating post-processing.

In addition, the printing accuracy and post-processing texture (baking at 200 °C) of cereal-based snacks (wheat) can be affected by variations in the printer settings: layer height and infill density. The former shows a proportional effect on the printed diameter, and an inverse effect on the shape height. The latter contributes highly to the changes in solid fraction and hardness of samples (Severini et al. 2016). Furthermore, a reduced solid fraction and increased void fraction throughout a 3D printed object create a foam structure, which can produce a larger object with variable textures upon post-processing, due to the incorporated porosity (Lipton and Lipson 2012).

However, besides varying the void fraction of the macrostructure through different levels of infill density and designs, the porosity can also be modified at the microstructure level, by coaxial extrusion of dough and bubbles (100 μm–1 mm diameter), affecting the mechanical properties of the cooked object (Noort et al. 2017), as shown in Figure 6.4. These findings provide an insight into the application of 3D printing technology for the design and textural customization of cereal-based products that require post-processing.

In addition to the 3D printing processing of doughs by extrusion, a combination of different constituents in the powder state can also be used to create multicomponent cereal-based materials by powder-based printing mechanisms. In this way, BJ, also known as powder-bed printing (PBP), can be used to 3D print a solid pasta that can be dried or boiled post-processing. The multi-component powder bed comprises flour, albumin, and guar gum (as filler, water-soluble protein, and hydrocolloid, respectively), and water as liquid spray (Diaz et al. 2015). In addition, SLS can be employed to create 2D and 3D bakery products using wheat flour-based powder with maltodextrin and palm oil acting as binding agents (Diaz et al. 2014). The fracture-ability of the samples can be modified by varying the cell size of 2D sheets of honeycomb pattern, which further yield increased hardness after steaming and baking. The compressibility of the samples

Figure 6.5 3D printed structures by Selective Laser Sintering (SLS). Top: 2D honeycomb-patterned sheets with varying cell sizes (macroporosity). Bottom: anisotropic 3D structure with varying compressibility behaviour based on direction. *Source*: Adapted from Noort et al. (2017), Figures 7, 9, pp. 275–276. Public domain.

can be affected by varying the cell wall thickness in different directions within the structure (Figure 6.5) (Noort et al. 2017).

6.4.3 Meat, Fish, and Seafood Products

Very little work has been done regarding the printability of meat, fish, and seafood products. It can be attributed to both the toughness of the materials due to the presence of fibres and connective tissue, which hinders the extrusion process (Dick et al. 2019a), and the cooling conditions required to keep a microbial safe environment during the printing process, which are not available in most 3D food printers. Some studies regard the use of viscosity enhancers to aid the rheological properties for 3D printing by extrusion, as well as the attuning of printing settings to enhance the printing performance. For instance, the addition of sodium chloride (Wang et al. 2018) and sweet potato starch (Dong et al. 2019) to fish surimi gel improved its physical properties for ease of extrusion, whereas the adjustment of nozzle size, print-move speed, and extrusion rate contributed to the creation of precise 3D geometries.

Other studies refer to the creation of composite printed meat products using dual-nozzle extrusion and their suitability for post-processing cooking (Dick et al. 2019b; Lipton et al. 2010). Furthermore, the production of sustainable products by seizing meat off-cuts for the development of value-added red meat was demonstrated as viable with the use of 3D printing technology (Sim 2017). In addition, meat and fish products have been used for 3D printing products with modified textures for people with special dietary needs, such as chewing and swallowing difficulties (Kouzani et al. 2017). In the latter study, fish in spring water was blended to form a slurry, requiring no post-processing. However, when 3D printing modified-texture pastes that will be further heated, studies regarding the textural changes after post-processing are required.

For end-products with regular textures, most studies report the printability of fibrous materials in the raw state, which are then either frozen or cooked following the printing step. Due to the intricate shapes that can be created with 3D printing technology, as compared to subtractive manufacturing, the post-processing feasibility of 3D printed meat products needs to be evaluated to ensure the shape stability throughout the manufacturing process. Recently, a composite multi-layer 3D printed meat product containing varying layers of lard (0–3 layers within the structure), at three different infill densities (50%, 75%, 100%, as shown in Figure 6.6), was fabricated and further subjected to post-processing sous-vide cooking (Dick et al. 2019b). The 3D meat products maintained their internal and external macrostructure during and after

Figure 6.6 Composite multi-layer meat products: (a) 3D printed raw and cooked samples (LF100: 1 layer of lard, 100% infill density); (b) Filament stream at 50, 75, and 100% infill density, for meat paste (2 mm nozzle diameter) and lard (1 mm nozzle diameter). *Source:* Dick et al. (2019b), Figure 3, p. 13. Reproduced with permission of Elsevier.

post-processing. However, the samples tended to shrink as the number of fat layers increased due to higher cooking loss, most likely attributed to the coalescence and displacement of unstable fat droplets during the cooking process, drifting towards the outer part of the porous structure and leaving room for the contraction of proteins. However, an opposite effect was observed by increasing the infill density (solid fraction) due to a more packed macrostructure, as suggested by the authors.

Hitherto, the extrusion mechanism seems to be the preferred technology for 3D printing of meat, fish, and seafood products, due to the viscoelastic properties of the formulated pastes. Nonetheless, it would be interesting to study the printability of meat powders and meat protein powders by PBP technologies, to contribute towards the production of value-added meat products from potential waste materials. Likewise, the use of such powdered products in formulated meat pastes for extrusion can be used as functional ingredients to increase the nutritional value of the pastes, and thus allowing the production of customized products, nutrition-wise.

6.4.4 Dairy Products and Derivatives

Milk products (e.g. cheese, butter, yoghurt) and milk derivatives (e.g. whey protein, sodium caseinate [NaCN]) display potential structural properties relevant for extrusion and shape stability in 3D printing processing, besides contributing functional components for customized nutrition (Ross et al. 2019). Nonetheless, a limited number of studies has been carried out to assess the printability of foods in this category.

Taking into account the complete process involved in 3D printing of dairy products, the suitability of the material's properties at each stage requires detailed evaluation. When 3D printing dairy products and derivatives by extrusion, the process can be divided into four steps: pre-processing, extrusion, deposition, and post-processing, depending on the food ink properties.

The pre-processing step refers to the preparation of the material before extrusion, which includes the assessment of a suitable formulation, melting, flowing, and viscoelastic behaviour. For the extrusion step, the effect of shearing and printing conditions on the microstructure, physical, and mechanical properties need to be accounted for, as this may affect the self-supporting ability upon deposition by solidification or stored energy, and thus the post-processing properties, which include the textural properties and the shape stability with time.

For instance, Liu et al. (2019b) studied the printability (rheological properties and printing performance) of milk protein composite gels by increasing the content of milk protein concentrate (MPC) in an NaCN dispersion (350, 400, 450, and 500 g/l of total MPC, and NaCN). Although all the gels with different protein combinations presented shear-thinning behaviour and seemed suitable for extrusion, significant differences in the apparent viscosity, yield stress, recovery rate, and elastic modulus affected the self-supporting ability of the 3D printed samples. The printing performance (shape fidelity and stability) increase with protein content up to 450 g/l, attributed to a transition from a concentrated solution to a gel structure, which presented lower deformation and higher recovery rate, and thus was able to maintain the given shape. However, with further protein addition, a hard and rough surfaced-gel was observed. At protein contents below 400 g/l, poor shear tolerance and low elastic modulus resulted in lack of self-supporting ability of the 3D printed material. Similarly, the best printing performance was observed when using semi-skimmed milk powder (SSMP) as a structural element in protein-rich pastes, as compared to pastes using skimmed-milk powder (SMP), starch, cellulose nanofiber, rye bran, oat, and/or fava bean protein concentrates (Lille et al. 2018). It was found that while a paste containing 60% of SSMP (9% fat) was easily extruded and able to hold the 3D structure, one containing 60% of SMP (0.4% fat) could not be extruded. As suggested by the authors, this may be attributed to the role of the fat acting as a plasticizer and aiding the extrusion, as well as the different carbohydrate content of the pastes (32% in SMP and 23% in SSMP), which may have significantly contributed to a higher storage (elastic) modulus (G') in 60% SMP, as shown in Figure 6.7.

In addition, the effect of 3D printing processing (melting, extrusion, solidification) on the textural and microstructural properties of processed cheese was analysed by Le Tohic et al. (2018). The authors compared samples of commercially processed cheese processed to different extents and reshaped into 20 mm diameter (aspect ratio:1) cylinders: untreated cheese (UC), melted cheese (MC), and melted and extruded at 4 ml/min (low-speed printed cheese, LSPC) and 12 ml/min (high-speed printed cheese, HSPC). The UC sample was core-extracted with a borer from the cheese block. The MC sample was melted in a water bath at 75 °C for 15 minutes and cooled at room temperature for 30 minutes. For the extruded sample, the processed cheese was filled into a syringe, melted in a water bath at 75 °C for 12 minutes, and further extruded into the desired shape. The results revealed decreased hardness and increased melt ability of MC as compared to UC, attributed to the coalescence of fat globules upon melting (Figures 6.8(a) and (b)).

However, printed samples (LSPC and HSPC) presented decreased hardness than UC and MC, with no differences found between LSPC and HSPC, suggesting no impact of extrusion speed. In addition, LSPC and HSPC displayed less stickiness and stiffness (i.e. lower complex modulus, G*) than UC. As suggested by the authors, these findings can be explained with the microstructural differences observed amongst the

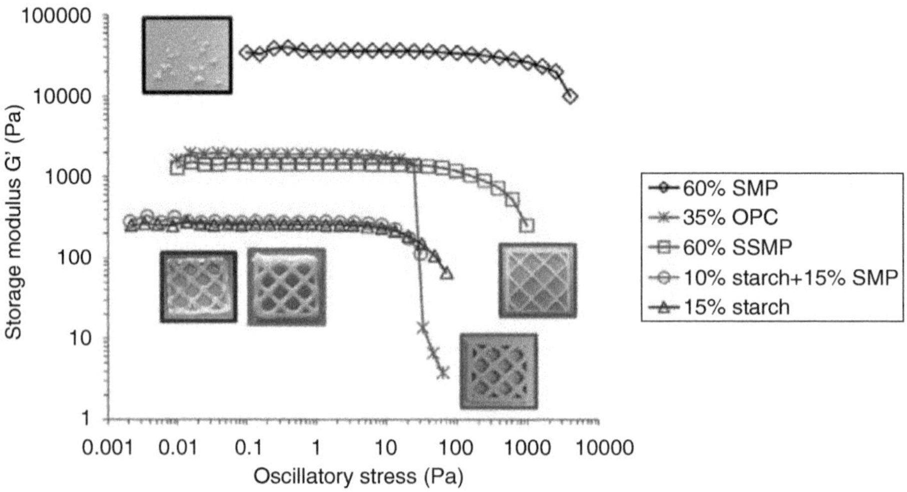

Figure 6.7 Storage modulus vs. oscillatory stress for protein-rich pastes containing SMP, semi skimmed-milk powder (SSMP), oat protein concentrate (OPC), and/or starch. *Source:* Lille et al. (2018), Figure 3, p. 25. Reproduced with permission of Elsevier.

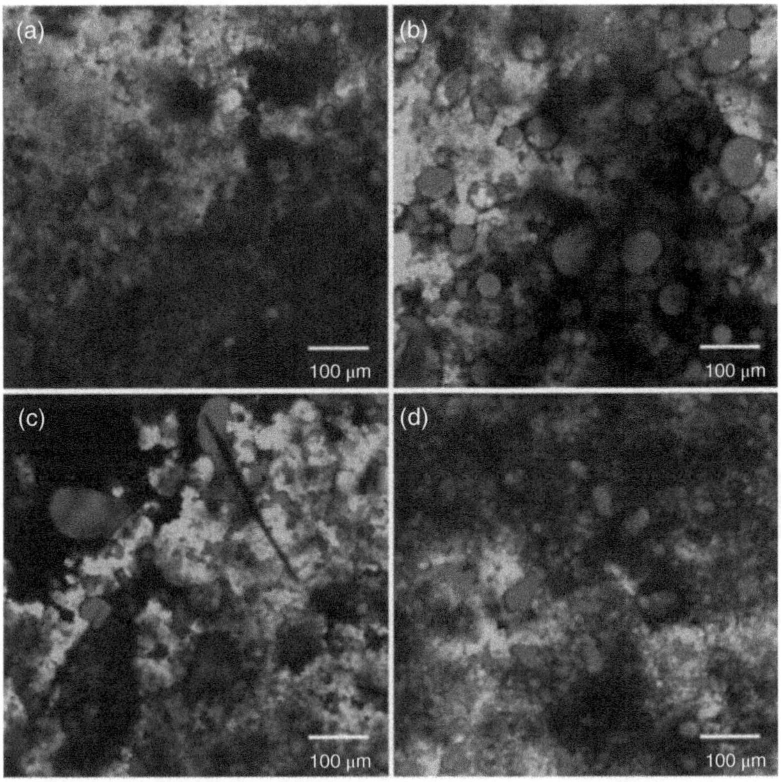

Figure 6.8 Confocal laser scanning micrographs of processed cheese samples (fat globules distributed within a protein phase). (a) UC: untreated cheese; (b) MC: melted cheese; (c) LSPC: melted and 3D printed at 4 ml/min; and (d) HSPC: melted and printed at 12 ml/min. *Source*: Le Tohic et al. (2018), Figure 5, p. 61. Reproduced with permission of Elsevier.

four samples. In UC sample (Figure 6.8(a)), an even distribution of spherical fat droplets emulsified within a protein phase can be seen, An increase in the size of the fat droplets is observed in the MC sample due to coalescence upon melting (Figure 6.8(b)). Contrastingly, non-spherical, disrupted fat droplets are visualized in both printed samples (Figures 6.8(c) and (d)), while smaller fat droplets, more evenly distributed size-wise, are observed in HSPC when compared to LSPC, attributed to a higher shear of extrusion that disrupted the fat globules. Furthermore, the 3D printing process affected the colour of the processes cheese samples. In this way, larger fat droplets formed upon melting relate to darker samples (MC, LSPC, HSPC), whereas the extrusion speed seemed to have an impact on the yellow-blue position of the samples on the colour space, attributed to variations in fat globule structure and protein phase (Le Tohic et al. 2018).

Besides the modification of physical and mechanical properties, 3D printing by extrusion allows the modification of nutritional quality and sensory perception through the combination of different materials in inhomogeneous spatial distribution. In this way, customization of food products for specific dietary requirements is possible by employing multi-nozzle extrusion or co-extrusion within one nozzle.

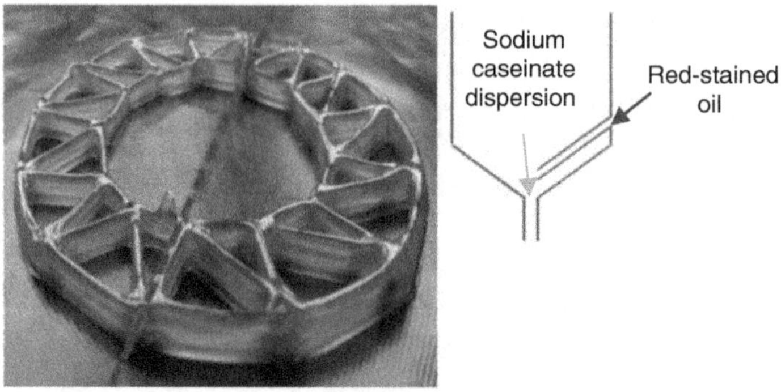

Figure 6.9 3D model printed using a dispenser with side-inlet. The continuous phase is made of a sodium caseinate dispersion, while the oil phase is dispensed in inhomogeneous distribution through the side-inlet. The left side of the structure was printed with sodium caseinate dispersion only, whereas the right side included red-stained oil droplets (darker spots near the junction zones). *Source*: Adapted from Schutyser et al. (2018), Figures 2, 8, pp. 51, 54. Reproduced with permission of Elsevier.

For instance, olive oil droplets were dispersed into a multi-layer structure of a continuous phase of NaCN dispersion (Schutyser et al. 2018). With a purposely-designed dispenser, which included a side-inlet, the authors were able to control the simultaneous dispensing of both materials to be delivered in specific locations within the structure, as shown in Figure 6.9.

These results demonstrate the relevant relationship of the rheological assessment of the food ink with the different steps involved in 3D printing processing, and the potential application of dual nozzle or co-extrusion to combine materials and their distribution within a product, and provide different nutritional values, mouthfeels, textures, eating experience, etc. of dairy products.

6.4.5 Confectionery and Desserts

In the category of confectionery and desserts, 3D printing technology allows the production of innovative, intricate product designs, which would be either time-consuming or rather unfeasible by traditional processes. Sugar-based products marked the beginning of 3D printed powdered foods in 2007 when The CandyFab Project created 3D printing machines that used a selective hot air sintering and melting (SHASAM) mechanism, employing a beam of hot air to melt targeted locations of a sugar grain bed (Oskay and Edman 2014). Similarly, flavoured sugar products (mint, sour cherry, vanilla) were printed using BJ mechanism by 3D Systems (Business Insider Australia 2014). Regardless of complex geometries created by using solely sugar, which seemed to be inexpensive and feasible, the nutritional value and application of the 3D printed products were not as promising.

On the other hand, the extrusion mechanism allows the usage of a broader range of food materials, in different states that can be adapted for extrusion, and thus has been widely used to 3D print confectionery and desserts, such as chocolate, gummy sweets (Magic Candy Factory 2017), and marzipan (Procusini®). For instance, the printability of chocolate has been widely studied, taking into account each stage of

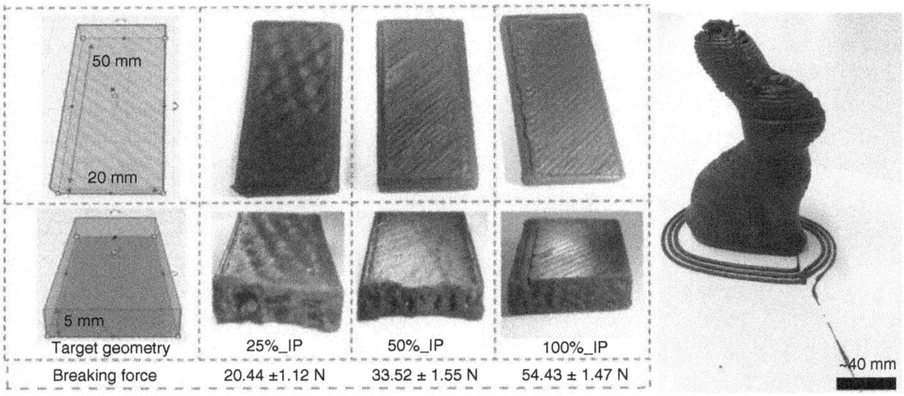

Figure 6.10 Left: 3D printed Callebaut dark chocolate with increasing infill density (IP): 25, 50, and 100%, and breaking force. *Source*: Adapted from Mantihal et al. (2019a), Figure 5, p. 6. Reproduced with permission of Elsevier. Right: 3D printed chocolate into a complex bunny structure with overhanging parts. *Source*: Adapted from Lanaro et al. (2017), Figure 9, p. 21. Reproduced with permission of Elsevier.

the 3D printing process, as detailed in Section 6.4.4. The rheological and thermal behaviour of chocolate provide valuable insights into its flowing behaviour during extrusion and help to set operational printing conditions for better quality of printing (Mantihal et al. 2017, 2019b). Likewise, the addition of processing and nutritional aids, such as magnesium stearate and plant sterols, can contribute to the flowing ability and nutritional quality of the chocolate ink, correspondingly, as reported by Mantihal et al. (2019b). Furthermore, the post-processing mechanical (e.g. snap ability) and textural properties (e.g. hardness) of the 3D printed chocolate can be modified through variations in the modelled internal structures (Mantihal et al. 2017, 2018). Figure 6.10 shows 3D printed chocolate with increasing infill density and breaking force.

On the other hand, by attuning the printing parameters, such as extrusion rate, nozzle speed, nozzle height, and cooling rates, more complex geometries comprising steep build angles and pendulous parts can also be created (Figure 6.10) (Hao et al. 2010; Lanaro et al. 2017).

6.4.6 Alternative Ingredients and End-Products

6.4.6.1 Sustainable Sources Aiming to support ethical consumption, food security, and environmentally sustainable sources, 3D printing technology has been applied to the production of laboratory-cultured meat and insect-based foods (Lupton and Turner 2018). Raw meat tissue can be grown from cultured stem cells using bioprinting technology, which employs the inkjet printing mechanism to dispense cultured stem cells from a syringe-based print head onto an agarose gel scaffold where the cells are fused to generate an engineered tissue, which is then matured into meat fibres (Forgacs et al. 2012; Sher and Tutó 2015). Nonetheless, the public acceptance and thus commercialization of laboratory-fabricated meat may not yet be underway, as cultural barriers may have to be overcome first since healthy, tasty, and natural foods seem to be the central drivers for consumer's food choices, while 3D-printed cultured meat may be considered unnatural, and thus not fresh, potentially harmful, tasteless, or unhealthy (Lupton and Turner 2018).

On the other hand, alternative ingredients, such as insect-based proteins and minerals, are non-conventional food ingredients that represent a novel source of nutrients (Sun et al. 2015b), and thus their use for 3D food printing applications has been triggered. For instance, Ground larvae of yellow mealworms were added to cereal-based snacks for protein enrichment, improving the essential amino acid content from 32 g to 41.3 g/100 g protein by increasing the enrichment from 0 to 20% ground insect, without displaying a negative impact on the technological quality of the printed foods (Severini et al. 2018a). In this way, the initial appearance of alternative ingredients, which can be a factor in discouraging their consumption (Soares and Forkes 2014), can be transformed into an aesthetic food product with high nutritional quality and sustainable imprint.

In addition, potential food waste such as dried bread, leftover-boiled rice, and odd-shaped or overripe fruits and vegetables have been used as non-conventional ingredients to create sweet and savoury appealing 2D and 3D food products with appetizing features (Dezeen 2018), suggesting new sustainable opportunities for the growing population.

Furthermore, meat-free steak and chicken breast formulated from vegetable sources like rice, peas, and seaweed are being developed by 3D printing the formulation into a desired shape and internal design to give the final product a fibrous texture like that of meat (NOVAMEAT 2020). Moreover, other non-meat derivatives with high protein value, such as egg-white protein, have been used for 3D printing processing, seizing their functional and nutritional properties. For instance, egg albumen protein powder was added to a mixture of gelatine, corn-starch, and sucrose to favour the heat-induced gelation of the mixture and provide it with enhanced rheological, frictional, and textural properties (Liu et al. 2019a), at the same time as favouring the protein content of the blend.

6.4.6.2 Tailor-made Products As an alternative for the ice cream-scooped pureed foods that people with chewing and swallowing difficulties are often served (Cichero 2015), 3D printing technology allows reshaping foods with modified textures into the original shape providing stimulating attributes that make the overall meal more appealing (Figure 6.11). In this way, the printability of broccoli puree (Tan et al. 2018), carrot puree (Huang 2018), pumpkin, beetroot, and fish purees (Kouzani et al. 2017) have been reported, besides the personalized texture-

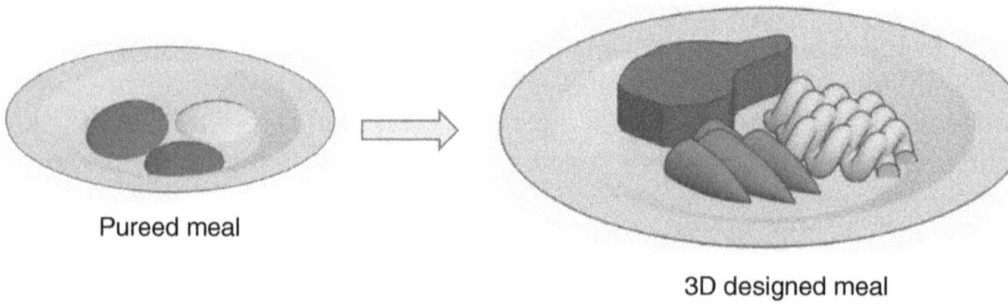

Figure 6.11 Scheme of modified-texture meal: from pureed to 3D designed. *Source*: Arianna Dick, Sangeeta Prakash, Bhesh Bhandari.

modified meals automated and manufactured as part of the PERFORMANCE project (CORDIS 2016).

Furthermore, through a combination of ingredients into the feeding material, tailored-foods for different dietary needs can be manufactured, such as customized fruit-based snacks for children of 3–10 years (Derossi et al. 2018). Whereas, by the addition of ingredients into targeted locations and quantities, foods with reduced unhealthy components such as sugar, salt, or fat that maintain the perceived taste and flavour can be produced (Prakash et al. 2019).

6.5 Scalability and Future Outlook

Currently, 3D printing technology is likely only applicable for in-house commodity, restaurants, or short-scale production due to the limitations of printable materials, and printers' dimensions and operating conditions. In order to reach an industrial scale, 3D food printing requires further research regarding the printability of a broader range of foods materials, with their corresponding printing mechanisms and operating conditions.

A printability classification system of 12 selected foods (jam, Greek yoghurt, ketchup, peanut butter, cheddar cheese, cream cheese, cookie dough, ground ham, mashed potato, bean paste, sugar paste, and marzipan) was proposed by Kim et al. (2017). The authors employed methylcellulose at varying concentrations (5–14%) as a reference material to simulate the rheological properties of the selected foods and thus established a guideline for printability based on the dimensional stability and degree of handling, and its relation to shear modulus to predict the deformation behaviour upon deposition. However, besides shear modulus, the printability can be explained by several rheological properties at each stage of the printing process (extrusion, recovery, and self-supporting stage), as proposed by (Liu et al. 2019c). Therefore, yield stress at the printing temperature, viscosity, and shear-thinning behaviour can give an insight into the suitability of the material for the extrusion stage. Similarly, shear recovery and temperature recovery can be linked to the recoverability of the material after exiting the nozzle, whereas yield stress and the elastic modulus recount for the self-supporting ability after deposition.

In this way, suitable rheological and mechanical properties could be demarcated to classify different types of food materials based on their printability, and thus the appropriate printing mechanism.

Furthermore, the scopes of 3D printers require more development. For the case of the extrusion mechanism, which is mostly used for food applications, the capacity of the hoppers and the difficulty of continuously feeding material limits the size of the printed objects. In addition, the speed–accuracy relationship needs advanced attuning to create more printed units in less time, thus increasing the cost-effectiveness of the process. Likewise, the temperature throughout the printer components – hopper, nozzle, and platform – should be controlled to ensure the microbiological safety of the products, despite the printing processing time (Dick et al. 2019a).

With the ongoing research on printing materials and printer's optimization, 3D printing technology for food applications is expected to reach an industrial scale within the next 10 years, to transform current, novel, and alternative food materials into value-added wholesome food products custom-made for individual aesthetic and dietary needs and preferences.

6.6 Conclusion

3D printing technology promises to be a tool to create novel food products from a range of common food materials, and to give value-added products to alternative food sources contributing to environmentally sustainable practices. Besides offering a lack of restrictions for appearance design, it allows the modification of the textural, mechanical, and nutritional qualities of 3D printed products in a way that would have been impractical through traditional subtractive manufacturing. However, further improvements of the technology are needed to allow its application for the manufacturing of a wider spectrum of foodstuffs.

References

Azam, S.M.R., Zhang, M., Mujumdar, A.S., and Yang, C. (2018). Study on 3D printing of orange concentrate and material characteristics. *Journal of Food Process Engineering* 41: e12689.

Business Insider Australia (2014). This is the best 3F food printer we've seen yet – and it makes stunning desserts. Pedestrian Group, BI Intelligence. www.businessinsider.com.au/3d-systems-chefjet-printers-2014-9 (accessed 13 August 2019).

Cichero, J.A.Y. (2015). Texture-modified meals for hospital patients. In: *Modifying Food Texture. Volume 2, Sensory Analysis, Consumer Requirements and Preferences* (eds. J. Chen and A. Rosenthal), 135–162. Amsterdam, Netherlands: Woodhead Publishing: Woodhead Publishing Series in Food Science, Technology and Nutrition.

CORDIS (The Community Research and Development Information Service) (2016). Development of Personalised Food using Rapid Manufacturing for the Nutrition of elderly Consumers. PERFORMANCE Project, Seventh Framework Program, European Commission. Luxembourg: The Publications Office of the European Union. https://cordis.europa.eu/project/id/312092 (accessed 15 October 2020)

Dankar, I., Pujolà, M., El Omar, F. et al. (2018). Impact of mechanical and microstructural properties of potato puree-food additive complexes on extrusion-based 3D printing. *Food and Bioprocess Technology* 11: 2021–2031.

Derossi, A., Caporizzi, R., Azzollini, D., and Severini, C. (2018). Application of 3D printing for customized food. A case on the development of a fruit-based snack for children. *Journal of Food Engineering* 220: 65–75.

Dezeen (2018). Elzelinde van Doleweerd creates 3D-printed snacks from food waste. www.dezeen.com/2018/10/03/upprinting-food-elzelinde-van-doleweerd-beijing-design-week-upprinting-food-design (accessed 13 August 2019).

Diaz, J.V., Van Bommel, K.J.C., Noort, M.W.-J. et al. (2014). Method for the production of edible objects using SLS and food products. US Patent Application 14/894,231, filed May 27, 2014, published April 14, 2016, and granted October 9, 2018.

Diaz, J.V., Noort, M.W.-J., and Van Bommel, K.J.C. (2015). Method for the production of an edible object by powder bed (3D) printing and food products obtainable therewith. US Patent Application 15/116,048, filed February 2, 2015 and published June 15, 2017.

Dick, A., Bhandari, B., and Prakash, S. (2019a). 3D printing of meat. *Meat Science* 153: 35–44.

Dick, A., Bhandari, B., and Prakash, S. (2019b). Post-processing feasibility of composite-layer 3D printed beef. *Meat Science* 153: 9–18.

Dong, X., Huang, Y., Pan, Y. et al. (2019). Investigation of sweet potato starch as a structural enhancer for three-dimensional printing of Scomberomorus niphonius surimi. *Journal of Texture Studies* 50 (4): 1–9.

Forgacs, G., Marga, F., and Jakab, K.R. (2012). Engineered comestible meat. US Patent Application 13/558,928, filed July 26, 2012 and granted April 22, 2014.

Godoi, F., Prakash, S., and Bhandari, B. (2016). 3D printing technologies applied for food design: status and prospects. *Journal of Food Engineering* 179: 44–54.

Hao, L., Mellor, S., Seaman, O. et al. (2010). Material characterisation and process development for chocolate additive layer manufacturing. *Virtual and Physical Prototyping* 5: 57–64.

Huang, C.Y. (2018). *Extrusion-Based 3D Printing and Characterization of Edible Materials*. University of Waterloo.

Kim, H.W., Bae, H., and Park, H.J. (2017). Classification of the printability of selected food for 3D printing: development of an assessment method using hydrocolloids as reference material. *Journal of Food Engineering* 215: 23–32.

Kim, H.W., Lee, J.H., Park, S.M. et al. (2018). Effect of hydrocolloids on rheological properties and printability of vegetable inks for 3d food printing. *Journal of Food Science* 83: 2923–2932.

Kim, H.W., Lee, I.J., Park, S.M. et al. (2019). Effect of hydrocolloid addition on dimensional stability in post-processing of 3D printable cookie dough. *LWT - Food Science and Technology* 101: 69–75.

Kouzani, A.Z., Adams, S., Whyte, D.J. et al. (2017). 3D printing of food for people with swallowing difficulties. Proceedings of The International Conference on Design and Technology (DesTech), Vol. 2017 (Geelong, Australia, 5–8 December 2016) (ed. P.K. Collins), 23–29. KnE Engineering. Dubai, UAE: Knowledge E Publishing. https://doi.org/10.18502/keg.v2i2.591

Lanaro, M., Forrestal, D.P., Scheurer, S. et al. (2017). 3D printing complex chocolate objects: platform design, optimization and evaluation. *Journal of Food Engineering* 215: 13–22.

Le Tohic, C., O'Sullivan, J.J., Drapala, K.P. et al. (2018). Effect of 3D printing on the structure and textural properties of processed cheese. *Journal of Food Engineering* 220: 56–64.

Lee, J.H., Won, D.J., Kim, H.W., and Park, H.J. (2019). Effect of particle size on 3D printing performance of the food-ink system with cellular food materials. *Journal of Food Engineering* 256: 1–8.

Lille, M., Nurmela, A., Nordlund, E. et al. (2018). Applicability of protein and fiber-rich food materials in extrusion-based 3D printing. *Journal of Food Engineering* 220: 20–27.

Lipton, J.I. and Lipson, H. (2012). Systems and methods for freeform fabrication of foamed structures. US Patent Application 13/356,194, filed January 23, 2012 and granted July 14, 2015.

Lipton, J., Arnold, D., Nigl, F. et al. (2010). Multi-material food printing with complex internal structure suitable for conventional post-processing. Proceedings of the Annual International Solid Freeform Fabrication Symposium, Austin, Texas (9-11 August 2010) (eds. D.L. Bourell, R.H. Crawford, C.C. Seepersad et al.) 809–815. Austin, Texas: The University of Texas at Austin.

Lipton, J.I., Cutler, M., Nigl, F. et al. (2015). Additive manufacturing for the food industry. *Trends in Food Science & Technology* 43: 114–123.

Liu, Z., Bhandari, B., Prakash, S., and Zhang, M. (2018a). Creation of internal structure of mashed potato construct by 3D printing and its textural properties. *Food Research International* 111: 534–543.

Liu, Z., Zhang, M., and Bhandari, B. (2018b). Effect of gums on the rheological, microstructural and extrusion printing characteristics of mashed potatoes. *International Journal of Biological Macromolecules* 117: 1179–1187.

Liu, Z., Zhang, M., and Yang, C. (2018c). Dual extrusion 3D printing of mashed potatoes/strawberry juice gel. *LWT – Food Science and Technology* 96: 589–596.

Liu, L., Meng, Y., Bhandari, B. et al. (2019a). Correction to: 3D printing complex egg white protein objects: properties and optimization. *Food and Bioprocess Technology* 12: 280–280.

Liu, Y., Yu, Y., Liu, C. et al. (2019b). Rheological and mechanical behavior of milk protein composite gel for extrusion-based 3D food printing. *LWT - Food Science and Technology* 102: 338–346.

Liu, Z., Bhandari, B., Prakash, S. et al. (2019c). Linking rheology and printability of a multicomponent gel system of carrageenan-xanthan-starch in extrusion based additive manufacturing. *Food Hydrocolloids* 87: 413–424.

Lupton, D. and Turner, B. (2018). Food of the future? Consumer responses to the idea of 3D-printed meat and insect-based foods. *Food and Foodways* 26: 269–289.

Magic Candy Factory (2017). 3D Magic Candy. http://magiccandyfactory.com/app (accessed 15 August 2019).

Malone, E. and Lipson, H. (2007). Fab@Home: the personal desktop fabricator kit. *Rapid Prototyping Journal* 13: 245–255.

Mantihal, S., Prakash, S., Godoi, F.C., and Bhandari, B. (2017). Optimization of chocolate 3D printing by correlating thermal and flow properties with 3D structure modeling. *Innovative Food Science & Emerging Technologies* 44: 21–29.

Mantihal, S., Prakash, S., and Bhandari, B. (2018). Textural modification of 3D printed dark chocolate by varying internal infill structure. *Food Research International* 121: 648–657.

Mantihal, S., Prakash, S., and Bhandari, B. (2019a). Texture-modified 3D printed dark chocolate: sensory evaluation and consumer perception study. *Journal of Texture Studies* 50 (5): 386–399.

Mantihal, S., Prakash, S., Godoi, F.C., and Bhandari, B. (2019b). Effect of additives on thermal, rheological and tribological properties of 3D printed dark chocolate. *Food Research International* 119: 161–169.

Noort, M., Van Bommel, K., and Renzetti, S. (2017). 3D-printed cereal foods. Cereal Foods World 62 (6): 272–277.

NOVAMEAT (2020). Plant-based meat, reinvented. www.novameat.com (accessed 21 August 2019).

Oskay, W. and Edman, L. (2014). The CandyFab Project. candyfab.org (accessed 13 August 2019).

Prakash, S., Bhandari, B.R., Godoi, F.C., and Zhang, M. (2019). Future outlook of 3D food printing. In: *Fundamentals of 3D Food Printing and Applications* (eds. F.C. Godoi, B.R. Bhandari, S. Prakash and M. Zhang), 373–381. Academic Press.

Ross, M.M., Kelly, A.L., and Crowley, S.V. (2019). Potential applications of dairy products, ingredients and formulations in 3D printing. In: *Fundamentals of 3D Food Printing and Applications* (eds. F.C. Godoi, B.R. Bhandari, S. Prakash and M. Zhang), 175–206. Academic Press.

Schutyser, M.A.I., Houlder, S., De Wit, M. et al. (2018). Fused deposition modelling of sodium caseinate dispersions. *Journal of Food Engineering* 220: 49–55.

Severini, C., Derossi, A., and Azzollini, D. (2016). Variables affecting the printability of foods: preliminary tests on cereal-based products. *Innovative Food Science & Emerging Technologies* 38: 281–291.

Severini, C., Azzollini, D., Albenzio, M., and Derossi, A. (2018a). On printability, quality and nutritional properties of 3D printed cereal based snacks enriched with edible insects. *Food Research International* 106: 666–676.

Severini, C., Derossi, A., Ricci, I. et al. (2018b). Printing a blend of fruit and vegetables. New advances on critical variables and shelf life of 3D edible objects. *Journal of Food Engineering* 220: 89–100.

Sher, D. and Tutó, X. (2015). *Review of 3D Food Printing*. Temes de Disseny.

Sim, T. (2017). Red meat printing initiative prompts 3D pop-up restaurant idea. Beef Central. https://www.beefcentral.com/trade/red-meat-printing-initiative-prompts-3d-pop-up-restaurant-idea/ (accessed 14 October 2020).

Soares, S. and Forkes, A. (2014). Insects Au gratin – an investigation into the experiences of developing a 3D printer that uses insect protein based flour as a building medium for the production of sustainable food. DS 78: Proceedings of the 16th International conference on Engineering and Product Design Education (E&PDE14), Design Education and Human Technology Relations, University of Twente, The Netherlands (4–5 September 2014) (eds E. Bohemia, A. Eger, W. Eggink et al.) Glasgow, UK: The Design Society, Wiltshire, UK: Institution of Engineering Designers.

Sun, J., Peng, Z., Zhou, W. et al. (2015a). A review on 3D printing for customized food fabrication. *Procedia Manufacturing* 1: 308–319.

Sun, J., Zhou, W., Huang, D. et al. (2015b). An overview of 3D printing technologies for food fabrication. *Food and Bioprocess Technology* 8: 1605–1615.

Tan, C., Chua, C.K., Li, L., and Wong, G. (2018). Enhancing 3D printability of pureed food by addition of hydrocolloids. Proceedings of the 3rd International Conference on Progress in Additive Manufacturing (Pro-AM 2018), Nanyang Technological University, Singapore (14–17 May 2018), 662–666. Singapore: Nanyang Technological University.

Wang, L., Zhang, M., Bhandari, B., and Yang, C. (2018). Investigation on fish surimi gel as promising food material for 3D printing. *Journal of Food Engineering* 220: 101–108.

Yang, F., Zhang, M., Bhandari, B., and Liu, Y. (2018a). Investigation on lemon juice gel as food material for 3D printing and optimization of printing parameters. *LWT - Food Science and Technology* 87: 67–76.

Yang, F., Zhang, M., Prakash, S., and Liu, Y. (2018b). Physical properties of 3D printed baking dough as affected by different compositions. *Innovative Food Science & Emerging Technologies* 49: 202–210.

Yang, F., Guo, C., Zhang, M. et al. (2019). Improving 3D printing process of lemon juice gel based on fluid flow numerical simulation. *LWT - Food Science and Technology* 102: 89–99.

7
Encapsulation Technologies Applied to Food Processing

Carlos Álvarez[1] and Daniel Pando[2]

[1] *Dept. Food Quality and Sensory Science, Teagasc Food Research Centre, Ashtown, Dublin, Ireland*
[2] *Nanovex Biotechnologies, Parque Tecnológico de Asturias, Asturias, Spain*

Summary

The food industry involves the use of many diverse ingredients that need to be incorporated into food products. Such ingredients may need to be preserved, dispersed, protected from the environment, or released under certain conditions. In order to address these challenges, encapsulation, which is a flexible tool, has been developed. This chapter will discuss the main encapsulation techniques and the main compounds that can be encapsulated for the food industry.

7.1 Introduction

According to many authors, encapsulation can be defined as the process of entrapping a substance into another material to generate capsules. In general, the encapsulated material is referred to as the core, fill, active, or payload phase; while encapsulants can be named as the membrane, coating, shell capsule, carrier, matrix, or external phase (Fang and Bhandari 2010). The nature of the materials employed for encapsulation, both the shell and the fill, are tremendously heterogeneous. Molecules such as proteins, lipids, vitamins, peptides, polyphenols, antimicrobials, carbohydrates, or probiotics can be encapsulated (McClements 2014). Meanwhile, diverse molecules such as carbohydrates, proteins, or lipids can be used as encapsulants. The size of the final particle, after the encapsulation process, can be at the range of nanometre, micrometre, or even millimetre scale and this forms the basis for them to be referred to as nano- or micro-encapsulation (Ray et al. 2016).

Encapsulation is useful in the fields of analytical chemistry, pharmaceuticals, nanoelectronics, metrology, cosmetics, research and, obviously, in food processing and

Food Formulation: Novel Ingredients and Processing Techniques, First Edition.
Edited by Shivani Pathania and Brijesh K. Tiwari.

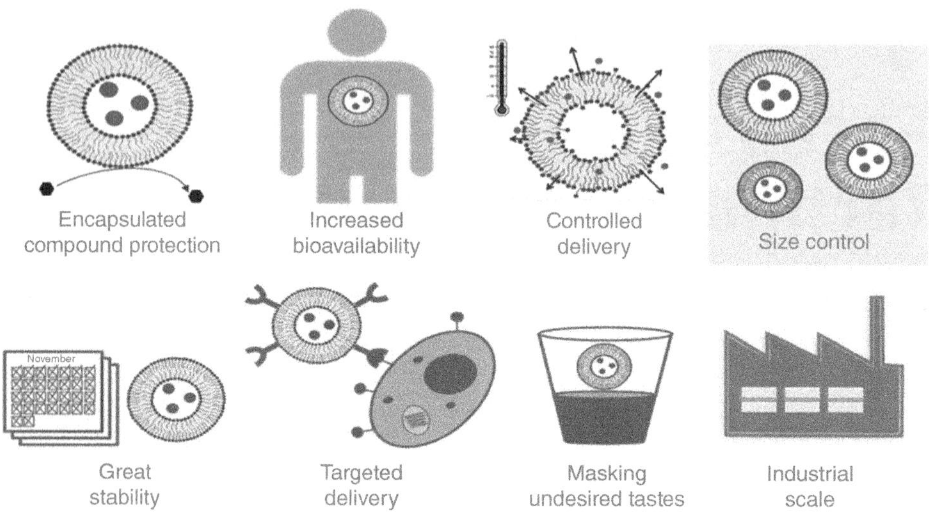

Figure 7.1 Main advantages of encapsulation in food products. *Source*: Based on Đorđević et al. (2015).

technology (Rodríguez et al. 2016). The purpose of encapsulation in the food area is broad and the range of applications include, but are not limited to, masking undesired flavours, protecting labile compounds from harsh environmental conditions (pH, light, oxygen, enzymes, moisture), inhibiting microbial growth, extending shelf life, triggering compound release, improving solubility, or even obtaining enhanced oral bioavailability, as reported in Figure 7.1 (Đorđević et al. 2015).

The role of encapsulation in protecting labile compounds in foods highlights its importance in industrial food production processes, which involve the addition of bioactive compounds for potential health benefits with antioxidant, antihypertensive, or heart protective properties (Fitzgerald et al. 2011), as well as inclusion of functional ingredients for tailored flavour, colour, or texture to the final product (McClements 2018b). It is worthy to note that the choice of encapsulating material will depend on the desired properties of the final product and the kind of application. For instance, when the encapsulation process is used in products for human consumption, encapsulating carriers must be food grade, biodegradable, and stable during processing, storage, and consumption. Furthermore, food properties such as solid, liquid or semisolid food, pH and ionic strength of the foodstuff, packaging conditions, availability or market price, usually influence the use of encapsulating material.

As encapsulated materials are added to perform a specific function, the importance of the encapsulation process cannot be emphasized enough, as each process is tailored for the load and final application. Additionally, regardless of the technology employed, as discussed later in this chapter, production costs can limit the scalability of this process to be used in industrial food production (Đorđević et al. 2016). However, the use of encapsulated materials in food products is increasing.

This chapter will discuss the main techniques employed for encapsulation (Section 7.2: Encapsulation Techniques) and the main applications within the food industry (Section 7.3: Applications in the Food Industry). In addition, the releasing mechanisms of encapsulating material have also been explained (Section 7.4: Factors Affecting Releasing Kinetics.

7.2 Encapsulation Techniques

Although encapsulation techniques can be classified based upon their size, application, hydrophobicity, or technique employed, a common and more general classification is based on how the particle is constructed. Two main approaches have been identified (Figure 7.2): bottom-up and top-down. In the bottom-up approach, the wall is formed by means of a self-assembling process of the molecules, which is driven by pH, ionic strength, or temperature. On the other hand, the top-down approach is defined by the application of energy in order to reduce the particle size.

The chosen technique will depend upon the hydrophilic or lipophilic nature of the core, as well as the desired characteristics of the particle. For example, emulsification is the right choice to obtain large particles to entrap any kind compound, while nanoprecipitation is the preferred method to encapsulate lipophilic compounds with a final small particle size.

7.2.1 Coacervation

The concept underpinning coacervation microencapsulation is the phase separation of one or many hydrocolloids from the initial solution and the subsequent deposition of the newly-formed coacervate phase around the active ingredient suspended or emulsified in the same reaction media (Gouin 2004). Coacervation can be simple or complex. Simple coacervation involves only one type of polymer with the addition of strongly hydrophilic agents to the colloidal solution, such as alginate and calcium. However, for complex coacervation, two or more types of polymers are used (Madene et al. 2006), such as gelatine and Arabic gum or chitosan and alginate.

Complex coacervation is a liquid–liquid phase separation phenomenon occurring between oppositely charged biopolymers through electrostatic interaction. Coacervates are generally produced from proteins and polysaccharides, thus they can have a wide range of functionality in various food products (Vankayala et al. 2018). Complex coacervation is usually associated with no definite forms.

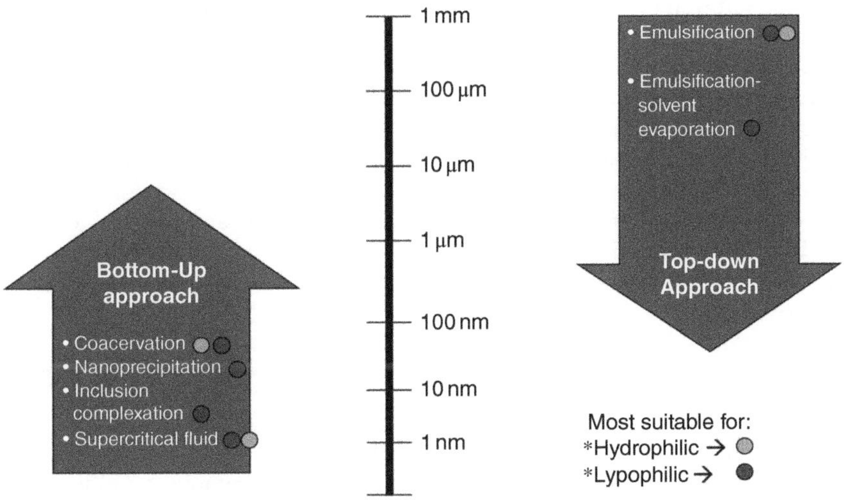

Figure 7.2 Classification of encapsulation techniques based on constructing approach.

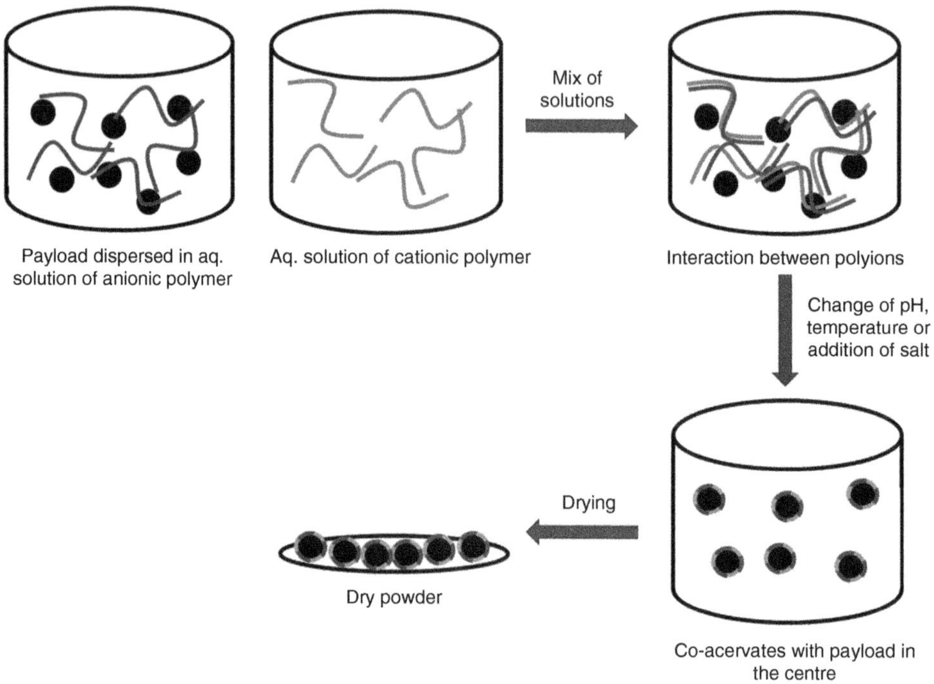

Figure 7.3 Scheme of the complex coacervation process (aq: aqueous). *Source*: Based on Timilsena et al. (2019).

The complex coacervation process is a three-phase system involving the solvent, the active material, and the coating material. In general, this process involves four steps (a simplified flow chart is shown in Figure 7.3): (i) preparation of an aqueous solution of two or more polymers. The solution is generally prepared above the gelling temperature and above the isoelectric point of the protein; (ii) mixing hydrophobic phase to the aqueous solution of one polymer, usually protein solution, and homogenizing the resultant mixture so that a stable emulsion is produced; (iii) change of pH and temperature to a certain required level to induce coacervation and phase separation; and (iv) hardening of the polymer matrices using elevated temperature, de-solvation agent, or cross-linker (Timilsena et al. 2019).

Some advantages of coacervation are a high payload and high encapsulation efficiency (up to 99%), relatively low processing cost, ability to use food-grade shell, and synthesis at ambient temperature; which makes coacervation an appropriate choice in food industries (Butstraen and Salaün 2014; Timilsena et al. 2019; Xiao et al. 2014), especially to high-value, labile functional ingredients, such as the encapsulation of polyphenols.

7.2.2 Co-Crystallization

Co-crystallization is an economic and flexible technique for the encapsulation of active compounds into powdered foods. In this process, it involves a modification of the crystalline structure of sucrose, from perfect to irregular agglomerated crystals,

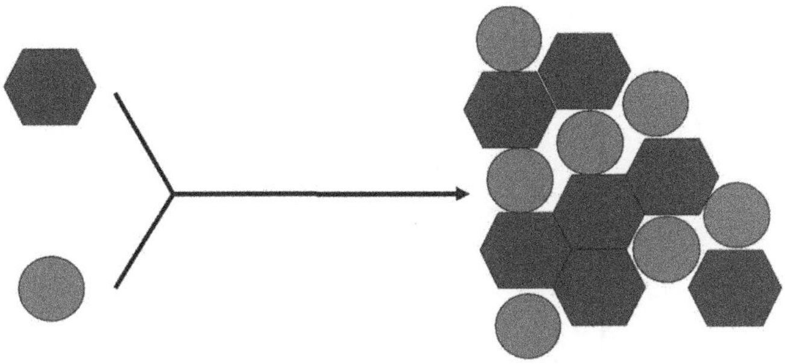

Figure 7.4 Scheme of the co-crystallization process.

providing a porous matrix in which a second active ingredient can be incorporated (Figure 7.4). The highly purified sucrose acts as a base to which a second ingredient is added to form a new structure, providing agglomerates with new functionalities.

Co-crystallization with sucrose could improve the solubility, dispersibility, wettability, anticaking, antidusting, antiseparation, homogeneity, flowability, and stability of food materials. Nevertheless, only a few studies have been carried out on this technique dealing with flavours, natural extracts, essential oils, honey, glucose, and fructose (Astolfi-Filho et al. 2005; Maulny et al. 2005; Sardar and Singhal 2013).

Spontaneous crystallization of supersaturated sucrose syrup is carried out at high temperature (above 120 °C) and low moisture (95–97 °Brix). If a second ingredient (active to encapsulate) is added at the same time, the spontaneous crystallization results in the incorporation of the second ingredient into the void spaces inside the agglomerates of the micronized crystals (Phanindra et al. 2018).

Co-crystallization is a technique able to obtain different co-crystallized products with a high concentration of the active compound. The resulting powders are free flowing and stable, even at high humidity conditions. Co-crystallized product properties like flowability, solubility, density, and size distribution mainly depended on the sucrose matrix.

7.2.3 Spray Drying

Spray drying is the oldest and the most widely used encapsulation technique in the food industry sector, used since the 1930s to prepare the first encapsulated flavours using gum acacia as wall material. It is a flexible, continuous, but more importantly, is an economical operation. Encapsulation by spray drying appears to be an effective way for those compounds in the powder to form because of its advantages (Ray et al. 2016). Moreover, it is useful for encapsulating heat-sensitive materials because of its short drying times (5–30 seconds) (Desai and Park 2005). This method produces powder micro-particles with low water activity which gives the product certain characteristics (Gouin 2004).

Different types of encapsulating agents have been used for spray drying, which include polysaccharides (starches, maltodextrins, corn syrups, and arabic gum), lipids (stearic acid, mono- and di-glycerides), and proteins (gelatin, casein, milk serum, soy, and wheat) (Saénz et al. 2009). An important step in the spray-drying technique is the selection of wall material that meets the required criteria such as mechanical strength, compatibility with the food product, compound to encapsulate, appropriate thermal

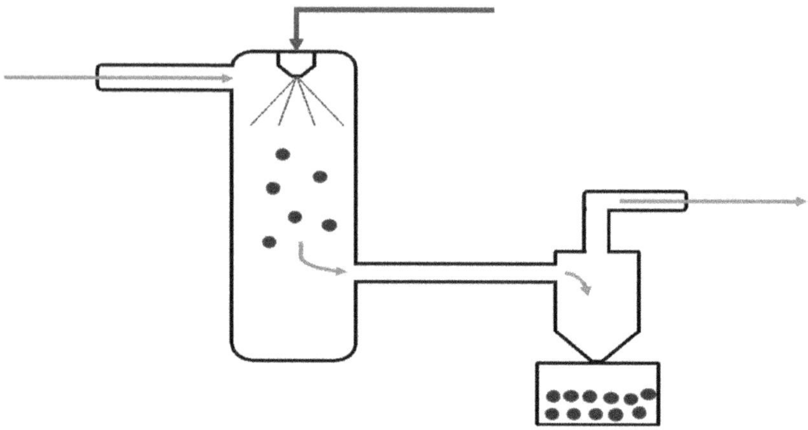

Figure 7.5 Diagram of the equipment and process of conventional spray drying.

or dissolution release, and particle size, amongst others (Ziaee et al. 2019). Spray drying is commonly used in the food industry to ensure microbiological stability, avoid chemical and/or biological degradation, and reduce the storage and shipping costs of the product by decreasing the water content. In addition, equipment is readily available and production costs are lower than other techniques; for instance, compared to freeze drying, spray drying is 30–50 times cheaper (Gharsallaoui et al. 2007).

Spray drying is an encapsulation technique related to the atomization of a liquid by means of an injector including a hot drying gas stream (Rattes and Oliveira 2007). The gas generally used is air or more rarely an inert gas such as nitrogen. The initial liquid feeding the sprayer can be a solution, an emulsion, or a suspension. Spray drying produces, depending on the starting feed material and operating conditions, a very fine powder (10–50 μm) or large size particles (2–3 mm) (Gharsallaoui et al. 2007). The technique is composed of three main steps: (i) homogenization of feed liquid; (ii) drying of feed solution by a hot gas (air or nitrogen to reduce the risk of explosion in the case of using certain solvents) with a temperature ranging usually between 150 and 220 °C to achieve the evaporation of the solvent, which under these conditions occurs almost instantaneously; and (iii) collection of dry particles by cyclones or a filter (Schafroth et al. 2012). A diagram of the process can be observed in Figure 7.5. Final particle size can be easily controlled, depending on the geometry of the nozzle, air speed, drying temperature, drying air flow rate, feed flow rate, speed of atomizer, and kind of carrier agent and concentration of carrier agent (Schoubben et al. 2010).

7.2.4 Spray Freeze Drying

Spray freeze drying is a relatively recent developed drying technique, involving a heterogeneous set of steps, which includes droplet formation, freezing, and sublimation. It has proven benefits over other drying methods in terms of producing products with improved structural integrity, superior quality, and better stability (Vishali et al. 2019). This method combines the narrow particle size distribution of an extrusion device and the freeze-drying process to prepare a dry powder of desired particle size, given as a result of narrow particle size distribution (Semyonov et al. 2010).

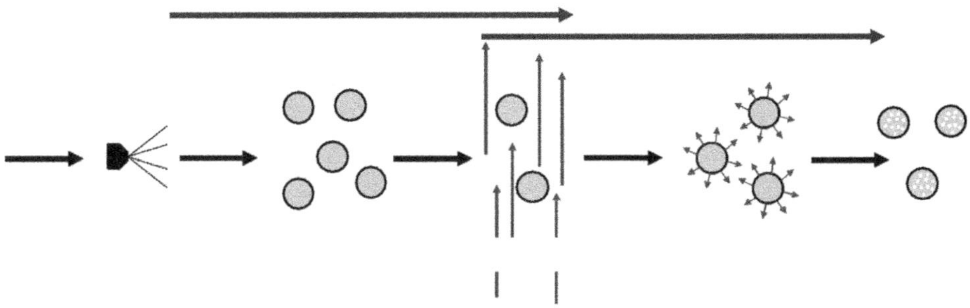

Figure 7.6 Main steps involved in a typical spray freeze drying process. *Source*: Based on Vishali et al. (2019).

Freeze drying is probably the most suitable technique for dehydration and microencapsulation of all heat-sensitive materials. The major disadvantages of freeze drying are the high energy input and long processing time; for this reason, this technique is only used to obtain very high-added value products.

As with spray drying, different types of encapsulating agents have been used for spray freeze drying, including Arabic gum, maltodextrin, and whey protein, amongst others (Ray et al. 2016). Nonetheless, high cryoprotectant concentrations and fast freezing rates are frequently used to achieve a successful formation of various polymeric and lipid nanostructures (Abdelwahed et al. 2006; Ali and Lamprecht 2017; Hirsjärvi et al. 2006).

The spray freeze drying technique is a three-step process which involves a liquid or solution being atomized into droplets, solidified by contact with a cold vapour phase of a cryogenic liquid, such as liquid nitrogen, so the droplets may start freezing during their passage through the cold vapour phase, and sublimed at low temperature and pressure in the final step (Ishwarya et al. 2015). A schematic flow chart can be seen in Figure 7.6.

Spray freeze drying powders have a controlled size, larger specific surface area, and a better porous character than other spray-dried powders (Semyonov et al. 2010). There are several factors affecting encapsulation efficiency and products functionality in the spray freeze drying process, such as feed flow rate, drying pressure and temperature, composition of the particles, and concentration of cryoprotectants, amongst others.

7.2.5 Emulsions

An emulsion is defined as a two-phase system consisting of two immiscible liquids of different composition, one in the shape of droplets (discontinuous phase) dispersed in the second one (continuous phase). Typically, the droplets diameter in emulsion systems ranges from 0.1 to 100 μm.

Emulsions can be classified according to the spatial organization of the oil and water phases. A system that consists of oil droplets dispersed in an aqueous phase is called an oil-in-water (o/w) emulsion, whereas a system that consists of water droplets dispersed in an oil phase is called a water-in-oil (w/o) emulsion (McClements et al. 2009).

Double emulsions (or multiple emulsions) are complex dispersion systems characterized by a low thermodynamic stability. They are 'emulsions of emulsions', e.g. water

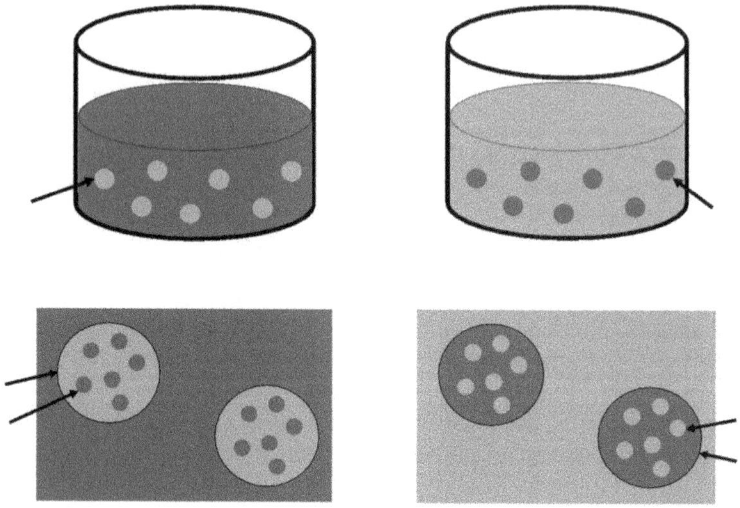

Figure 7.7 Simple emulsions: O/W emulsion (left-top) and W/O emulsion (right-top). Double or multiple emulsions: W/O/W emulsion (left-bottom) and O/W/O emulsion (right-bottom).

droplets-in-oil droplets-in-water (water-in-oil-in-water, w/o/w) or oil droplets-in-water droplets-in-oil (oil-in-water-in-oil, o/w/o). A diagram of the phases is shown in Figure 7.7. These systems can be used in different fields such as food, medicine, pharmaceuticals, or cosmetics, because the release of a substance from the inner to the outer phase is highly controllable (Muschiolik 2007). They also have a high potential for application in the food processing industry to encapsulate or protect sensitive and active food components from degradation, both in the aqueous and in the oily phase. They also control the release of the encapsulated compounds, masking flavours, improve sensory characteristics of foods, or to produce low calorie and reduced fat products, amongst others.

In recent years, nanoemulsions have gained importance in food applications. Nanoemulsions are metastable submicron oil-in-water dispersions with droplet diameter in the range of 10–100 nm (McClements 2011). Potential advantages of nanoemulsions over conventional emulsions are higher physical stability and bioavailability; higher entrapment efficiency and lower turbidity make them attractive systems for application in food (Ghosh et al. 2013).

However, such emulsions are thermodynamically unstable and yield droplets with a tendency to coalesce, leading to phase separation. In order to obtain stable emulsions, emulsifiers must be used. Emulsifiers work by forming physical barriers that prevent droplets from coalescing. Emulsifiers contain both a hydrophilic head group and a hydrophobic tail. Therefore, emulsifiers are attracted to both polar and nonpolar compounds. When added to an o/w emulsion, emulsifiers surround the oil droplet with their nonpolar tails extending into the oil, and their polar head groups facing the water. For a w/o emulsion, the emulsifiers surround the water droplet with their polar heads inside the aqueous droplet, while nonpolar tails extend into the oil phase.

Food emulsions are compositionally complex. Their droplets are stabilized to differing extents by proteins, small-molecule surfactants (emulsifiers) and, in certain cases, polysaccharides (hydrocolloids) (Dickinson 2010).

The most common methods to prepare emulsions, multiple emulsions, or nanoemulsions are: membrane emulsification, ultrasonic emulsification, high-speed homogenization, or phase inversion, etc. The purpose is to facilitate the interaction between emulsion phases and the emulsifiers, at the same time that the droplet size is controlled.

Membrane emulsification offers compact devices for preparation of nanoemulsions with low energy consumption, tuneable droplet size, monomodal distribution, and high encapsulation efficiency of entrapped functional components without shear or thermal degradation (Vladisavljević 2019). Emulsion preparation by ultrasonic is an effective method for decreasing the mean droplet diameter size. Frequency, ultrasonication time, and power of sonication are the most important parameters (Maali and Hamed Mosavian 2013). High-speed homogenization is one of the low-energy emulsification methods where rotor/stator mixers or high-speed stirrers are used at speeds of around 6000– 4 000 rpm (Wang et al. 2008).

There are some other spontaneous emulsification methods within the low-energy category, such as phase inversion temperature (PIT) and phase inversion composition (PIC), both of which are based on changes in interfacial properties (Anton and Vandamme 2009). These changes can be achieved by keeping the composition fixed while the temperature is changed (PIT) or keeping the temperature constant while the composition is changed (PIC) (Maestro et al. 2008).

7.2.6 Liposomes

Liposomes are structurally and functionally some of the most versatile supra-molecular assemblies in existence. Since the beginning of active research on lipid vesicles in 1965, the field has progressed enormously and applications are well established in several areas, such as medicine, cosmetics, or food, etc. (Jesorka and Orwar 2008).

Liposomes are basically vesicles composed by one or more phospholipid bilayers encapsulating a volume of aqueous media. The mechanism of liposome formation is essentially based on the unfavourable interactions occurring between phospholipids and water molecules, where the polar head groups of phospholipids are exposed to the aqueous phases (inner and outer), and the hydrophobic hydrocarbon tails are forced to face each other in a bilayer, as represented in Figure 7.9 (Jesorka and Orwar 2008).

Several food-grade materials can be used as building blocks for different types of liposomes, for example from eggs, sunflower, lecithin, and free fatty acids. They can encapsulate hydrophilic compounds in their core while simultaneously entrapping lipophilic compounds in the lamellar bilayers amongst the hydrophobic tails (Nowak et al. 2019). Other compounds can be used as stabilizers to improve the physicochemical properties and stability of liposomes. One of the most used stabilizers of liposomes is cholesterol, since it is able to increase the packing of phospholipid molecules, reduce bilayer permeability, improve vesicle resistance to aggregation, and change the fluidity of the lipid bilayer. However, it is possible to use phytosterols instead of cholesterol for food applications (Chan et al. 2004).

Size of the liposome can differ from very small (25 nm) to large (more than 1 μm) vesicles. Additionally, on the basis of their size and number of bilayers, liposomes can be classified into two types: (i) unilamellar vesicles; and (ii) multilamellar vesicles (MLV). In unilamellar vesicles, the structure is composed of a single phospholipid bilayer sphere surrounding the aqueous solution, while in MLV, there are several concentric bilayers (Figure 7.8).

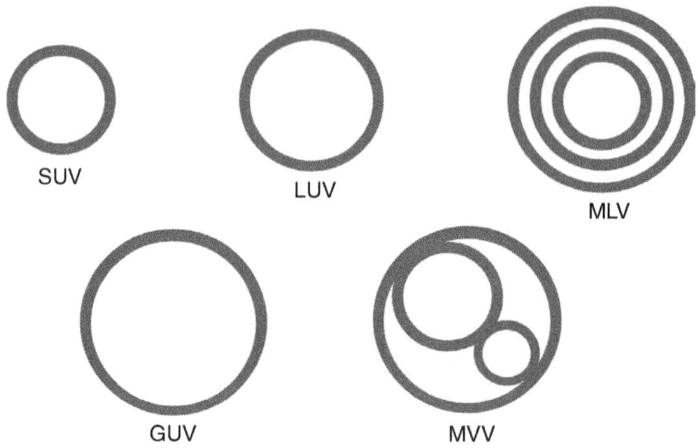

Figure 7.8 Most common structures of liposomes. SUV (small unilamellar vesicles); LUV (large unilamellar vesicles); MLV (multilamellar vesicles); GUV (giant unilamellar vesicles); and MVV (multivesicular vesicles).

Unilamellar vesicles can also be classified into two subgroups: (1a) small unilamellar vesicles (SUV); and (1b) large unilamellar vesicles (LUV). There are also other structures such as giant unilamellar vesicles (GUV, size >1 μm) and multivesicular vesicles (MVV, size >1 μm). The selection of the liposome size and morphology will depend mainly on the compound to be encapsulated and its final application.

Liposomes have several advantages, such as protection of encapsulated compound, increased stability, high bioavailability and absorption, can encapsulate both hydrophilic and lipophilic compounds, and mask undesired odours and tastes (Mozafari et al. 2008; Shade 2016). Several methods are available for the manufacture of liposomes on the small or large scale (Akbarzadeh et al. 2013), as for example thin-film layer hydration, reverse phase evaporation, or sonication. Table 7.1 shows the advantages and disadvantages of different liposome preparation methods (Amoabediny et al. 2018; Huang et al. 2014; Maherani et al. 2011; Mozafari 2010).

7.2.7 Niosomes

Niosomes are vesicles of non-ionic amphiphilic molecules, which form closed bilayer structures (Figure 7.9) (Paecharoenchai et al. 2013). Niosomes are similar, in terms of structure and physical properties, to liposomes. They are also prepared, as unilamellar or MLVs, following the same procedures and under the same conditions (Marianecci et al. 2014).

Niosomes, compared to liposomes or other encapsulation methods, offer numerous advantages as they are osmotically active, chemically stable, and have longer storage time compared to liposomes. Their surface formation and modifications are easy because of the functional groups on their hydrophilic heads exhibiting high compatibility with biological systems and low toxicity because of their non-ionic nature. Niosomes are biodegradable and non-immunogenic, and can entrap lipophilic drugs into vesicular bilayer membranes and hydrophilic drugs in aqueous compartments. Interestingly, they can improve the therapeutic performance of the encapsulated molecules by protecting the drug from the biological environment, resulting in better availability and controlled drug delivery. In addition, access to raw materials is convenient; unlike phospholipids, the handling of surfactants requires no special precautions and conditions; an

Table 7.1 Advantages and disadvantages of liposome preparation methods.

Method	Advantages	Disadvantages
Thin-film hydration method	Relatively simple method	Use of potentially toxic solvents; time consuming; difficult to scale-up
Ethanol/ether injection	Simple procedure	Organic solvent residue; nozzle blockage in ether system; time consuming; sterilization issue
Reverse phase evaporation	Simple design; acceptable encapsulation efficiency	Not suitable for the encapsulation of sensitive material due to large quantity of organic solvent use; time consuming; sterilization issue
Microfluidization	Controlled particle size; large volume manufacture in a continuous and reproducible manner	Employment of high pressures (up to l0 000 psi)
Supercritical Fluid Process	Controlled particle size; possible in situ sterilization; low organic solvent consumption	High cost; low yield; high pressure up to 350 bar used
Dual asymmetric centrifugation	Simple method, yields products with narrow size distribution; high encapsulation efficiency	Not suitable for bulk production; high pressure and high shear force
Sonication	Simple and fast technique	Overheating of the lipid suspension causing degradation; sonicator tips release metal particles into the product
Membrane contactor technology	Homogeneous vesicles with high encapsulation efficiency for lipophilic drugs can be obtained; easy to scale up.	Encapsulation of hydrophilic material still needs further investigation
Cross-flow filtration detergent depletion method	Vesicles of defined size, homogeneity and high stability can be prepared in a short time; sterile products can be obtained	Requires high amount of detergent use; multistep procedure
Freeze drying double emulsions method	Sterile product with good storage stability can be obtained	Use of carbohydrates as cryoprotectants limits the applications for diabetic patients

Source: Adapted from Huang et al. (2014).

increase in the oral bioavailability is observed; and, finally, the variable characteristics of the niosomes can be controlled. Characteristics such as the type of the niosomes according to their size, entrapment efficiency, and stability can be controlled by the type of preparation method, nature of surfactant, stabilizer content, size, surface charge, and suspension concentration (Moghassemi and Hadjizadeh 2014).

The research interest in niosomes is growing because these systems are able to overcome some disadvantages associated with liposomes, surfactants are easily derivatized and give a higher versatility to the vesicular structure; and moreover, they

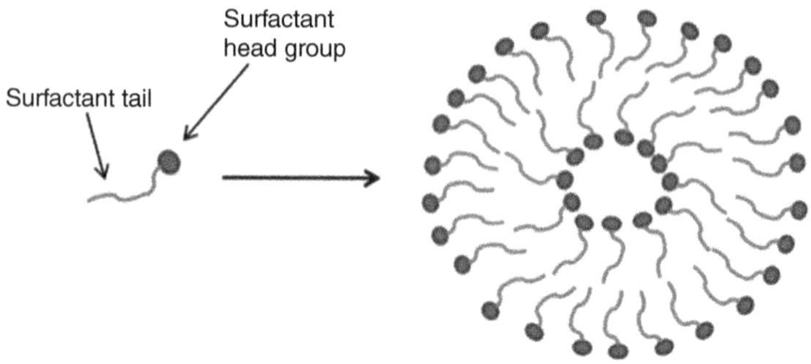

Figure 7.9 Niosome structure. *Source*: Based on Pando (2014).

have lower costs than the phospholipids employed in liposomes. Niosomes are a potential vehicle for bioactive delivery systems such as in food applications, since they can be prepared by natural products with low costs compared with other types of vesicles (Matos et al. 2019).

There are several works using niosomes for the encapsulation of compounds for food applications (Gutiérrez et al. 2016; Pando et al. 2015). Niosomes uses for dairy products have been studied for nearly two decades, but only recently, food scientists have begun to utilize them for the controlled delivery of functional components such as peptides, enzymes, vitamins, and flavours, aiming to fortify dairy products, increasing their nutraceutical and organoleptic qualities, meaning they show great promise in this field (Manconi et al. 2010).

7.2.8 Inclusion Complexation

This inclusion complexation encapsulation technique is based on the use of cyclodextrins (CDs), which are a family of cyclic oligosaccharides derived from starch, with a particular space configuration. Such molecules are composed of a truncated cone-shaped hydrophobic cavity that can entrap other molecules; whilst the outer surface has a hydrophilic nature (Recharla et al. 2017). In fact, the most characteristic treat of CDs is their capacity to form inclusion complexes with entrapped compounds through host–guest interactions (Crini et al. 2018). Such interactions are mainly driven by hydrogen bonding, van de Waals' forces, or hydrophobic effects. Several conformations, molecule-to-molecule ratios, can be found for the final particle, as illustrated in Figure 7.10. The main advantage of this technique is their capacity to improve the solubility of entrapped compounds, or preserve aromatic compounds from evaporation.

According to Recharla et al. (2017), different methods are available to prepare an inclusion complex: (i) simple mechanical blending; (ii) kneading method, in which a paste of cyclodextrin is created in a kneading machine, after the target compound is added to the paste, homogenized, and further dried; (iii) co-precipitation, in which the bioactive is dissolved in a proper solvent, and afterwards this solution is added to an aqueous CD solution and the obtained precipitated is recovered; (iv) spray drying, in which a solution of CD and the entrapped molecule are processed; (v) freeze drying; and (vi) supercritical antisolvent technique, in which CO_2 is employed as an alternative to organic solvents.

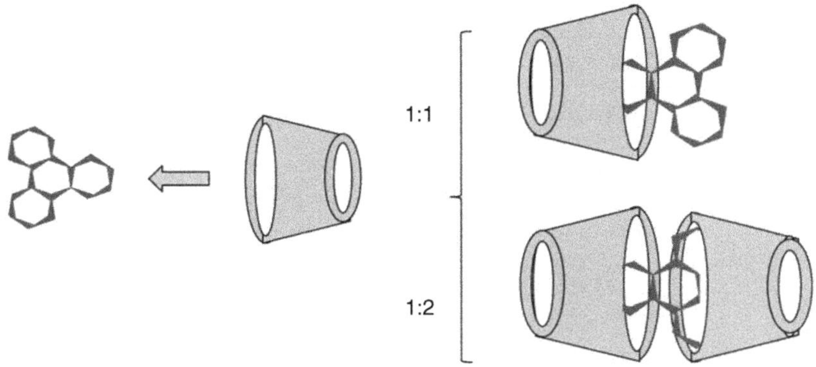

Figure 7.10 Schematic representation of inclusion complexation using cyclodextrins at different molecule/molecule ratio.

7.3 Applications in the Food Industry

Consumers' demands within the food sector are increasing more than ever, wherein modern living trends are converging on tastier, healthier, and convenience products. This has steered research and development in the food industry, to fulfil consumer demands. Without doubt, and as highlighted by the increasing number of research articles on the topic, encapsulation is becoming one of the hot topics in food research (Đorđević et al. 2016). As reported by McClements (2018b), the number of papers in this area has grown from 3 paper in 1990 up to almost 11 000 in 2017; and these articles have focused specifically on targeted delivery of materials. A systematic approach to encapsulation application to the food industry can be made in different ways, i.e. based on the technology employed, based on wall material, based on the size of the final particle, the final application or purpose, or as it happens in this chapter, based on the particle load. Table 7.2 summarizes the main challenges that can be overcome by encapsulation technology.

7.3.1 Aromatic and Volatile Compounds

Undoubtedly, aromatic compounds contribute greatly to the overall eating quality of the food products; therefore, their preservation is essential to preserve the final product quality (Reineccius 2017). Encapsulation has been employed to overcome many of the limitations related to the use of volatile compounds such as controlled fast release, decreased evaporation rate, increased bioavailability, improved thermal and oxidative stability; and consequently, expanding their applications in food systems (Saifullah et al. 2019). Some examples of compounds successfully encapsulated are peppermint oil, limonene, eugenol, menthol, thymol, geraniol, caffeine, vanillin, citral, orange terpenes, or cocoa aroma (Saifullah et al. 2019). Meanwhile, most common carrier agents are those based on carbohydrates (amylose, corn starch, modified starch, maltodextrin, alginates, chitosan, or sucrose) and proteins (rice flour, whey protein isolate, zein, soybean protein isolate, and gelatin).

Aromatic and volatile compounds are composed of low molecular weight organic molecules; such compounds are volatile and sensitive to external factors such as heat,

Table 7.2 Main challenges addressed by encapsulation based on the type of load.

Type of load	Challenges	Type of load	Challenges
Volatile compounds	Low stability Fast release High evaporation rate Low bioavailability Prone to oxidation	**Enzymes**	Short shelf life Lack of stability Easy to recover Sensible to pH Thermo labile
Protein/peptides	Hygroscopicity Degraded in the gastric tract Bitter taste	**Polyphenols**	Bitter taste Thermo labile Sensitive to extreme pH Prone to oxidation
Vitamins	Low stability Low solubility Off-flavours Prone to oxidation Thermolabile	**Lipids**	Prone to oxidation Low bioavailability Sensory acceptance Short shelf life

light, air, or moisture, and so far more than 8000 different aroma compounds have been identified in food products (Reineccius 2017). The same author highlighted that the chemical properties of such molecules differ enormously in terms of solubility, boiling point, or stability; this fact, combined with the multiple final applications and vast variety of types of final product, makes designing encapsulation techniques a challenging task. We recommend the above-mentioned author as further reading to obtain a better understanding of the reactions leading to aroma degradation and evaporations as mechanisms of flavour loss (Reineccius 2017).

The most employed techniques for aroma encapsulation are spray drying (assisted or not by homogenization), spray chilling, extrusion, electrostatic extrusion, molecular inclusion, coacervation, complex coacervation, co-precipitation, freeze drying, electrospinning, electrospraying, fluidized bed coating, high-pressure homogenization, or sonication. (Gupta et al. 2016). However, only a few techniques are industrially available and the vast majority is encapsulated using spray drying (80–90%), spray chilling (5–10%), and extrusion (2–3%). It is worthwhile noting that spray drying involves high temperatures (although for a very short time), which can drastically reduce the encapsulation efficacy and increase the loss of the payload by evaporation, whereas freeze drying will ensure higher process efficiency. However, the cost of spray drying is much lower and this is the main driver when selecting an industrial encapsulation process. Therefore, the choice of the encapsulation process thus depends on the encapsulating material, process efficiency, desired product quality, and overall cost implications

Factors affecting the encapsulation rate are numerous and vary with the encapsulation technique employed. For example, when using spray drying, inlet and outlet temperature, ratio of matrix and core, nozzle pressure, and wall material are of the highest relevance. In order to simplify this task, statistical approaches such as RSM (response surface methodology) can be employed, making this statement valid for any of the core compounds discussed here. For instance, using this approach, Getachew and Chun (2016) found that the encapsulation efficacy of coffee aroma using polyethylene glycol (PEG) ranged from 41 to 80% by modifying the temperature, pressure, and core/matrix ratio; optimum points were determined using RSM tools. A similar example was reported by Murali et al. (2017), in which spray drying was optimized for the encapsulation of rice oil using jackfruit starch.

7.3.2 Vitamins

Vitamins are a group of bioactive molecules, obtained by the human body through a balanced diet and play a crucial role in maintaining our body metabolism and promoting good health (Wildman 2016). Vitamins are susceptible to degradation and therefore preservation methods are employed to protect them against oxidation, UV light, or thermal degradation. Encapsulation has been applied to extend the shelf life and stability of vitamins. Also, the use of encapsulation increases the efficiency of administration, therefore the doses required are lower and the risk of hypervitamonis, and the related side effects, are minimized (Gonnet et al. 2010)

Nano- and microencapsulation have been used for vitamins encapsulation. According to Katouzian and Jafari (2016), both techniques have advantages on their own. Microencapsulation is advantageous in controlled release and precise determination of vitamin delivery, cost-effectiveness, masking off-flavours, and entrapment of a mixture of vitamins. On the other hand, nanoencapsulation provides faster dissociation, higher surface area, high intracellular uptake, targeting efficiency, less use of wall material, and longer shelf life. The particle size and characteristics need to be adapted to the intended administration pathway, which could be oral, topical, parental, or even local injection.

In addition, when designing an encapsulation protocol for vitamins, the first consideration is to determine the solubility of the vitamin to be encapsulated. Liposoluble vitamins are A, D, E, K, and carotenoids, which are not strictly a vitamin, but at least 50 of them are identified as pro-vitmain A elements and hydrosoluble are all other vitamins, and their encapsulation presents different challenges. The first vitamins present a very limited solubility in aqueous solutions, which poses a challenge when they are to be incorporated in water-based products. Within the water-soluble vitamins, vitamin C is the most popular in the food industry; it is stable as a powder, but such stability decreases remarkably when dissolved in water (Wilson and Shah 2007). Special attention is worthy for vitamin B_{12}, which presents a particular challenge because is the largest of all vitamins and the most chemically complex; this compound, although stable in water in a pH range from 4 to 7, is lacking in human diets with limited or null intake of animal origin products leading to a potential risk of deficiency. Food fortified with encapsulated Vitamin B_{12} is an increasing topic of research (Estevinho et al. 2016).

In an excellent review recently published (Katouzian and Jafari 2016), the main techniques applied to the encapsulation of lipophilic and hydrophilic vitamins were discussed. Several techniques for macro-encapsulation can be applied (spray drying, liposomes, coacervation, or simple emulsion techniques); and many others have been researched when reducing the particle size to the nanoscale (nanoemulsification, nanoliposomes, nanoprecipitation, coacervation, CDs, or electrospinning).

7.3.3 Polyphenols

Polyphenols are a wide variety of secondary metabolites that can be found in all vascular plants. Regardless of the complexity or the different molecular weight that all of these substances have, the presence of one or several benzenic cycles is common

(Munin and Edwards-Lévy 2011). Polyphenols have the physiological role of acting as antioxidants to prevent the oxidative stress originated by the accumulation of reactive oxygen species. Mechanism of action is proposed to be the complexation with pro-oxidant proteins, by chelating pro-oxidant metals (Fe^{3+}, Al^{3+}, or Cu^{2+}) or by direct scavenging of reactive species. However, it has been reported that, in spite of their health benefits, the levels required to be effective *in vitro* are at least 10 times higher than those found *in vivo* (Fang and Bhandari 2010). Polyphenols are highly sensitive (requires protection) to temperature, oxidation, light, and extreme pH; and even more, polyphenols usually present a bitter taste, so requiring flavour masking. Therefore, by means of encapsulation, these concerns can be addressed while increasing their bioavailability.

Amongst all the available technologies for polyphenols encapsulation, spray drying is the most employed technique. Wall materials as maltodextrin or sodium alginate have been satisfactorily applied to encapsulate anthocyanins, spent coffee extracts, or olive leaf extract respectively (Ballesteros et al. 2017, Ersus and Yurdagel 2007, González et al. 2019). Special attention has been given to the use of CDs as materials for inclusion encapsulation; after its inclusion inside the structure of the CDs the water solubility is remarkably improved at the same time that the antioxidant effect is enhanced and cross-linking with proteins is avoided (Roy et al. 2018). Nevertheless, other protocols such as encapsulation coacervation, liposomes, thin film evaporation, or freeze drying have also been employed (Ozkan et al. 2019). However, as highlighted by Ozkan et al. (2019), the parameters affecting the encapsulation process should be optimized for each molecule of interest, wall material, and encapsulation technique; the same authors reported that selection of the encapsulation technique is mainly associated with the thermal stability and solubility of the active compounds.

7.3.4 Peptides, Proteins, and Hydrolysates

Functional and bioactive properties of protein peptides and hydrolysates has been extensively reported in the literature (Bhat et al. 2017; Daliri et al. 2017; Sánchez and Vázquez 2017). The main challenge for bioactive peptides is their susceptibility to being transformed or further hydrolyzed in the gastric tract by enzymes such as trypsin or pepsin, leading to an undesirable loss of bioactivity. Due to their health-imparting properties, proteins are intentionally hydrolyzed at industrial scale to obtain hydrolysates and peptides. However, their sensory as well as physical properties are negatively affected; therefore, their inclusion in encapsulated forms is preferred to mask the bitter taste and reduce their hygroscopicity for stability and extended shelf life (Mohan et al. 2018).

Mohan et al. (2015) reported that proteins, polysaccharides, and lipids are the main carriers employed for peptide encapsulation. Each one of them presents pros and cons, for example, the good emulsification properties of proteins, the abundance and cost-effective sourcing of carbohydrates, or the suitability of lipids for peptides with different charge. On the other hand, high-energy processing is required for the first two, while the latter can suffer oxidation over long storage times. The most frequently used industrial techniques for peptide encapsulation are liposomes, spray drying, and coacervation. Liposomes are an effective and simple way to encapsulate; however, it involves the use of solvents that need to be removed, with the consequent extra cost. Spray drying is effective to entrap the target compounds; however, the high temperatures employed can lead to protein denaturation or Maillard reactions. Coacervation has been reported to have high encapsulation efficiency (Black et al. 2014), but it needs to be observed that the interaction between core and encapsulants can be too strong, making further release complicated.

7.3.5 Enzymes

A particular type of protein that requires special attention is enzymes. The knowledge and technology about them are constantly increasing; therefore, an increasing number of enzymes are becoming available for industrial applications in the food sector (dairy, baking, alcoholic fermentation, juice processing, or hydrolysates production). However, industrial application of enzymes can be limited by short shelf life and lack of stability. Immobilization of enzymes can help to remove these barriers, which can make an enzyme robust and resilient to environmental changes (Homaei et al. 2013). In contrast with previous sections, where encapsulation techniques were developed aiming to facilitate product incorporation into food stuffs, to increase bioavailability and/or to design control release mechanisms, immobilization of the enzyme is mostly used at the processing level. As discussed below, techniques and approaches might be different to that which has been reviewed so far; however, this is a very interesting research topic of capital importance for the food industry, and in essence it deals with the idea of protecting an active molecule.

By definition, an immobilized enzyme is 'an enzyme which is attached to an inert, organic or inorganic, insoluble material such as calcium alginate or silica' (Homaei 2015). By means of this attachment, the enzyme's resistance is amplified, thus preserving its activity to external factors such as pH or temperature. In addition, the fact that enzymes are conjugated with insoluble particles, facilitates its further separation from the enzymatic products (centrifugation or filtration), allowing the generation of a final product free of the enzyme.

Materials employed to entrap enzymes are required to have high permeability, provide the sufficient functional groups for enzyme attachment under non-denaturing conditions, large surface area, water insolubility, and high stability at different external conditions, amongst others. Although many different techniques have been developed for enzyme immobilization, those falling into the category of physical entrapment use encapsulation techniques such as liposomes (mostly used in cheese ripening where control release can be tailored) (Mohammadi et al. 2015).

7.3.6 Lipids

Essential fatty acids, specifically polyunsaturated fatty acids (PUFAs), play an important role in the human body and help to prevent various diseases (Chang and Nickerson 2018). The most relevant essential fatty acids are α-linolenic acid (ALA), eicosapentanoic acid (EPA), docosahexanoic acid (DHA), docosapentanoic acid (DPA), or oleic acid, which can be classified as omega-3, 6, or 9, depending on the location of the first double bond in the carbon chain. The main sources of these compounds are vegetable and marine oils. However, in spite of all these beneficial properties, PUFAs have been found to have low bioavailability, poor sensory acceptance, and are prone to oxidation, which has a negative impact on the final product such as off-flavours, colour changes, and shortened shelf life (Chang and Nickerson 2018).

Therefore, strategies have been developed to preserve lipid integrity, allowing their broader use in the food industry (Assadpour and Jafari 2019).

Although several methods have been reported for lipids encapsulation (complex coacervation, electrospraying, ultrasonic atomization, freeze drying) (Kaushik et al. 2015), bibliographical results confirmed that spray drying is the preferable technique for its high efficiency, an easy-to-handle powder is generated, and a large variety of wall materials are available, flexible, and scalable. As mentioned earlier, even though the contact time is short in spray drying, the high temperatures employed in the outlet can lead to lipid oxidation (Dias et al. 2017). In general terms, regardless of the encapsulation technique employed, an emulsification step is useful in obtaining a high-quality final product.

7.4 Factors Affecting Releasing Kinetics

Another aspect that needs to be carefully considered is the release mechanism, which can be defined as the migration of a target compound from one environment to another over a defined period of time (Lakkis 2007). Many factors exert and effect the kinetics release of the encapsulated material, and most relevant is the nature of the load itself, the composition and properties of the wall materials, the particle size, the surrounding environment (pH, temperature, ionic strength, etc.), and conjugation or complexation between load and encapsulants. In general terms, the encapsulated compounds are initially released at high rates (burst phase), followed by a phase with a more sustained pace. Furthermore, delivery systems are frequently designed to retain the load until a triggering external factor initiates the release (change in pH, temperature, ionic strength, or enzymatic activity) (McClements 2014, 2018a). Careful consideration must be given to the composition and properties of the food product within which the capsules are being incorporated, especially food stuffs with relatively high ionic strength, extreme pH, or containing enzymes that could initiate the release before it is intended.

Some mechanisms for compounds release have been described, as for instance simple diffusion, changes in the wall pore size, capsule disintegration, or change in the molecular interactions (McClements 2018a). In diffusion mechanisms, the size of the molecule and more specifically, the hydrodynamic radius, compared to the pore size are of high importance; large molecules or small pore size will decrease the release speed. Pore size can be modified by altering the surrounding environment, for example capsule walls composed by polyelectrolytes (alginate, cellulose, or pectins) are prone to swell at low ionic strengths, while shrinking is observed when electrostatic repulsion between the polymer chains are reduced by increasing ionic strength (Bannikova et al. 2017). Controlling the interaction (i.e. attractive or repulsive forces) between the load and the capsule molecules might be another approach to tailor the release mechanisms. As mentioned in the previous example, modifying pH and ionic strength are the most common techniques to control such interactions, and subsequently, to regulate the release (Rahmani et al. 2017).

The last mechanism proposed was the disruption of the capsule by external factors as endogenous enzymes: amylases in saliva, proteases in gastric tract, or lipases in the intestine, that will attack specific components of the wall: starches, proteins, lipids, and dietary fibres, respectively. In the same way that interaction between load and encapsulants can be modified by pH or ionic strength, interactions between the wall components can be modulated as well. Alternatively, temperature can be used to destabilize the capsule, allowing the release of the entrapped molecules, also known as release by melting. The ability to control the release under different pH conditions is extremely useful when designing capsules to be able to resist the acidic pH of mouth and stomach, and to release its content on reaching the intestine, where pH is more alkaline.

When dealing with aromatic and volatile substances, two factors control the release rate. The first one is a thermodynamic factor (comparative volatility of the load in the matrix and the air phases under equilibrium), and a kinetic factor (the resistance to mass transfer from product to atmosphere). In addition, the presence of other molecules such as salt, lipids, proteins, or small molecules can modify the retention or release of the flavour from the capsule (Madene et al. 2006).

Vitamin release is driven by a passive (degradation of particle wall), as well as an active mechanism (vitamin dispersion throughout the matrix). The diffusion rate of the particles plays an important role once the vitamin has been released; since they have to diffuse in a hydrophilic environment (Dan 2016). Main factors affecting the release kinetics of vitamins are related to the vitamin type and its concentration: higher load concentrations increase the release kinetics, meanwhile different release profiles were observed for vitamins B_2 (water-soluble) and E (lipid soluble) in a simulated gastric digestion after being encapsulated in a double emulsion (w/o/w) system (Li et al. 2012).

In the case of peptides, the hydrophobicity of the peptide is important to determine the velocity of release. It has been reported that in water-based environments, the release of peptides was inversely proportional to their hydrophobicity (O'Neill et al. 2014). Proteins are susceptible to conformational changes and different aggregation states by modifying their net charge under different conditions of pH and ionic strength; these phenomena can be used to modify their hydrodynamic radius, interaction with the wall components, or even their solubility.

7.5 Conclusions

New functional and bioactive compounds, with applications in the fields of food, pharmaceuticals and cosmetics, are being isolated and identified, each one of them with different chemical and physical properties. However, simply adding the new ingredients into the food products is not the most efficient or preferable way to integrate them into new food formulations. Such active ingredients can degrade over time, are prone to oxidative processes and thermal degradation, or compounds are released in the wrong time/space. Besides, some of them present very low bioavailability or solubility, which makes it more challenging for their incorporation into food products. In order to address these challenges, encapsulation is a well-established tool which has been proved to have a positive effect on the shelf life and stability of entrapped compounds, can be tailored for triggered release of the core compounds under specific stimulus (pH or ionic strength), improves the incorporation and dispersibility into the food matrix, and enhance the bioavailability.

Since encapsulation techniques and wall materials are so diverse, it is imperative to design specific protocols based on the compound to be encapsulated and the final application. This chapter reviewed the main techniques for encapsulation currently employed at both industrial and research levels, as well as the main applications based on the properties of the core material.

References

Abdelwahed, W., Degobert, G., and Fessi, H. (2006). Investigation of nanocapsules stabilization by amorphous excipients during freeze-drying and storage. *European Journal of Pharmaceutics and Biopharmaceutics* 63 (2): 87–94. https://doi.org/10.1016/j.ejpb.2006.01.015.

Akbarzadeh, A., Rezaei-Sadabady, R., Davaran, S. et al. (2013). Liposome: classification, preparation, and applications. *Nanoscale Research Letters* 8 (1): 102. https://doi.org/10.1186/1556-276X-8-102.

Ali, M.E. and Lamprecht, A. (2017). Spray freeze drying as an alternative technique for lyophilization of polymeric and lipid-based nanoparticles. *International Journal of Pharmaceutics* 516 (1): 170–177. https://doi.org/10.1016/j.ijpharm.2016.11.023.

Amoabediny, G., Haghiralsadat, F., Naderinezhad, S. et al. (2018). Overview of preparation methods of polymeric and lipid-based (niosome, solid lipid, liposome) nanoparticles: a comprehensive review. *International Journal of Polymeric Materials and Polymeric Biomaterials* 67 (6): 383–400.

Anton, N. and Vandamme, T.F. (2009). The universality of low-energy nano-emulsification. *International Journal of Pharmaceutics* 377 (1): 142–147. https://doi.org/10.1016/j.ijpharm.2009.05.014.

Assadpour, E. and Jafari, S.M. (2019). Advances in spray-drying encapsulation of food bioactive ingredients: from microcapsules to nanocapsules. *Annual Review of Food Science and Technology* 10: 103–131.

Astolfi-Filho, Z., Souza, A.C., Reipert, É.C.D., and Telis, V.R.N. (2005). Encapsulação de suco de maracujá por co-cristalização com sacarose: cinética de cristalização e propriedades físicas. *Ciência e Tecnologia de Alimentos* 25 (4): 795–801.

Ballesteros, L.F., Ramirez, M.J., Orrego, C.E. et al. (2017). Encapsulation of antioxidant phenolic compounds extracted from spent coffee grounds by freeze-drying and spray-drying using different coating materials. *Food Chemistry* 237: 623–631.

Bannikova, A., Rasumova, L., Evteev, A. et al. (2017). Protein-loaded sodium alginate and carboxymethyl cellulose beads for controlled release under simulated gastrointestinal conditions. *International Journal of Food Science & Technology* 52 (10): 2171–2179.

Bhat, Z.F., Kumar, S., and Bhat, H.F. (2017). Antihypertensive peptides of animal origin: a review. *Critical Reviews in Food Science and Nutrition* 57 (3): 566–578.

Black, K.A., Priftis, D., Perry, S.L. et al. (2014). Protein encapsulation via polypeptide complex coacervation. *ACS Macro Letters* 3 (10): 1088–1091.

Butstraen, C. and Salaün, F. (2014). Preparation of microcapsules by complex coacervation of gum Arabic and chitosan. *Carbohydrate Polymers*: 608–616. https://doi.org/10.1016/j.carbpol.2013.09.006.

Chan, Y.-H., Chen, B.-H., Chiu, C.P., and Yi-Fa, L. (2004). The influence of Phytosterols on the encapsulation efficiency of cholesterol liposomes. *International Journal of Food Science & Technology* 39 (9): 985–995. https://doi.org/10.1111/j.1365-2621.2004.00867.x.

Chang, C. and Nickerson, M.T. (2018). Encapsulation of omega 3-6-9 fatty acids-rich oils using protein-based emulsions with spray drying. *Journal of Food Science and Technology* 55 (8): 2850–2861.

Crini, G., Fourmentin, S., Fenyvesi, É. et al. (2018). Fundamentals and applications of cyclodextrins. In: *Cyclodextrin Fundamentals, Reactivity and Analysis* (eds. S. Fourmentin, G. Crini and E. Lichtfouse), 1–55. Cham: Springer.

Daliri, E.B.-M., Deog, H.O., and Lee, B.H. (2017). Bioactive Peptides. *Food* 6 (5): 32.

Dan, N. (2016). Transport and release in nano-carriers for food applications. *Journal of Food Engineering* 175: 136–144.

Desai, K.G.H. and Park, H.J. (2005). Recent developments in microencapsulation of food ingredients. *Drying Technology* 23 (7): 1361–1394. https://doi.org/10.1081/DRT-200063478.

Dias, D.R., Botrel, D.A., Fernandes, R.V.d.B., and Borges, S.V. (2017). Encapsulation as a tool for bioprocessing of functional foods. *Current Opinion in Food Science* 13: 31–37.

Dickinson, E. (2010). Food emulsions and foams: stabilization by particles. *Current Opinion in Colloid & Interface Science* 15 (1): 40–49. https://doi.org/10.1016/j.cocis.2009.11.001.

Đorđević, V., Balanč, B., Belščak-Cvitanović, A. et al. (2015). Trends in encapsulation technologies for delivery of food bioactive compounds. *Food Engineering Reviews* 7 (4): 452–490.

Đorđević, V., Paraskevopoulou, A., Mantzouridou, F. et al. (2016). Encapsulation technologies for food industry. In: *Emerging and Traditional Technologies for Safe, Healthy and Quality Food* (eds. V. Nedović, P. Raspor, J. Lević, et al.), 329–382. Springer.

Ersus, S. and Yurdagel, U. (2007). Microencapsulation of anthocyanin pigments of black carrot (*Daucus carota* L.) by spray drier. *Journal of Food Engineering* 80 (3): 805–812.

Estevinho, B.N., Carlan, I., Blaga, A., and Rocha, F. (2016). Soluble vitamins (vitamin B_{12} and vitamin C) microencapsulated with different biopolymers by a spray drying process. *Powder Technology* 289: 71–78.

Fang, Z. and Bhandari, B. (2010). Encapsulation of polyphenols–a review. *Trends in Food Science & Technology* 21 (10): 510–523.

Fitzgerald, C., Gallagher, E., Tasdemir, D., and Hayes, M. (2011). Heart health peptides from macroalgae and their potential use in functional foods. *Journal of Agricultural and Food Chemistry* 59 (13): 6829–6836.

Getachew, A.T. and Chun, B.-S. (2016). Optimization of coffee oil flavor encapsulation using response surface methodology. *LWT – Food Science and Technology* 70: 126–134.

Gharsallaoui, A., Roudaut, G., Chambin, O. et al. (2007). Applications of spray-drying in microencapsulation of food ingredients: an overview. *Food Research International* 40 (9): 1107–1121. https://doi.org/10.1016/j.foodres.2007.07.004.

Ghosh, V., Mukherjee, A., and Chandrasekaran, N. (2013). Ultrasonic emulsification of food-grade nanoemulsion formulation and evaluation of its bactericidal activity. *Ultrasonics Sonochemistry* 20 (1): 338–344. https://doi.org/10.1016/j.ultsonch.2012.08.010.

Gonnet, M., Lethuaut, L., and Boury, F. (2010). New trends in encapsulation of liposoluble vitamins. *Journal of Controlled Release* 146 (3): 276–290.

González, E., Gómez-Caravaca, A.M., Giménez, B. et al. (2019). Evolution of the phenolic compounds profile of olive leaf extract encapsulated by spray-drying during in vitro gastrointestinal digestion. *Food Chemistry* 279: 40–48.

Gouin, S. (2004). Microencapsulation: industrial appraisal of existing technologies and trends. *Trends in Food Science & Technology* 15 (7): 330–347. https://doi.org/10.1016/j.tifs.2003.10.005.

Gupta, S., Khan, S., Muzafar, M. et al. (2016). Encapsulation: entrapping essential oil/flavors/aromas in food. In: *Encapsulations, Vol. 2 of Nanotechnology in the Agri-Food Industry* (ed. A.M. Grumezescu), 229–268. Academic Press.

Gutiérrez, G., Matos, M., Barrero, P. et al. (2016). Iron-entrapped niosomes and their potential application for yogurt fortification. *LWT– Food Science and Technology* 74 (December): 550–556. https://doi.org/10.1016/j.lwt.2016.08.025.

Hirsjärvi, S., Peltonen, L., Kainu, L., and Hirvonen, J. (2006). Freeze-drying of low molecular weight poly(L-lactic acid) nanoparticles: effect of cryo- and lyoprotectants.

Journal of Nanoscience and Nanotechnology 6 (9–10): 3110–3117. https://doi.org/10.1166/jnn.2006.439.

Homaei, A. (2015). Enzyme immobilization and its application in the food industry. *Advances in Food Biotechnology* 9: 145–164.

Homaei, A.A., Sariri, R., Vianello, F., and Stevanato, R. (2013). Enzyme immobilization: an update. *Journal of Chemical Biology* 6 (4): 185–205.

Huang, Z., Li, X., Zhang, T. et al. (2014). Progress involving new techniques for liposome preparation. *Asian Journal of Pharmaceutical Sciences* 9 (4): 176–182. https://doi.org/10.1016/j.ajps.2014.06.001.

Ishwarya, S., Anandharamakrishnan, C., and Stapley, A.G.F. (2015). Spray-freeze-drying: a novel process for the drying of foods and bioproducts. *Trends in Food Science & Technology* 41 (2): 161–181. https://doi.org/10.1016/j.tifs.2014.10.008.

Jesorka, A. and Orwar, O. (2008). Liposomes: technologies and analytical applications. *Annual Review of Analytical Chemistry* 1 (1): 801–832. https://doi.org/10.1146/annurev.anchem.1.031207.112747.

Katouzian, I. and Jafari, S.M. (2016). Nano-encapsulation as a promising approach for targeted delivery and controlled release of vitamins. *Trends in Food Science & Technology* 53: 34–48.

Kaushik, P., Dowling, K., Barrow, C.J., and Adhikari, B. (2015). Microencapsulation of omega-3 fatty acids: a review of microencapsulation and characterization methods. *Journal of Functional Foods* 19: 868–881.

Lakkis, J.M. (ed.) (2007). *Encapsulation and Controlled Release Technologies in Food Systems*. Wiley Online Library.

Li, B., Jiang, Y., Liu, F. et al. (2012). Synergistic effects of whey protein–polysaccharide complexes on the controlled release of lipid-soluble and water-soluble vitamins in W1/O/W2 double emulsion systems. *International Journal of Food Science & Technology* 47 (2): 248–254.

Maali, A. and Hamed Mosavian, M.T. (2013). Preparation and application of nanoemulsions in the last decade (2000–2010). *Journal of Dispersion Science and Technology* 34 (1): 92–105.

Madene, A., Jacquot, M., Scher, J., and Desobry, S. (2006). Flavour encapsulation and controlled release–a review. *International Journal of Food Science & Technology* 41 (1): 1–21.

Maestro, A., Solè, I., González, C. et al. (2008). Influence of the phase behavior on the properties of ionic nanoemulsions prepared by the phase inversion composition method. *Journal of Colloid and Interface Science* 327 (2): 433–439. https://doi.org/10.1016/j.jcis.2008.07.059.

Maherani, B., Arab-Tehrany, E., Mozafari, M.R. et al. (2011). Liposomes: a review of manufacturing techniques and targeting strategies. *Current Nanoscience* 7 (3): 436–452. https://doi.org/10.2174/157341311795542453.

Manconi, M., Mura, S., Manca, M.L. et al. (2010). Chitosomes as drug delivery systems for C-phycocyanin: preparation and characterization. *International Journal of Pharmaceutics* 392 (1): 92.

Marianecci, C., Di Marzio, L., Rinaldi, F. et al. (2014). Niosomes from 80s to present: the state of the art. Advances in Colloid and Interface Science. *Special Issue in honor of Bjorn Lindman* 205: 187–206.

Matos, M., Pando, D., and Gutiérrez, G. (2019). Nanoencapsulation of food ingredients by niosomes. In: *Lipid-Based Nanostructures for Food Encapsulation Purposes, Vol. 2 of Nanoencapsulation in the Food Industry* (ed. S.M. Jafari), 447–481. Academic Press.

Maulny, A.P.E., Beckett, S.T., and Mackenzie, G. (2005). Physical properties of co-crystalline sugar and honey. *Journal of Food Science* 70 (9): E567–E572.

McClements, D.J. (2011). Edible nanoemulsions: fabrication, properties, and functional performance. *Soft Matter* 7 (6): 2297–2316.

McClements, D.J. (2014). *Nanoparticle-and Microparticle-Based Delivery Systems: Encapsulation, Protection and Release of Active*. Compounds: CRC press.

McClements, D.J. (2018a). Encapsulation, protection, and delivery of bioactive proteins and peptides using nanoparticle and microparticle systems: a review. *Advances in Colloid and Interface Science* 253: 1–22.

McClements, D.J. (2018b). Recent developments in encapsulation and release of functional food ingredients: delivery by design. *Current Opinion in Food Science* 23: 80–84.

McClements, D.J., Decker, E.A., Park, Y., and Weiss, J. (2009). Structural design principles for delivery of bioactive components in nutraceuticals and functional foods. *Critical Reviews in Food Science and Nutrition* 49 (6): 577–606.

Moghassemi, S. and Hadjizadeh, A. (2014). Nano-niosomes as nanoscale drug delivery systems: an illustrated review. *Journal of Controlled Release* 185 (July): 22–36.

Mohammadi, R., Mahmoudzadeh, M., Atefi, M. et al. (2015). Applications of nanoliposomes in cheese technology. *International Journal of Dairy Technology* 68 (1): 11–23.

Mohan, A., Rajendran, S.R.C.K., He, Q.S. et al. (2015). Encapsulation of food protein hydrolysates and peptides: a review. *RSC Advances* 5 (97): 79270–79278.

Mohan, A., Rajendran, S.R.C.K., Thibodeau, J. et al. (2018). Liposome encapsulation of anionic and cationic whey peptides: influence of peptide net charge on properties of the nanovesicles. *LWT – Food Science and Technology* 87: 40–46.

Mozafari, M.R. (2010). Nanoliposomes: preparation and analysis. In: *Pharmaceutical Nanocarriers, Vol. 1 of Liposomes: Methods and Protocols*, Methods in Molecular Biology (ed. V. Weissig), 29–50. Totowa, NJ: Humana Press.

Mozafari, M., Johnson, C., Hatziantoniou, S., and Demetzos, C. (2008). Nanoliposomes and their applications in food nanotechnology. *Journal of Liposome Research* 18 (November): 309–327.

Munin, A. and Edwards-Lévy, F. (2011). Encapsulation of natural polyphenolic compounds; a review. *Pharmaceutics* 3 (4): 793–829.

Murali, S., Kar, A., Patel, A.S. et al. (2017). Optimization of rice bran oil encapsulation using jackfruit seed starch–whey protein isolate blend as wall material and its characterization. *International Journal of Food Engineering* 13 (4): 25–35. https://doi.org/10.1515/ijfe-2016-0409.

Muschiolik, G. (2007). Multiple emulsions for food use. *Current Opinion in Colloid & Interface Science* 12 (4): 213–220.

Nowak, E., Livney, Y.D., Niu, Z., and Singh, H. (2019). Delivery of bioactives in food for optimal efficacy: what inspirations and insights can be gained from pharmaceutics? *Trends in Food Science & Technology* 91 (September): 557–573.

O'Neill, G.J., Egan, T., Jacquier, J.C. et al. (2014). Whey microbeads as a matrix for the encapsulation and immobilisation of riboflavin and peptides. *Food Chemistry* 160: 46–52.

Ozkan, G., Franco, P., De Marco, I. et al. (2019). A review of microencapsulation methods for food antioxidants: principles, advantages, drawbacks and applications. *Food Chemistry* 272: 494–506.

Paecharoenchai, O., Teng, L., Yung, B.C. et al. (2013). Nonionic surfactant vesicles for delivery of RNAi therapeutics. *Nanomedicine* 8 (11): 1865–1873.

Pando Rodríguez, D. (2014). Formulation and preparation of niosomes containing bioactive compounds. PhD thesis. Universidad de Oviedo, Spain. http://digibuo.uniovi.es/dspace/handle/10651/30525

Pando, D., Beltrán, M., Gerone, I. et al. (2015). Resveratrol entrapped niosomes as yoghurt additive. *Food Chemistry* 170 (March): 281–287. https://doi.org/10.1016/j.foodchem.2014.08.082.

Phanindra, P., Ramesh, P., Poshadri, A., and Beeram, V.V.R. (2018). Micro encapsulation of bioactive compounds through co crystallization. *International Journal of Pure & Applied Bioscience* 6 (2): 1366–1371. https://doi.org/10.18782/2320-7051.6582.

Rahmani, V., Elshereef, R., and Sheardown, H. (2017). Optimizing electrostatic interactions for controlling the release of proteins from anionic and cationically modified alginate. *European Journal of Pharmaceutics and Biopharmaceutics* 117: 232–243.

Rattes, A.L.R. and Oliveira, W.P. (2007). Spray drying conditions and encapsulating composition effects on formation and properties of sodium diclofenac microparticles. *Powder Technology* 171 (1): 7–14.

Ray, S., Raychaudhuri, U., and Chakraborty, R. (2016). An overview of encapsulation of active compounds used in food products by drying technology. *Food Bioscience* 13: 76–83.

Recharla, N., Riaz, M., Ko, S., and Park, S. (2017). Novel technologies to enhance solubility of food-derived bioactive compounds: a review. *Journal of Functional Foods* 39: 63–73.

Reineccius, G. (2017). Aroma encapsulation and controlled delivery. In: *Springer Handbook of Odor* (ed. A. Buettner), 33–34. Springer Handbooks. Springer.

Rodríguez, J., Martín, M.J., Ruiz, M.A., and Clares, B. (2016). Current encapsulation strategies for bioactive oils: from alimentary to pharmaceutical perspectives. *Food Research International* 83: 41–59.

Roy, P., Dinda, A.K., Chaudhury, S., and Dasgupta, S. (2018). β-Cyclodextrin encapsulated polyphenols as effective antioxidants. *Biopolymers* 109 (1): e23084.

Saénz, C., Tapia, S., Chávez, J., and Robert, P. (2009). Microencapsulation by spray drying of bioactive compounds from cactus pear (Opuntia ficus-indica). *Food Chemistry* 114 (2): 616–622.

Saifullah, M., Shishir, M.R.I., Ferdowsi, R. et al. (2019). Micro and nano encapsulation, retention and controlled release of flavor and aroma compounds: a critical review. *Trends in Food Science & Technology* 86: 230–251.

Sánchez, A. and Vázquez, A. (2017). Bioactive peptides: a review. *Food Quality and Safety* 1 (1): 29–46.

Sardar, B.R. and Singhal, R.S. (2013). Characterization of co-crystallized sucrose entrapped with cardamom oleoresin. *Journal of Food Engineering, SI: Extraction and Encapsulation* 117 (4): 521–529. https://doi.org/10.1016/j.jfoodeng.2012.12.011.

Schafroth, N., Arpagaus, C., Jadhav, U.Y. et al. (2012). Nano and microparticle engineering of water insoluble drugs using a novel spray-drying process. *Colloids and Surfaces B: Biointerfaces* 90 (February): 8–15. https://doi.org/10.1016/j.colsurfb.2011.09.038.

Schoubben, A., Blasi, P., Giovagnoli, S. et al. (2010). Development of a scalable procedure for fine calcium alginate particle preparation. *Chemical Engineering Journal* 160 (1): 363–369. https://doi.org/10.1016/j.cej.2010.02.062.

Semyonov, D., Ramon, O., Kaplun, Z. et al. (2010). Microencapsulation of lactobacillus paracasei by spray freeze drying. *Food Research International* 43 (1): 193–202. https://doi.org/10.1016/j.foodres.2009.09.028.

Shade, C.W. (2016). Liposomes as advanced delivery Systems for Nutraceuticals. *Integrative Medicine: A Clinician's Journal* 15 (1): 33–36.

Timilsena, Y.P., Akanbi, T.O., Khalid, N. et al. (2019). Complex coacervation: principles, mechanisms and applications in microencapsulation. *International Journal of Biological Macromolecules* 121 (January):1276–1286.https://doi.org/10.1016/j.ijbiomac.2018.10.144.

Vankayala, J.S., Battula, S.N., Kandasamy, R. et al. (2018). Surfactants and fatty alcohol based novel nanovesicles for resveratrol: process optimization, characterization and evaluation of functional properties in RAW 264.7 macrophage cells. *Journal of Molecular Liquids* 261 (July): 387–396. https://doi.org/10.1016/j.molliq.2018.04.058.

Vishali, D.A., Monisha, J., Sivakamasundari, S.K. et al. (2019). Spray freeze drying: emerging applications in drug delivery. *Journal of Controlled Release* 300 (April): 93–101. https://doi.org/10.1016/j.jconrel.2019.02.044.

Vladisavljević, G.T. (2019). Preparation of microemulsions and nanoemulsions by membrane emulsification. *Colloids and Surfaces A: Physicochemical and Engineering Aspects* 579 (October): 123709.

Wang, X., Jiang, Y., Wang, Y.-W. et al. (2008). Enhancing anti-inflammation activity of curcumin through O/W nanoemulsions. *Food Chemistry* 108 (2): 419–424.

Wildman, R.E.C. (2016). *Handbook of Nutraceuticals and Functional Foods*. CRC press.

Wilson, N. and Shah, N.P. (2007). Microencapsulation of vitamins. *ASEAN Food Journal* 14 (1): 1.

Xiao, Z., Liu, W., Zhu, G. et al. (2014). A review of the preparation and application of flavour and essential oils microcapsules based on complex Coacervation technology. *Journal of the Science of Food and Agriculture* 94 (8): 1482–1494.

Ziaee, A., Albadarin, A.B., Padrela, L. et al. (2019). Spray drying of pharmaceuticals and biopharmaceuticals: critical parameters and experimental process optimization approaches. *European Journal of Pharmaceutical Sciences* 127: 300–318.

8 Advances in Extrusion Technologies

Manzoor Ahmad Shah[1], Shabir Ahmad Mir[2], and B.N. Dar[3]

[1]*Department of Food Science & Technology, Government PG College for Women, Gandhi Nagar, Jammu, JK, India*
[2]*Department of Food Science & Technology, Government College for Women, M. A. Road, Srinagar, JK, India*
[3]*Department of Food Technology, IUST-Kashmir, JK, India*

Summary

Extrusion technology has a great impact on the food industry with its numerous benefits over other processing methods. Different products from various food groups can be prepared by modifying minor or major ingredients and processing conditions using extrusion technology. The recent advances in extrusion technology have made it possible to control the process parameters so as to achieve the desired characteristics of the final product. This technology has been used to produce numerous products like breakfast cereals, savory snacks, pre-cooked flours, cereal based baby food and textural proteins. It is a very practical and economical processing method to manufacture diverse types of food products with high consumer acceptability.

8.1 Introduction

The word 'extrusion' refers to a process by which a liquid to a partially liquid product is forced through a desired die in order to obtain a desired shape and structure. According to Rossen and Miller (1973), 'Food Extrusion is a process in which a food material is forced to flow, under one or more varieties of conditions of mixing, heating and shear, through a die which is designed to form and/or puff dry the ingredients'. The basic science and technology underlying food extrusion systems has been the same since its inception; however, new applications, efficiency of use and design have made food processing using extrusion much easier, reliable, less of an art form depending on the operator interference, and more of a science and technology and self-correcting continuous process.

Food Formulation: Novel Ingredients and Processing Techniques, First Edition.
Edited by Shivani Pathania and Brijesh K. Tiwari.

Extrusion is mainly dependent on multitude of factors; however, the most important parameter, as like any other food processing technology, would be the raw material. Physical and chemical composition (moisture content, starch, proteins, fats and sugar content and pH) of the raw material are critical to achieve final product properties. A slight change in raw material properties can yield undesirable products leading to time, energy and capital loss. In terms of process parameters, barrel temperature and pressure, screw configuration and rotation speed, die diameter and shear force can significantly impact the product properties (Steel et al. 2012). Various food groups, ranging from cereals, pulses and oilseeds, meat and fish products, roots and tubers, fruits and vegetables have found useful application in extrusion processing. The food extruder is a short bioreactor capable of continuous conversion of raw ingredients into a wide range of final products. Changes in raw material properties, processing parameters, extrusion variables can yield a range of food products with diverse structures as well as functional properties (Figure 8.1). Extrusion-based product range diversity (Table 8.1)

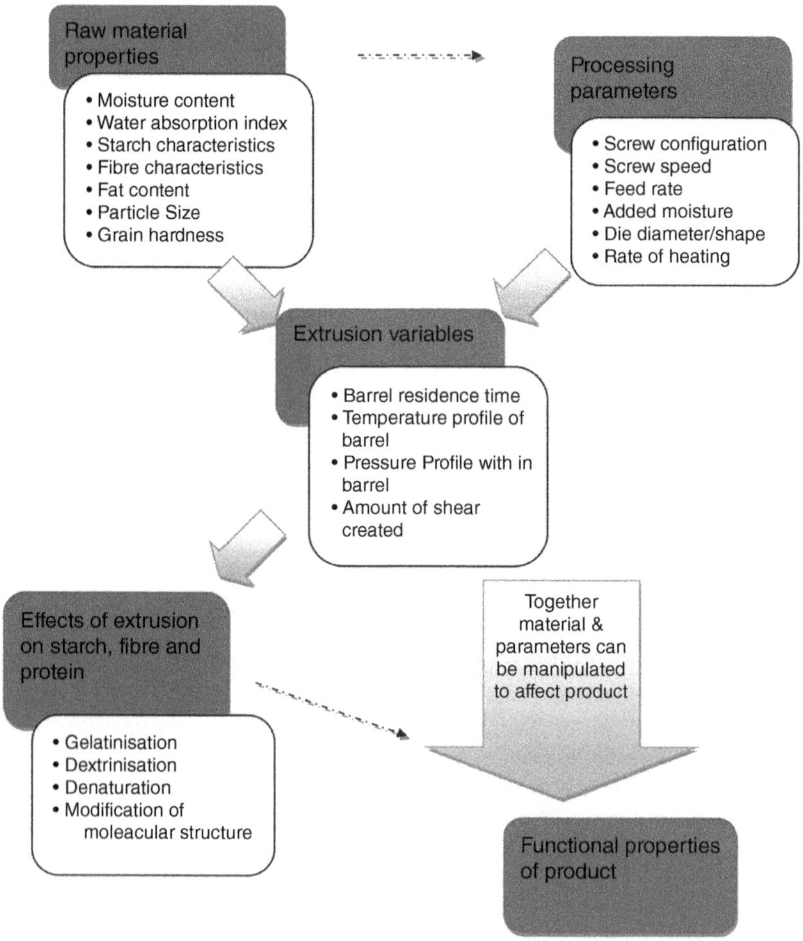

Figure 8.1 Potential interactions of raw materials, process variables and extruder to form product. *Source*: Brennan et al. (2013), Reproduced with permission from Wiley.

Table 8.1 Extruded rice and rice-based products produced under various extrusion process conditions.

Products	Ingredients	Extrusion conditions		
		Barrel/Die temperature (°C)	Feed moisture (%)	Screw speed (rpm)
Single-screw extruded products				
Floating fish pellets	Rice, Taro	79–100	40	150
Extruded flour	Rice, whey protein concentrate	106.36–173.64	16–23	140
Gluten-free pasta	Brown rice	63.3–96.8	26.6–43.4	80
Rice pasta	Rice	120	40	–
Amaranth-enriched, gluten-free rice pasta	Parboiled milled rice, Amaranth seed	25–120	40	–
Instant rice	Rice	30–70	28–36	150–300
Extruded product	Whole corn, brown rice	175	14	150
Snack	Soybean flour, purple rice	150–190	15–19	350–450
Rice-grit snack	Rice grit, durum flour, partially defatted hazelnut flour and fruit waste	150–175	12–18	200–280
Adlay-based extrudates	Adlay, rice	140	–	50
Cactus pear rice-based extrudate	Orange yellow and red cactus pear fruit, rice	100–160	16	250
Rice vermicelli	Rice flour and modified starch	70–110	30	20–40
Rice grit extrudate	Rice grit, pea	130–170	18–24	100
Chestnut-rice flour-based snack	Chestnut flour, rice flour	90–120	35	80
Black rice pasta	Black rice, joha rice, egg white powder	35	35	100
Mineral fortified snack	Wheat bran and broken rice flour	93–97	–	300
Twin-screw extruded products				
Extruded black rice	Black rice	90–120	–	150
Pea fortified rice snack	Rice, pea protein isolate	130	19–23	400–600
Extruded snacks	Brown rice grits	130–170	14–18	400–550
Rice noodles	Rice flour	90	35	200

(*Continued*)

Table 8.1 (Continued)

Products	Ingredients	Extrusion conditions		
		Barrel/Die temperature (°C)	Feed moisture (%)	Screw speed (rpm)
Brown rice pasta	Brown japonica and indica rice flour, soy protein isolate, gaur gum, sodium alginate, xanthan gum, monoglyceride	100–120	28	80–120
Rice-pinto bean extrudate	Brown rice, pinto bean flour	80–110	17.2–20.1	300
Soy-fortified millet-based snack	Rice, soybean, finger millet	184	18	285
Pea-rice-based extruded product	Pea, rice, gums	118	Water feed rate 0.39 l/h	306
Energy bar	Hempseed powder and rice flour	60–130	20	200
Gluten-free snack	Rice, soy, teff flour, milk powder, potato, corn starch, cran berry, beetroot, apple, carrot	80–150	-	200–350
Rice flour extrudate	Rice flour, sucrose	70–100	27.9–42.1	150
Low GI extruded product	Rice, chick pea, fenugreek	118	Water feed rate 0.42 kg/h	300
Gluten-free snack	Rice, maize, chickpea	110	13.5	220–340
Rice-based expanded snack	Rice	86.4–153.6	11.27–24.73	132.3–367.7
Ready-to-eat snack	Corn meal, rice or potato flour, sweet whey solids or whey protein concentrate	125	Water feed rate 0.5–1.0 l/h	300

Source: Dalbhagat et al. (2019). Reproduced with permission from Elsevier.

was exemplified by Dalbhagat et al. (2019) wherein they illustrated how rice flour – as the main feed material, when extruded using different co-ingredients and process parameters – impacts the organization and transformation of the melt, yielding absolutely different food products.

Extrusion technology allows fast and efficient conversion of various raw materials into diverse food products by changing feed formulation and operating conditions on the same extrusion equipment (Riaz 2006). Since its introduction, extrusion has replaced traditional batch-scale food processing. It allows for flexibility, thus leading to process and new product development; automatic online process modifications have been made to achieve required product characteristics, and several different products ranging from highly viscous or wet to relatively dry materials, using the same equipment. Additionally, extrusion is frequently able to process foods in one manufacturing step. It is a widely used technology in food processing, producing hundreds of food products, including traditional snacks, breakfast cereals, ingredients, pasta, texturized vegetable protein, cookie dough, baby foods, processed cheese, bread and many more. The success of extrusion technology in the food industry is mainly due to its ability to continuously manoeuvre very viscous and complex fluids and powders. As food is being injected through the extruder, foods are cooked by the high pressure, high shear and high temperature created by the screws, enclosed in the barrel. Upon exiting, materials often puff due to the release of pressure and conversion of water into steam. The entire process is continuous and has short residence time of less than a minute. This chapter will discuss the state of the art of extrusion process and advances in the technology leading to new food structures and functions.

8.2 History and State of the Art

Greek mathematician and physicist Archimedes (287–212 BCE) is credited for the simple design of a screw within a barrel chamber; he developed this wooden apparatus to move water from lower level to a higher level with the turn of screw within a round chamber. This simple design later became the foundation of many different industries including food and feed industries. Modern-day extruder consists of one or more screws encased in a metal barrel, attached to a drive motor. Extruders have come a long way starting from rubber processing and then applied to plastics processing and very recently to food and feed industry. The first extrusion patent was granted in 1797 in the name of Joseph Bramah for making a lead pipe by forcing the raw material through a die to form a continuous profile (Kazemzadeh 2012). Extruders were developed in late nineteenth century to manufacture sausage. Application of single-screw extruder evolved during the 1930s, when it was used to produce pasta from semolina and breakfast cereal pellet production. During the late 1930s and 1940s, corn curls were made using extruders. In the 1950s, it was applied for pet food manufacturing. The first patent on application of twin-screw extrusion was filed in mid-1950s, after that, the application of food extrusion technology has widened and grown considerably and has been developed as an alternative method for texturization, mixing, forming, reacting and cooking of food.

As far as design of extruder is concerned, several designs are possible for food extrusion, with single and twin screw being the most common. Simply explained, the extruder is a continuous processing machine consisting of one or two screws that are mounted on shafts and rotate in a fixed closed housing called a barrel. Single-screw extruders are installed with one screw and twin-screw extruders have two intermeshing screws. Both types of extruders operate in a similar manner, providing transport, compression, mixing, cooking, shearing, heating, cooling and shaping of raw material final products but with different efficiencies and quality.

8.2.1 Process/Working

A general illustration of extrusion processing line is shown in Figure 8.2. Raw material characteristics significantly impact the material transfer in the extruder, melt viscosity and final product properties; therefore, the importance of raw material optimization

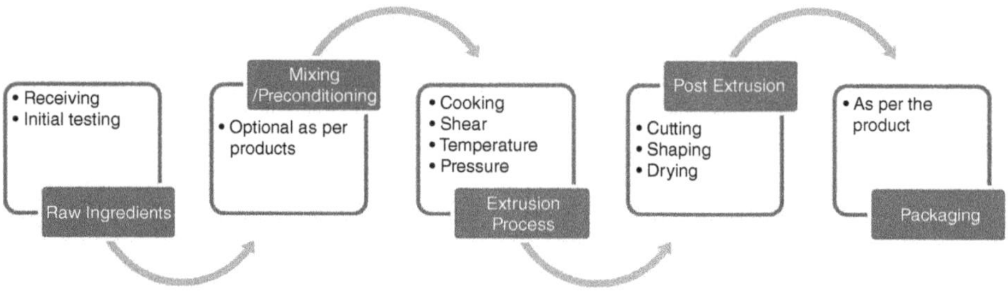

Figure 8.2 Flow chart of a typical extrusion processing line.

cannot be stressed further. Raw ingredients are suggested to be premixed using blenders and/or preconditioned to ensure uniformity in particle size as well as optimum moisture levels before they enter the extruder.

Extrusion processing is the main cooking step where the raw ingredients are changed from solid powders to a hot dough inside the extruder with high shear and temperature for a short period. This facilitates to transform these raw ingredients into the cooked and formed products once this hot dough is forced outside through a die at the end of the extruder. The hot dough coming out of the extruder encounters a sudden drop in pressure, resulting in rapid expansion and a decrease in temperature, helping it to transform into a cooked product (Alvarez-Martinez et al. 1988).

After extrusion, several other operations known as post-extrusion processing operations are used to alter the properties of the extruded products. These operations include cutting, drying, seasoning or coating. Cutting is done to produce the product into appropriate sizes, drying to reduce the moisture to the desired level, seasoning or coating to provide the desired flavour and taste to the extruded products prior to packaging.

The final quality parameters of the extruded product depend on several factors, such as ingredient mixture and its properties, extrusion conditions and post-extrusion processing conditions. As extrusion processing offers a great flexibility, it has found very diverse applications in food industry.

8.3 Application of Extrusion Technology

Extruder is a multifunctional, versatile machine that allows it to be used for wide range of applications across food and feed industry. Some of the functions identified by Riaz (2000) are agglomeration, degassing, dehydration, expansion, gelatinization, grinding, homogenization, mixing, pasteurization, sterilization, protein denaturation, shaping, shearing, texture alteration, thermal cooking and unitizing.

Extrusion is a widely used processing method for manufacturing diverse food items with various ingredients requiring different processing conditions and includes starch, protein, lipids, water and additives. Extrusion process resulted in gelatinization of starch, denaturation of protein and formation of complexes between starch:lipids and between proteins:lipids. Changes in the properties of foods rich in starch caused by the addition of lipids are mainly due to the formation of complexes between amylose and lipids.

Extrusion cooking technology is a high temperature short time (HTST) processing being used increasingly in the food industries for the development of various products such as cereal-based snacks, including dietary fibre, baby foods, breakfast cereals and modified starch from cereals (Alam et al. 2016). As it is a HTST process, it reduces microbial load and inactivates enzymes, but the main mode of preservation of both hot and cold extruded foods is due to low water activity of product (0.1–0.4) (Bordoloi and Ganguly 2014). The thermal energy generated by viscous dissipation extruded with combination of shearing effect cooks quickly the raw mixture so that the properties of the materials are modified by physicochemical changes of the biopolymers (Chanvrier et al. 2015).

Extrusion cooking is a continuous process with high production capacity, versatility and low cost per product unit. The extrusion cooking, especially used in the production of precooked and modified starches, ready to eat cereals, infant formulae and snack foods, has increased recently. During extrusion cooking, the native starch undergoes substantial changes leading to a greater molecular disorganization. The factors such as shear, temperature, moisture and feed composition are important in the transformation of starch by the extrusion process. It has become an important technique in increasing variety of food products. The benefits of this technology are: a low-cost technique, variability of product shapes, high-quality production of new foods, higher productivity, inactivation of anti-nutritional factors, etc. (Chakraborty and Banerjee 2009).

8.4 Recent Advances in Extrusion Process

Extrusion technology is widely used for manufacturing of a variety of extruded products. It is used to produce different products by simply changing the raw materials and processing conditions. The common raw materials for preparing extruded products are cereal flours, but some other ingredients derived from other food sources may also be used, provided they meet the requirement of the equipment used. Other raw materials used are pseudo-cereals, fruits and vegetables, legumes, pulses, oilseeds, roots and tubers, nuts and seeds and meats. Various combinations of these raw materials are used to produce products varying in shapes, textures, colours and appearances. Diversity in these products can be achieved by varying raw material mix, equipment assembly and operating conditions (Riaz 2000).

The ingredients used in the manufacture should be of high quality, and only permitted additives of required purity must be used. The properties of raw materials which are important are cohesiveness, hardness and surface friction. The most commonly used ingredients are flours and granules from maize, rice, wheat and potatoes. The other raw materials used are sorghum, cassava, oats, barley, rye and buckwheat. The basic product structure and texture is produced by changing and manipulating the biopolymers (starches and proteins). In products where proteins play the major role, raw materials are selected from oilseed proteins like soybean or from fractionated cereal proteins. The materials when incorporated into a formulation tend to change the process, thereby modifying the resulting product (Offiah et al. 2019). Bhat et al. (2019) prepared whole wheat flour-based snacks supplemented with crude lycopene, tomato powder and saffron extracts (Figure 8.3).

Extrusion cooking has been the widely used processing method for food and feed products, and it is rapidly evolving from an art into a science (Riaz et al. 2009). It has been used to produce a variety of speciality foods including pasta products and ready-to-eat breakfast cereals, baby foods, snack foods, texturized vegetable protein, dried soups and dry beverage mixes, as it not only improves digestibility (Singh et al. 2010)

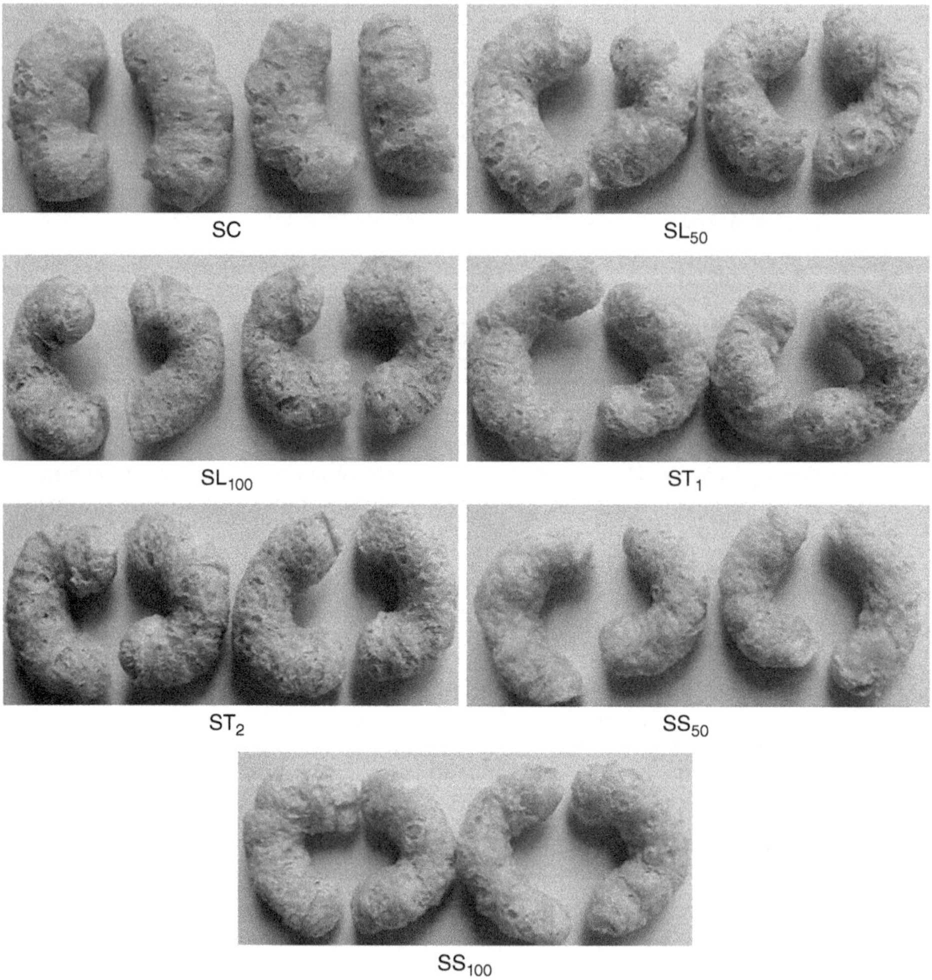

Figure 8.3 Whole wheat flour-based snacks supplemented with crude lycopene, tomato powder and saffron extracts. SC: Control snacks without any supplement; SL_{50}: Snacks with added crude lycopene (50 mg/100 g of flour); SL_{100}: Snacks with added crude lycopene (100 mg/100 g of flour); ST_1: Snacks with added tomato powder (2 g/100 g); ST_2: Snacks with added tomato powder (4 g/100 g); SS_{50}: Snacks with SE of 50 mg/100 g of flour; SS_{100}: Snacks with SE of 100 mg/100 g of flour. *Source*: Bhat et al. (2019). Reproduced with permission of Elsevier.

but also improves nutrients' bioavailability (Gu et al. 2008) compared to conventional cooking. In addition to these properties, extrusion cooking is preferred over conventional cooking/processing techniques because of its ability to produce a range of products with distinct textural properties including expansion, crispiness and general mouthfeel. It is a versatile technique with high productivity, low operating costs, energy efficiency and shorter cooking times (Brennan et al. 2011).

Extrusion process is used to produce a gelatinized, chewy product such as fruit gums and liquorice, from a mixture of sugar, glucose and starch. The starch is gelatinized by heat, sugar is dissolved and excess water is vaporized and released from the extruder. Different colours and flavours are added to produce different products. The product texture can be adjusted from soft to elastic by changing the formulation and processing conditions. The shape can be changed by changing the die through which the product comes out of the extruder. Different combinations of these factors allow the production of a wide range of products including liquorice, toffee, fudge, boiled sweets, creams and chocolate (Fellows 2017).

Cereal products like crispbread and breakfast cereals have been produced by the extrusion process. In the manufacture of crispbread, ingredients (wheat flour, milk powder, corn starch, sugar and water) are mixed and subjected to extrusion at a high temperature and pressure. The crispbread is then toasted to decrease the moisture content and to develop brown colour on the surface. This results in reduced energy consumption as compared to oven baking. The corn flakes are manufactured from dough pellets produced in a low-pressure extruder. The size of the corn flakes can be controlled by using different size of the pellets. Then, these are flaked, toasted and sprayed with a vitamin solution (Fellows 2017).

Extrusion cooking has been used in manufacture of protein-based foods such as texturized vegetable protein and meat and fish products. It has been found that extrusion cooking results in the destruction of the enzymes present in soybeans. It destroys urease which decreases the shelf life, lipoxidase which produces off flavours by oxidation of soya oil and a trypsin inhibitor which decreases protein digestibility. Thus, extrusion increases the acceptability, digestibility and shelf life of the extruded product (Fellows 2017).

Texturized vegetable proteins are expanded sponge-like textures developed through low moisture extrusion and dried before packaging. They are generally hydrated before consumption and are often used to improve the functional properties of processed meat products (beef patties and sausages) in order to reduce the shrinkage and improve their water holding capacity (Gnanasambandam and Zayas 1994). High-moisture meat analogues, are not only meat extenders, can be directly used to develop meat-like products such as burgers, patties, sausages and mince due to their highly fibrous textures mimicking meat or muscle (Cheftel et al. 1992). The recent interest in high moisture extrusion is to develop meat-like structures for flexitarian population, to encourage the consumption of plant proteins. High-moisture meat analogues can, texturally, mimic meat fibres and along with right combination of flavors, taste like meat. Soy protein and wheat gluten are the main components of the meat analogues as they can mimic not only the texture, appearance and functionality but also the nutritional value of the meat-based products. Soy protein isolate, wheat gluten and starch have also been used in a different combination to develop meat analogues (Lin et al. 2000). Studies have shown that soy protein isolate and starch in the ratio of 90 : 10 is the preferred formulation to achieve high-quality meat substitutes.

Protein content of the feed material, screw profile, screw speed, moisture content and cooking temperature are the parameters that impact the texturization process in development of high-moisture meat analogues. Lin et al. (2002) extruded soy protein isolate and wheat starch at a fixed feed rate of 6.8 kg/h and screw speed of 150 rpm and varied the moisture content from 60 to 70% and cooking temperature from 138 to 160 °C; they found that moisture content, as compared to cooking temperature, controlled the protein denaturation and texturization thereby influencing the sensory characteristics. Formulation can impact final product properties. Samard et al. (2019) studied that wheat gluten incorporation in soy protein isolate can contribute to a spongy texture in a low-moisture extruded product which without its inclusion yields an expanded protein structure. Additionally, wheat gluten inclusion in high-moisture extruded product yielded meat-like, fibrous and dense structure. However, soy and wheat are known to be common food allergens, protein concentrates of pea, chickpea and faba bean have been extruded to develop meat analogues but more work is needed on these new materials to achieve the textures and functional properties similar to soy-based products (Figure 8.4).

The other recent applications of extrusion technology include the use of extruders as enzymatic reactors to modify starches using thermostable α-amylase. Extruders have been used to partially hydrolyse caseinates using proteases to improve their colour, flavour, water absorption and microbial stability. The oligosaccharide mixtures have been produced from potato starch in an extruder without using enzymes. These types of products are suitable for dietetic and infant food. It has also been used to produce high-quality oil from soybean. The oil was extracted by using dry extrusion (10–14% moisture) at 135 °C followed immediately by pressing in a continuous screw press. Extruders are also being used for decontamination of spices. The sterilization of cocoa nibs prior to roasting for chocolate manufacture has been done by using extruders. Extrusion decreases microbial by 1000 times and removes off-flavours and thus that reduces the need for time consuming and expensive stages during processing (Fellows 2017).

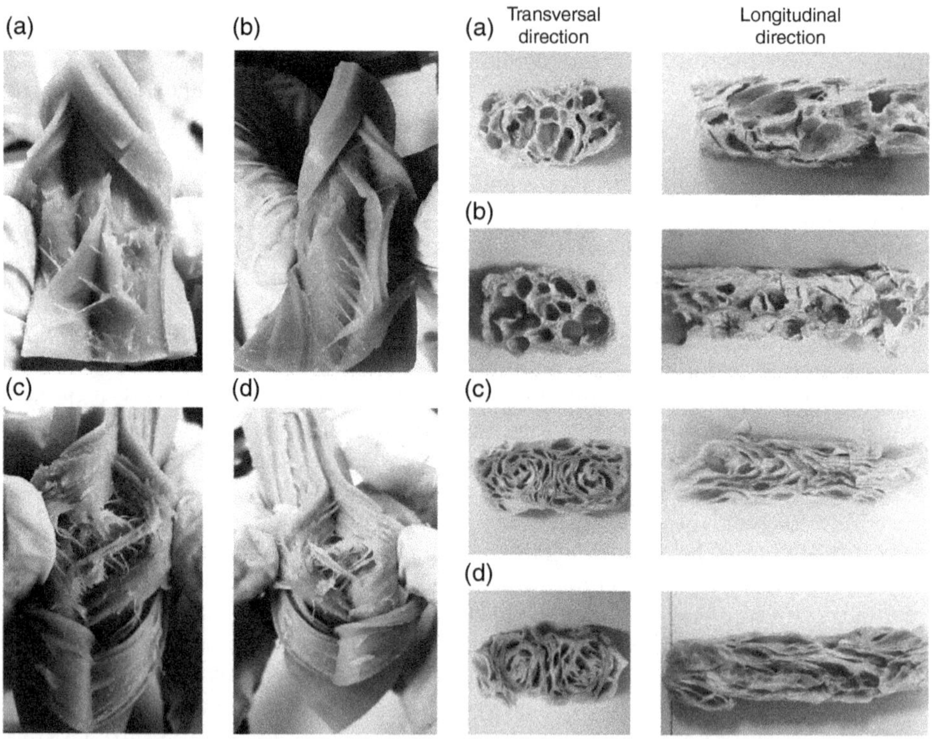

Figure 8.4 Left side (a to d): Photographs of high-moisture meat analogues (left) without [(a) and (b)] and with added [(c) and (d)] wheat gluten extruded at screw speeds of 150 and 200 rpm, respectively. Right side (a to d): Photographs with low-moisture meat analogues without wheat gluten extruded at screw speeds of 150 (a) and 200 rpm (b), and of analogues with added wheat gluten at the same screw speeds [150 (c) and 200 rpm (c)]. *Source*: Samard et al. (2019). Reproduced with permission of Wiley.

Extrusion assisted by supercritical CO_2 is an innovative food processing technology exploited in food industry. This technology offers a sub 100 °C expansion utilizing direct supercritical CO_2 injection during extrusion, providing many additional advantages over conventional extrusion processing. Supercritical CO_2 changes the rheological properties of the material inside the extruder and also acts as an expansion agent. The decrease of viscosity will result in the limitation of mechanical and shear stresses and will allow a decrease in operating temperatures. Therefore, the materials with limited thermal stability can be easier to handle. Also, its dissolution in the polymer under pressure is accompanied by large volume expansion during the return to atmospheric pressure (Sauceau et al. 2011; Chauvet et al. 2017).

Reactive extrusion is a processing method that is a combination of the traditionally separated chemical processes (polymer synthesis and/or modification) and extrusion (melting, blending, structuring, devolatilization and shaping) processes; carried out as a single process using an extruder. This process is usually used in food industry as it offers a more flexible alternative to polymerization in the presence of a solution, suspension or emulsion and subsequent compounding. A typical reactive extrusion process involves the reactants to be fed through the main hopper into the twin-screw extruder, where the reactants get heated up to start the chemical reaction. This method offers many possibilities for flexible and economic production of materials with specific characteristics. Additionally, the continuous nature of the process, the small residence time distribution and the high flexibility also facilitate the economic production of specialty polymers (Kalambur and Rizvi 2004; Moad 2011).

8.5 Effect of Processing Conditions on Food Formulation

Extrusion is used to produce a number of effects in different food systems, because of the different levels of starch, proteins, moisture and other components which affect the structure and texture of the resulting products. The operating conditions and raw materials influence the extent of changes in starch which result in a number of different textures in the final product. The extrusion cooking has a slight effect on the natural colour and flavour of foods because of the HTST conditions. However, the colour of the products can be altered by addition of synthetic colorants in various forms (Fellows 2017).

The combined effect of high pressure, high temperature and shear leads to several changes in food components, nutritional composition and functional properties of the food (Dalbhagat et al. 2019). The changes occurring during the extrusion process may range from simple to complex, and include gelatinization, solubilization and dextrinization or complex formation of starches, denaturation, polymerization or crosslinking and texturization of proteins, partial or complete deactivation of enzymes, browning reactions, denaturation of vitamins and inactivation of anti-nutritional factors (Wang et al. 2016). The quality of shear applied during the process also has a significant effect on the food materials. High shear results in disruption of starch and protein molecules and thus affects their functionality. The high-pressure zone of the extruder barrel is the site where most of the chemical reactions occur, thus the heat-sensitive components like flavours and vitamins may be fed into the extruder immediately before the die so as to reduce their exposure to heat and shear (Steel et al. 2012).

Extrusion cooking leads to the alteration of food biopolymers at low moisture levels and the formation of plasticized dough or fluid systems. As the temperature reaches about 110–200 °C within the dough, water present is superheated but remains in a liquid state (Ajita and Jha 2017). This facilitates changes in the forms and structures of the natural biopolymers. Due to the recent advances in the extrusion technology, it has been made possible to control the thermo-mechanical changes during the extrusion process so as to achieve the required product properties (Offiah et al. 2019).

The gelatinization of starch results in expanded products and with increased gas holding capacity. Gelatinization leads to the breakage of intermolecular hydrogen bonds resulting in the disruption of crystalline structure and increase in water absorbing capacity. Moreover, the starch granules swell resulting in release of amylose molecules (Wang et al. 2017, 2018; Dalbhagat et al. 2019).

The effect of amylose–amylopectin ratio on extruded products has been extensively studied (Chanvrier et al. 2007; Tacer-Caba et al. 2014). Amylopectin results in harder extrudates with less expansion, while amylose produces lightness and elasticity in the products, although with a sticky surface (Offiah et al. 2019). About 5–20% amylopectin content is recommended in raw materials to produce food products with good texture and hardness (Moscicki et al. 2013) while waxy starches need high gelatinization temperature (Alcazar-Alay and Meireles 2015). The swelling capacity is considerably reduced by the lipids associated with amylose fractions (Moscicki et al. 2013).

Proteins are appreciably influenced by the extrusion process because of their heat-labile nature (Dalbhagat et al. 2019). Several changes occur in proteins during the extrusion process, and protein denaturation is the most important one (Offiah et al. 2019). Proteins lose their native structure during extrusion cooking under the influence of thermo-mechanical energy and form a viscoelastic mass. Denaturation reduces the protein solubility and improves digestibility. The enzymes also lose their biological activity due to high temperature and shear during the process (Ramachandra and Thejaswini 2015). The inactivation of enzymes by extrusion cooking reduces the risk of lipid oxidation and other enzyme-mediated adverse reactions in the food products (Zhu et al. 1996). Extrusion is used to produce texturized protein products (Omohimi et al. 2014; Jiddere and Filli 2015) that resemble the structure and texture of meats (Joshi and Kumar 2015).

The mechanical shear developed during the extrusion process leads to the disruption of protein molecules that leads to the improved protein digestibility. Arribas et al. (2017) reported that the in vitro protein digestibility was significantly increased in rice pea and carob flour blend. This may be due to denaturation of proteins and inactivation of anti-nutritional factors such as protease inhibitors (Dalbhagat et al. 2019). Rathod and Annapure (2016) reported that extrusion was most useful method for elimination of the anti-nutritional factors such as phytic acid (99.30%), trypsin inhibitors (99.54%) and tannins (98.83%) in lentils, without changing protein content. Digestibility of the protein was also increased by 89%.

The extrusion process may lead to the production of reducing sugars which in turn may react with amino acids and thus may result in development of coloured compounds via Maillard reactions favoured by low moisture and high temperature. These reactions lower the protein quality which depends on the types of raw material, composition and processing conditions used during the extrusion process. Csapo et al. (2008) reported that the amount of several amino acids was decreased by the extrusion process carried out at 220 °C in dry full fat soy. Singh et al. (2007) assessed the effect of extrusion on egg and milk proteins, and the reducing sugars (galactose and fructose) using pre-gelatinized flour (wheat) and reported that lysine showed the lowest retention, while retention values for the other amino acids in most products were in the range of 80–100%.

Lipids serve an important role in extrusion cooking because of their role as plasticizers or emulsifiers and greatly affect the texture and stickiness of the extruded products. The effects of extrusion processing parameters on the lipids have been studied by many researchers. At temperatures greater than 40 °C, lipids change into a liquid state and disperse within other materials as small oil droplets. The dispersed droplets change the melt rheological behaviour (viscosity and flow behaviour) of the feed mix through the extruder, and is affected by the quantity of shear inside the extruder (Ilo et al. 2000). Tumuluru et al. (2013) observed that the lipid content was reduced by 15% in fish and rice flour extrudates with the increase in temperature from 100 to 200 °C. This may be due to the formation of lipid–protein complexes. The high temperature results in oxidation of unsaturated fatty acids to lipid hydroperoxides that lead to a decrease in lipids in the final extruded product (Tumuluru et al. 2013). According to Sandrin et al. (2018), the increase in the extrusion temperature changes the lipids into liquid form and increases the migration of the oil. The combined effect of shear stress and high pressure results in the expulsion of oil.

Lipids interact with amylase and form amylose–lipid complexes during the extrusion cooking process. This results in the modification of physicochemical properties (expansion ratio, bulk density and water solubility index) of the products. Formation of amylose–lipid complexes depends on the nature of starch and lipids present in the feed. Lipid levels greater than 5% may significantly decrease expansion, mainly due to its effects on starch gelatinization, and levels less than 3% did not show any significant effect. High lipid concentration decreases extruder performance by decreasing shear within the extruder barrel. For some snack products, decreasing moisture content may compensate for expansion in formulations with high lipid content (Offiah et al. 2019).

Dietary fibre is an important component in the human diet as it reduces the risk of many diseases. It leads to the reduction of calorific value by incorporating in different food formulations (Arribas et al. 2017). Dietary fibres can be categorized into insoluble or soluble fibres, based on their solubility in water. Insoluble fibres may be converted to soluble fibres because of the modifications in their physicochemical and structural properties due to extrusion processing (Martınez-Bustos et al. 2011). This may be due to breakage of non-covalent and covalent bonds between fibres and other molecules, resulting in smaller molecules with greater solubility. Some authors have reported that extrusion resulted in the conversion of insoluble fibres in cauliflower trimming powder into soluble fibre (Stojceska et al. 2008; Alam et al. 2016). The extent of degradation of fibres depends on shear stress intensity. Several researchers have reported that increase in dietary fibre content resulted in a decrease in expansion in extruded products. This may probably be due to inhibition of starch gelatinization (Yanniotis et al. 2007).

Vitamins as heat-labile compounds and are susceptible to thermal degradation during extrusion cooking. Their stability varies the process because of their variation in their chemical composition and structure. Heat treatment at 100 °C and above is expected to reduce the heat-sensitive vitamins, especially water-soluble vitamins (Moscicki et al. 2013). However, due to the short duration of the high temperature during extrusion cooking and rapid cooling after extrusion, these losses are usually low as compared to other traditional thermal treatments (Offiah et al. 2019). Research has been carried out especially on retention of B-vitamins in cereals, particularly thiamine, riboflavin and niacin (Riaz et al. 2009). Retention of such vitamins is often affected by temperature, moisture content and screw speed. Vitamins A and E have been reported to be highly stable during extrusion cooking. Athar et al. (2006) reported that the retention of B vitamins was about 44–62% in cereal-based snack.

Careful control of extrusion parameters is necessary to obtain products with good nutritional quality (Moscicki and van Zuilichem 2011).

Minerals are less susceptible to thermal processing operations and have been found to be stable during the processing of various food matrices (Singh et al. 2007). However, some studies have shown that the mineral content of the pasta prepared from brown rice and corn flour increased. The extrusion cooking increased the iron content in these products (da Silva et al. 2016). Singh et al. (2000) observed a significant increase in iron, calcium, phosphorous and copper in a blend of wheat bran and rice flour.

8.6 Conclusions

Extrusion technology has gained importance due to the development of various types of traditional and novel products. Extrusion technology has proven to be a very practical and economical processing tool to produce various types of food products that the consumers enjoy. The selection of ingredients and optimum processing conditions are useful for the development of high-quality products with better sensory properties and consumer acceptability.

References

Ajita, T. and Jha, S.K. (2017). Extrusion cooking technology: principal mechanism and effect on direct expanded snacks: an overview. *International Journal of Food Studies* 6: 113–128.

Alam, M.S., Pathania, S., and Sharma, A. (2016). Optimization of the extrusion process for development of high fibre soybean-rice ready-to-eat snacks using carrot pomace and cauliflower trimmings. *LWT – Food Science and Technology* 74: 135–144.

Alcazar-Alay, S.C. and Meireles, M.A.A. (2015). Physicochemical properties, modifications and applications of starches from different botanical sources. *Food Science and Technology (Campinas)* 35: 215–236.

Alvarez-Martinez, L., Kondury, K., and Harper, J. (1988). A general model for expansion of extruded products. *Journal of Food Science* 53 (2): 609–615.

Arribas, C., Cabellos, B., Sanchez, C. et al. (2017). The impact of extrusion on the nutritional composition, dietary fiber and in vitro digestibility of gluten-free snacks based on rice, pea and carob flour blends. *Food and Function* 8 (10): 3654–3663.

Athar, N., Hardacre, A., Taylor, G. et al. (2006). Vitamin retention in extruded food products. *Journal of Food Composition and Analysis* 19 (4): 379–383.

Bhat, N.A., Wani, I.A., Hamdani, A.M., and Gani, A. (2019). Effect of extrusion on the physicochemical and antioxidant properties of value added snacks from whole wheat (Triticum aestivum L.) flour. *Food Chemistry* 276: 22–32.

Bordoloi, R. and Ganguly, S. (2014). Extrusion technique in food processing and a review on its various technological parameters. *Indian Journal of Scientific Research and Technology* 2: 1–3.

Brennan, C., Brennan, M., Derbyshire, E., and Tiwari, B.K. (2011). Effects of extrusion on the polyphenols, vitamins and antioxidant activity of foods. *Trends in Food Science and Technology* 22: 570–575.

Brennan, M.A., Derbyshire, E., Tiwari, B.K., and Brennan, C.S. (2013). Ready-to-eat snack products: the role of extrusion technology in developing consumer acceptable and nutritious snacks. *International Journal of Food Science and Technology* 48: 893–902.

Chakraborty, P. and Banerjee, S. (2009). Optimization of extrusion process for productionof expanded product from green gram and rice by response surface methodology. *Journal of Scientific and Industrial Research* 68 (2): 140–148.

Chanvrier, H., Appelqvist, I.A.M., Bird, A.R. et al. (2007). Processing of novel elevated amylose wheats: functional properties and starch digestibility of extruded products. *Journal of Agricultural and Food Chemistry* 55: 10248–10257.

Chanvrier, H., Nordstrom Pillin, C., Vandeputte, G. et al. (2015). Impact of extrusion parameters on the properties of rice products: a physicochemical and X-ray tomography study. *Food Structure* 6: 29–40.

Chauvet, M., Sauceau, M., and Fages, J. (2017). Extrusion assisted by supercritical CO_2: a review on its application to biopolymers. *The Journal of Supercritical Fluids* 120: 408–420.

Cheftel, J.C., Kitagawa, M., and Queguiner, C. (1992). New protein texturization processes by extrusion cooking at high moisture levels. *Food Reviews International* 8: 235–275.

Csapo, J., Varga-Visi, E., Loki, K. et al. (2008). The influence of extrusion on loss and racemization of amino acids. *Amino Acids* 34: 287–292.

Dalbhagat, C.G., Mahato, D.P., and Mishra, H.N. (2019). Effect of extrusion processing on physicochemical, functional and nutritional characteristics of rice and rice-based products: a review. *Trends in Food Science and Technology* 85: 226–240.

Fellows, P.J. (2017). *Food Processing Technology: Principles and Practices*, 4ee. Woodhead Publishing.

Gnanasambandam, R. and Zayas, J.F. (1994). Chemical and bacteriological stability of frankfurters extended with wheat germ, corn germ and soy proteins. *Journal of Food Processing and Preservation* 18: 159–171.

Gu, L., House, S.E., Rooney, L.W., and Prior, R.L. (2008). Sorghum extrusion increases bioavailability of catechins in weanling pigs. *Journal of Agricultural and Food Chemistry* 56: 1283–1288.

Ilo, S., Schoenlechner, R., and Berghofe, E. (2000). Role of lipids in the extrusion cooking processes. *Grasas Y Aceites* 51: 97–110.

Jiddere, G. and Filli, K.B. (2015). The effect of feed moisture and barrel temperature on the essential amino acids profile of sorghum malt and bambara groundnut based extrudates. *Journal of Food Processing and Technology* 6: 6–11.

Joshi, V.K. and Kumar, S. (2015). Meat analogues: plant based alternatives to meat products – a review. *International Journal of Food and Fermentation Technology* 5: 107–119.

Kalambur, S.B. and Rizvi, S.S. (2004). Starch-based nanocomposites by reactive extrusion processing. *Polymer International* 53 (10): 1413–1416.

Kazemzadeh, M. (2012). Introduction to extrusion technology. In: *Advances in Food Extrusion Technology* (eds. M. Maskan and A. Altan), 1–22. Boca Raton: CRC Press.

Lin, S., Huff, H.E., and Hsieh, F. (2000). Texture and chemical characteristics of soy protein meat analog extruded at high moisture. *Journal of Food Science* 65 (2): 264–269.

Lin, S., Huff, H.E., and Hsieh, F. (2002). Extrusion process parameters, sensory characteristics, and structural properties of a high moisture soy protein meat analog. *Journal of Food Science* 67 (3): 1066–1072.

Martınez-Bustos, F., Viveros-Contreras, R., Galicia-Garcıa, T. et al. (2011). Some functional characteristics of extruded blends of fiber from sugarcane bagasse, whey protein concentrate, and corn starch. *Ciência e Tecnologia de Alimentos* 31: 870–878.

Moad, G. (2011). Chemical modification of starch by reactive extrusion. *Progress in Polymer Science* 36 (2): 218–237.

Moscicki, L. and van Zuilichem, D.J. (2011). Extrusion-cooking and related technique. In: *Extrusion-Cooking techniques: Applications, Theory and Sustainability* (ed. L. Moscicki), 1–24. Wiley-VCH.

Moscicki, L., Mitrus, M., Wojtowicz, A. et al. (2013). Extrusion cooking of starch. In: *Advances in Agrophysical Research* (eds. S. Grundas and A. Stepniewski), 319–346. Intech.

Offiah, V., Kontogiorgos, V., and Falade, K.O. (2019). Extrusion processing of raw food materials and by products-a review. *Critical Reviews in Food Science and Nutrition* 59: 2979–2998.

Omohimi, C.I., Sobukola, O.P., Sarafadeen, K.O., and Sanni, L.O. (2014). Effect of thermo-extrusion process parameters on selected quality attributes of meat analogue from mucuna bean seed flour. *Nigerian Food Journal* 32: 21–30.

Ramachandra, H.G. and Thejaswini, M.L. (2015). Extrusion technology: a novel method of food processing. *International Journal of Innovative Research in Science Engineering and Technology* 2: 358–369.

Rathod, R.P. and Annapure, U.S. (2016). Effect of extrusion process on antinutritional factors and protein and starch digestibility of lentil splits. *LWT – Food Science and Technology* 66: 114–123.

Riaz, M.N. (2000). *Extruders in Food Applications*. Boca Raton: CRC Press.

Riaz, M.N. (2006). Extruded snacks. In: *Handbook of Food science, Technology and Engineering* (eds. Y.H. Hui, E. Castell-Perez, L.M. Cunha, et al.), 168. Boca Raton: CRC Press.

Riaz, M., Asif, M., and Ali, R. (2009). Stability of vitamins during extrusion. *Critical Reviews in Food Science and Nutrition* 49: 361–368.

Rossen, J.L. and Miller, R.C. (1973). Food extrusion. *Food Technology* 27 (8): 46–53.

Samard, S., Gu, B.Y., and Ryu, G.H. (2019). Effects of extrusion types, screw speed and addition of wheat gluten on physicochemical characteristics and cooking stability of meat analogues. *Journal of the Science of Food and Agriculture* 99 (11): 4922–4931.

Sandrin, R., Caon, T., Zibetti, A.W., and de Francisco, A. (2018). Effect of extrusion temperature and screw speed on properties of oat and rice flour extrudates. *Journal of the Science of Food and Agriculture* 98 (9): 3427–3436.

Sauceau, M., Fages, J., Common, A. et al. (2011). New challenges in polymer foaming: a review of extrusion processes assisted by supercritical carbon dioxide. *Progress in Polymer Science* 36 (6): 749–766.

da Silva, E.M.M., Ascheri, J.L.R., and Ascheri, D.P.R. (2016). Quality assessment of gluten-free pasta prepared with a brown rice and corn meal blend via thermoplastic extrusion. *LWT – Food Science and Technology* 68: 698–706.

Singh, D., Chauhan, G.S., Suresh, I., and Tyagi, S.M. (2000). Nutritional quality of extruded-vsnakes developed from composite of rice broken and wheat bran. *International Journal of Food Properties* 3: 421–431.

Singh, S., Gamlath, S., and Wakeling, L. (2007). Nutritional aspects of food extrusion: a review. *International Journal of Food Science and Technology* 42 (8): 916–929.

Singh, J., Dartois, A., and Kaur, L. (2010). Starch digestibility in food matrix: a review. *Trends in Food Science and Technology* 21: 168–180.

Steel, C., Leoro, M., Schmiele, M. et al. (2012). Thermoplastic extrusion in food processing. In: *Thermoplastic Elastomers* (ed. E.-S. Adel), 265–290. Intech.

Stojceska, V., Ainsworth, P., Plunkett, A. et al. (2008). Cauliflower by-products as a new source of dietary fibre, antioxidants and proteins in cereal based ready-to-eat expanded snacks. *Journal of Food Engineering* 87 (4): 554–563.

Tacer-Caba, Z., Nilufer-Erdil, D., Boyacioglu, M.H., and Ng, P.K.W. (2014). Evaluating the effects of amylose and Concord grape extract powder substitution on physicochemical properties of wheat flour extrudates produced at different temperatures. *Food Chemistry* 157: 476–484.

Tumuluru, J.S., Sokhansanj, S., Bandyopadhyay, S., and Bawa, A.S. (2013). Changes in moisture, protein, and fat content of fish and rice flour coextrudates during single-screw extrusion cooking. *Food and Bioprocess Technology* 6 (2): 403–415.

Wang, L., Duan, W., Zhou, S. et al. (2016). Effects of extrusionconditions on the extrusion responses and the quality of brown rice pasta. *Food Chemistry* 204: 320–325.

Wang, P., Fu, Y., Wang, L. et al. (2017). Effect of enrichment with stabilized rice bran and extrusion process on gelatinization and retrogradation properties of rice starch. *Starch Staerke* 69: 1600201.

Wang, S., Chao, C., Xiang, F. et al. (2018). New insights into gelatinization mechanisms of cereal endosperm starches. *Scientific Reports* 8 (1): 3011.

Yanniotis, S., Petraki, A., and Soumpasi, E. (2007). Effect of pectin and wheat fibers on quality attributes of extruded cornstarch. *Journal of Food Engineering* 80: 594–599.

Zhu, S., Riaz, M.N., and Lusas, E.W. (1996). Effect of different extrusion temperatures and moisture content on lipoxygenase inactivation and protein solubility in soybeans. *Journal of Agricultural and Food Chemistry* 44: 3315–3318.

9
Thermal Processing Technologies

Tomás Lafarga[1,2], Anna Vallverdú Queralt[3,4], Gloria Bobo[1], Maribel Abadias[1], and Ingrid Aguiló-Aguayo[1,*]

[1] *Institute of Agrifood Research and Technology (IRTA), Lleida, Spain*
[2] *Department of Chemical Engineering, University of Almería, Almería, Spain*
[3] *Food Science and Gastronomy XaRTA, University of Barcelona, Institute of Nutrition and Food Safety (INSA-UB), Barcelona, Spain*
[4] *Consorcio CIBER, M.P. Fisiopatología de la Obseidad y Nutrición (CIBEROBn), Instituto de Salud Carlos III (ISCIII), Madrid, Spain*

Summary

Traditional thermal treatments are of key importance in the food industry, to provide products that are both safe and stable during storage. However, conventional thermal processing can lead to negative sensorial or organoleptic attributes, as well as degradation of thermolabile health-promoting compounds such as vitamins. This has led to the development of a number of novel thermal technologies to meet the required food safety or shelf life demands, but minimizing the negative effects of temperature on the organoleptic and nutritional quality of the products, including ohmic heating, microwave heating, and combinations of mild temperatures with technologies. These technologies have gained industrial interest and have potential to replace, or partially replace, traditional preservation processes. This chapter will discuss conventional and a range of novel thermal technologies.

9.1 Introduction

Thermal technologies have been an important tool for food preservation and production. The industry has widely used heat as a preservation mechanism (sterilization, pasteurization, drying, concentration, etc.) and/or as a method to confer textural and structural changes in products (e.g. baking, vacuum cooking, frying, etc.) (Richardson 2001).

* Corresponding author.

Food Formulation: Novel Ingredients and Processing Techniques, First Edition.
Edited by Shivani Pathania and Brijesh K. Tiwari.

Conventional technologies are based on the generation of heat outside the product to be treated using conduction and convection mechanisms. Limitations of heat processes might include non-uniform heat transfer efficiency, thermal damage by overheating, or a high cost of energy, which are important considerations when selecting the process method (Fellows 2017; Pereira and Vicente 2010). For these reasons, many researchers have included novel thermal processing (ohmic heating, pulsed electric heating, radiant technologies, etc.), or even combinations of thermal and non-thermal technologies like high-pressure or ultrasound treatments, into their extensive research.

Application of the treatment in food could be by direct or indirect contact with the matrix. For instance, the application of electrical energy is directly induced in pulsed electric field processing and indirect electric heating is induced in infrared heaters. In the case of conventional thermal methods, indirect heating systems are the most common in steam or hot water generated by a boiler. In novel thermal processing, dielectric and ohmic heating are used to preserve foods, whereas infrared radiation is mainly used to alter the eating qualities of foods by changing the surface colour, flavour, and aroma (Fellows 2017).

In thermal processing, a complete knowledge of the thermophysical (e.g. heat penetration coefficient, thermal conductivity, diffusivity) and thermal (e.g. chemical composition, structure of the product, temperature or, e.g. dielectric constant or dielectric loss factor) properties of food is essential to determine how a product will behave when applying a heat treatment (Delgado et al. 2012).

In this chapter, the advantages and disadvantages of some of these novel technologies and their main applications are explored. On the one hand, these new technologies could present some limitations to their full industrial application. On the other hand, some of them could present some interesting advantages such as working at sublethal temperatures (avoiding thermolabile degradation) or being environmentally friendly; due to, for example, energy and water savings, eliminating microorganisms without the use of chemical compounds, reducing the amount of solids wastes, or not generating organic volatiles (Pereira and Vicente 2010).

9.2 Conventional Thermal Processing Technologies

Thermal processing techniques, such as pasteurization and sterilization, have been used in food preservation for millennia to process and extend the shelf life of food, due to their ability to inactivate microorganisms and enzymes. A disadvantage of this approach, however, is that degradation reactions are accelerated at higher temperatures, so heat treatments can adversely affect nutritional components of food with health-related properties, such as vitamins, pigments, polyphenols, and antioxidant compounds (Lafarga et al. 2018; Patras et al. 2009). However, the cell damage caused by high temperatures in food products are not always a drawback, and there is evidence that thermal treatment can make bioactive compounds more available for absorption (Fernández-García et al. 2008). When applying thermal processing methods to food, several factors should be taken into consideration, including the type of microorganisms or spores targeted for inactivation, and the pH and thermo-physical properties of the food.

9.2.1 Effect of Thermal Treatments on Polyphenols

9.2.1.1 Blanching The most commonly used blanching method in the food industry is water blanching (typically 75–95 °C for 1–10 minutes) owing to its lower costs relative to other options (Rawson et al. 2011). Generally, blanching is employed to inactivate enzymes prior to further processing, but its effect on the properties of food is not clear, as reflected by the frequent contradictions in published data. The impact of blanching on polyphenol contents has been studied in grapes, apples, aonla shreds, mandarins, potatoes, blueberry purees, and red cabbage (Brownmiller et al. 2008; Cabrera and Moon 2015; Chu et al. 2000; Heras-Ramírez et al. 2012; Ojha et al. 2016; Prajapati et al. 2011; Volden et al. 2008; Wolfe and Liu 2003).

In whole Campbell grapes, the flavonoid content was found to increase after one minute of blanching before juice processing (Cabrera and Moon 2015), which could be attributed to compound release from cell walls promoted by the heat treatment. Similarly, it was reported that blanching apple peel in boiling water for 10–20 seconds increased the total phenolic and flavonoid contents compared to untreated samples (Wolfe and Liu 2003). However, in another study, samples of apple pomace maintained their original nutrient composition whether blanched or not before drying (Heras-Ramírez et al. 2012). Therefore, the level of bioactive compounds varies according to the matrix in which they are embedded and their structure.

In aonla shreds, significantly more tannin was retained after blanching with hot water than with potassium meta bisulphate (Prajapati et al. 2011), whereas in mandarin peel, water blanching reduced the tannin content (Ojha et al. 2016). In another study, in which the effect of blanching time was determined in green leaves of sweet potatoes, antioxidant activity was unaltered after one minute in boiling water (Chu et al. 2000).

In contrast with the previous studies, blanching (95 °C for 3 minutes) in combination with pasteurization during the preparation of blueberry puree reduced total monomeric anthocyanins by 43% compared to fresh fruit (Brownmiller et al. 2008). Similarly, a 59% loss of anthocyanin content was observed in red cabbage after blanching (Volden et al. 2008). Thus, no agreement was established in the literature regarding the effects of blanching on polyphenols, which may vary from insignificant to significant increase or decrease.

9.2.1.2 Pasteurization and Sterilization Another well-established thermal treatment is pasteurization, during which thermal degradation, polymerization, and depolymerization might cause the degradation of food bioactive compounds. However, as mentioned above, some compounds may increase due to cell disruption by thermal treatments (Fuleki and Ricardo-da-Silva 2003). The effect of pasteurization has been studied in a range of beverages (plum, mulberry, grape, strawberry, orange, pineapple and apple juices, and fruit juice-milk and fruit juice-soya milk drinks) (Agcam et al. 2014; Davidov-Pardo et al. 2011; Fuleki and Ricardo-da-Silva 2003; Goh et al. 2012; Medina-Meza and Barbosa-Cánovas 2015; Morales-De La Peña et al. 2010, 2011; Morales-de la Peña et al. 2017; Odriozola-Serrano et al. 2008, 2009; Velázquez-Estrada 2011).

After pasteurization (70 °C for 1 hour), the total flavonoid content in grape peel tended to increase, yet in plum peel it decreased (Medina-Meza and Barbosa-Cánovas 2015). Grape seed was also found to contain higher amounts of tannin, catechin and derivatives, and procyanidins when the temperature of pasteurization was increased (from 65 °C for 30 minutes to 75 °C for 20 minutes) (Davidov-Pardo et al. 2011). Thus, temperature is a key factor to modulate during pasteurization. In cold-pressed grape juice, the concentration of catechins and procyanidins was also increased by thermal treatment, although it decreased in hot-pressed grape juice (Fuleki and Ricardo-da-Silva 2003).

In grape, orange, and apple juices, pasteurization (80 °C for 30 seconds) produced an increase in certain polyphenols such as flavonoids, but at a higher temperature (90 °C for 30 seconds), the level of other polyphenols decreased (He et al. 2016) due to oxidative degradation or polymerization. In another study, the total flavonoid and quercetin aglycone content in mulberry juice decreased during pasteurization (Wang et al. 2017; Yu et al. 2015). This differential effect might be because some polyphenol forms undergo hydrolysis under heat treatment (Morales-De La Peña et al. 2010).

A high pasteurization temperature (90 °C for 1 min) increased the concentration of polyphenols such as hesperidin, rutin, and quercetin in a fruit juice-milk beverage (Morales-de la Peña et al. 2017), as well as the hesperidin, rutin, narirutin, quercetin, and apigenin content in fruit juice-soy milk (Morales-de La Peña et al. 2011) in comparison with untreated samples. Another study of a fruit juice-soy milk beverage revealed that the total aglycone content increased significantly after pasteurization (90 °C for 60 seconds) compared to the untreated control, whereas no significant differences were found in the total glucoside content. Again, the differences might be because some polyphenol forms undergo hydrolysis under heat treatment (Morales-De La Peña et al. 2010).

Higher hesperidin concentration was found in thermally treated orange juice (90 °C for 1 minute) compared to fresh juice, whereas eriocitrin, naringenin, narirutin, eriodictyol, hesperetin, and kaempferol concentrations were lower in thermally treated juice (Velázquez-Estrada 2011). Also in orange juice, bioactive compounds such as naringin, kaempferol, luteolin, and apigenin increased after thermal treatment (90 °C for 10 second), whereas quercetin, isoquercetin, rutin, eriocitrin, neoeriocitrin, and naringenin decreased relative to the control (Agcam et al. 2014). Thus, the structure of the polyphenols and whether they are present in glycoside form may affect the final amount in the processed food. In contrast, a study with strawberry juice found no significant difference in bioactive compounds such as kaempferol, quercetin, and myricetin between fresh and thermally treated samples (90 °C for 30 and 60 seconds) (Odriozola-Serrano et al. 2008), whereas the level of other polyphenols in tomato juice was negatively affected by thermal treatment (90 °C for 30 and 60 seconds) (Odriozola-Serrano et al. 2009). Also, thermal treatment (97 °C for 15 minute) increased the total flavonoid content in pineapple juice compared to fresh juice (Goh et al. 2012).

Sterilization, another thermal process, uses higher temperatures and longer times than pasteurization, and its effect has been studied in berry products, jams, cherries and plums, tomato and sweet corn (Dewanto et al. 2002; Howard et al. 2010; Jiménez-Aguilar et al. 2015; Kim and Padilla-Zakour 2004; Nayak et al. 2015; Vallverdú-Queralt et al. 2012).

Thermal degradation of polyphenols during sterilization was found to decrease the antioxidant capacity of berry products (Nayak et al. 2015). Similarly, heat sterilization (131 ° C for 2 seconds) caused a loss of quercetin, kaempferol, and isorhamnetin in pear juice compared to fresh juice (Jiménez-Aguilar et al. 2015). However, a higher retention of anthocyanins was observed in blueberry jam after processing at elevated temperatures (101–105 °C and pasteurization in boiling water for 10 minutes) (Howard et al. 2010).

In another study, heat processing during jam-making (at 104–105 °C) reduced the total phenolic content of some varieties of cherries and plums, whereas no significant change ($p < 0.05$) occurred in raspberries and other varieties of plums and cherries (Kim and Padilla-Zakour 2004). On the whole, high temperatures and extended processing times are not recommended to preserve the phenolic content of fruits (Nayak et al. 2015). At high temperatures of 90–120 °C, phenolic antioxidants undergo chemical oxidation to quinones and their polymers, as well as caramelization (of reducing sugars and ascorbic acid), Maillard reactions (between reducing sugars and amino acids), Strecker degradation (of dicarbonyl compounds with amino acids), and hydrolysis (of esters and glucosides of antioxidants).

Despite a loss of vitamin C content, industrially-processed tomato and sweet corn exhibited higher antioxidant activities than fresh products, due to an enhanced release of phenolic compounds bound in the food matrices (Dewanto et al. 2002). However, a decrease in total polyphenol content and antioxidant activity in tomato was detected after the sterilization step in industrial processing (Vallverdú-Queralt et al. 2012).

9.2.2 Effect of Thermal Treatments on Carotenoids

Studies on how thermal treatments affect carotenoids have provided conflicting results. Some report that heat treatments favour the release of lycopene from chromoplasts by rupturing cell walls and thus enhance lycopene bioavailability (Benlloch-Tinoco et al. 2015; Karakaya and Yılmaz 2007; Ryan et al. 2008). However, another study found that cooking tomatoes for 15 minutes at 100 °C did not change lycopene bioavailability (Bugianesi et al. 2004). Moreover, in tomato pulp with variable moisture levels, lycopene content decreased during heating at different temperatures. Similarly, heating tomato products under different processing conditions to manufacture tomato juice, pulp, powder, etc. caused the degradation of lycopene (Goula et al. 2006). A consensus on the effects of thermal treatments on carotenoid bio-accessibility, which reportedly vary from no change to a significant increase or decrease, has therefore not yet been reached.

The bioavailability of β-carotene in tomatoes was generally found not to be significantly changed by thermal processing (Bugianesi et al. 2004). In contrast, a significant increase in bio-accessibility of β-carotene in carrots was observed *in vitro* when temperature and processing time were augmented (Lemmens et al. 2011). Similarly, thermal processing at high temperatures and long treatment times increased the bio-accessibility of lycopene in tomato products (Colle et al. 2010a, b, c). However, this improvement in bio-accessibility was only significant at temperatures of about 130 and 140 °C, and these processing conditions may cause the degradation of other thermolabile compounds. No changes in carotenoid bio-accessibility were found after conventional thermal pasteurization (97 °C, 30 seconds).

Thus, bio-accessibility of carotenoids also varies according to the matrix in which they are embedded and their structure. For example, xanthophylls are generally more bio-accessible than carotenes. This is generally caused by the greater polarity of xanthophylls, which improves transfer to the mixed micelles (Goñi et al. 2006).

The results of these studies confirm the hypothesis that the type of microorganisms or spores targeted for inactivation, the pH and thermophysical properties of food, as well as the processing temperature, have an influence on the amounts of polyphenols and other bioactive compounds in the final product.

9.2.3 Effect of Thermal Treatments on Vitamins

Thermal treatments cause variable losses of vitamins, depending on the cooking method and the type of food (Lafarga et al. 2018). The most important factors that influence the degradation of vitamins are temperature, presence of oxygen, light, moisture, pH, and, of course, heat treatment duration. The most labile vitamins during thermal treatments are retinol, vitamin C, folate, and thiamine, whereas niacin, biotin, and pantothenic acid are stable (Lešková et al. 2006).

9.3 Novel Thermal Technologies

9.3.1 Ohmic Heating

As highlighted in the previous section, conventional thermal processing, namely pasteurization and sterilization, have certain disadvantages including overheating, negative sensorial modifications, and loss of nutritional compounds, amongst others. Ohmic heating, also known as Joule heating, consists of heating by passing an alternating electric current through the food product. The electrical resistance of the food promotes the rapid generation of heat inside the product (Jermann et al. 2015) (Figure 9.1). One of the main advantages of this technology is that although it is based on the use of electricity, which is becoming increasingly less appreciated nowadays, the electrical conversion into heat is nearly 100% (Roux et al. 2016). This technology is generally included within the group of novel processing technologies or novel alternatives to conventional thermal processing. However, this technology has been known since the nineteenth century, when it was applied to pasteurize milk (Cappato et al. 2017). Since then, several ohmic heating processes that use electrical current for heating food have been patented and developed. This technology has been also widely studied and reviewed in scientific publications (Cappato et al. 2017; Gavahian et al. 2019; Knirsch et al. 2010; Sakr and Liu 2014; Varghese et al. 2014; Yildiz-Turp et al. 2013). Overall, most of the recent scientific publications that studied the effect of ohmic heating on foods operated at either 20 kHz (Ito et al. 2014; Kim and Kang 2015a; Park and Kang 2013) or 50 kHz (Engchuan et al. 2014; Jakób et al. 2010). Somavat et al. (2012) studied the effect of ohmic heating frequency (10 or 60 kHz) on the inactivation kinetics of *Geobacillus stearothermophilus* spores and suggested that higher frequencies showed an improved antimicrobial efficiency at higher temperatures. Similar results were reported by Somavat et al. (2013). Electric field strengths commonly range between 20 and 30 V/cm, although some authors operated at milder or more intense conditions. For example, El Darra et al.

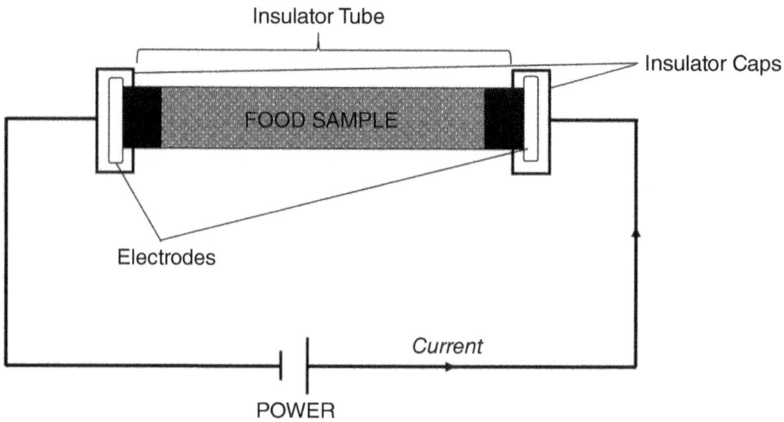

Figure 9.1 Ohmic heating concept.

(2013) improved the extraction of polyphenols from grape pomace by processing with pulsed electric fields within the range 100–800V/cm. Most commonly studied food matrices were meat or processed meat products (Engchuan et al. 2014; Ito et al. 2014; Lascorz et al. 2016; Zell et al. 2010) and liquid foods: fruit and vegetable juices, nectars, and purees (Achir et al. 2016; Brochier et al. 2016; Jaeschke et al. 2016; Kim and Kang 2015b; Kim and Kang 2017; Kim et al. 2018; Makroo et al. 2017; Mercali et al. 2013, 2014; Park and Kang 2013; Sarkis et al. 2013; Somavat et al. 2012, 2013), and milk or milk-analogues (Kim and Kang 2015b; Li et al. 2018; Parmar et al. 2018). The main reason for this is that ohmic treatment can alter the textural properties of foods, although a correct optimization of the ohmic process conditions, namely electrical specification, processing time, temperature, and product formulation, can help to minimize negative textural modifications (Varghese et al. 2014).

Ohmic heating could be potentially used for cooking meat or processed meat products. Ohmic heating of meat and processed meat products led to lower processing times and more uniform cooking and colour when compared to conventional cooking methods (Lascorz ct al. 2016; Zell et al. 2010). Moreover, previous studies reported no major differences between the quality and sensorial attributes of ohmically heated and conventionally cooked chicken meat (Ito et al. 2014). Ohmic heating also shows potential for inactivating microorganisms, including human pathogens such as *Escherichia coli* O157:H7, *Salmonella enterica* Serovar Typhimurium, and *Listeria monocytogenes*, as well as spores of *G. stearothermophilus* and *Bacillus coagulans* (Kim et al. 2018; Park and Kang 2013; Somavat et al. 2012, 2013). Overall, ohmic heating resulted in higher microbial inactivation efficiency and lower processing times when compared to conventional heat treatments. Other advantage of ohmic heating when compared to conventional thermal processing methods is that this technology can potentially minimize the loss of valuable compounds suffered during processing. However, to minimize the degradation of healthy food compounds it is important to optimize the process as, for example, Sarkis et al. (2013) reported that the level of degradation of anthocyanins in blueberry pulp, which was measured as 7.2% after conventional processing, varied from 5.7 to 14.7% after ohmic heating. Achir et al. (2016) reported a higher retention of carotenoids after ohmic heating of juice when compared to conventional pasteurization. However, Jaeschke et al. (2016) observed no differences in the degradation kinetics of ascorbic acid and carotenoids when comparing ohmic and conventional heating. The authors of that study reported that oxygen was the limiting reagent of the degradation reactions. Similar results were reported by other research groups (Mercali et al. 2013, 2014). Ohmic treatments also showed potential for inactivating enzymes (Brochier et al. 2016; El Darra et al. 2013; Makroo et al. 2017) and for improving the extraction yield of valuable compounds such as phytochemicals with health-promoting properties (Loypimai et al. 2015; Pereira et al. 2016).

Despite the high capital investment and some safety concerns raised by several researchers, the future of ohmic heating in the food industry is promising. Indeed, Jermann et al. (2015) designed and conducted two surveys to identify the present and future potential of different technologies for being commercialized in either 5 or 10 years and respondents (which were professionals from industry, academia, or government) suggested that ohmic heating could be a useful technology for sauces, soups, and purees. In that same study, the authors reported that in North America, ohmic heating scored the highest in terms of potential commercial importance in the next decade (together with other technologies including high-pressure processing and ozone treatments, amongst others).

9.3.2 Microwave Processing

Microwaves (MW) are electromagnetic waves within 300 MHz–300 GHz of frequency and wavelengths ranging from 1 mm to 1 m. While domestic microwaves generally operate at a frequency of 2.45 GHz, industrial microwave systems operate at frequencies of 915 MHz and 2.45 GHz (Datta and Anantheswaran 2000). MW heating is caused by the ability of the materials to absorb MW energy and convert it into heat. Microwave heating of food materials mainly occurs due to dipolar and ionic mechanisms. Heating takes place due to the interaction of electromagnetic radiation at certain frequencies with dielectric materials; this is also called volumetric heating due to, in contrast with convection or conduction mechanisms, microwave radiation directly penetrating the material causing volumetric heat generation in the material, resulting in high-energy efficiency and lower heating times (Zhu et al. 2007). From the physical point of view, microwave heating in foods is a complex process which depends upon the propagation of microwaves governed by Maxwell's equations for electromagnetic waves, the interactions between microwaves and foods determined by dielectric properties, and heat dissipation governed by basic heat and mass transfer theories (Tang et al. 2002). Dielectric properties of product and penetration depths are the main factors that affect MW heating. This technology has found many applications in the food industry, as it has important advantages over conventional heating, such as reduced processing time, high-energy efficiency, safe handling, ease of operation, low maintenance, and improved food quality. These applications include cooking, baking, blanching, drying, pasteurization, sterilization, thawing, tempering, baking, etc. (Chandrasekaran et al. 2013). However, one of the main disadvantages recognized in microwave heating is the non-uniform temperature distribution resulting in hot and cold spots in microwave-heated products (Vadivambal and Jayas 2010).

One of the uses of MW heating include pasteurization and sterilization. Pasteurization is a process in which pathogenic microorganisms such as bacteria in the vegetative form are destroyed by thermal treatment to enhance food safety and shelf life and also involves inactivation of undesirable enzymes on certain products such as juices. Destruction of microbes or enzymes by MW at sub-lethal temperatures was explained by one or more of the following theories: selective heating, electroporation, cell membrane rupture, and magnetic field coupling (Kozempel et al. 1998). Salazar-González et al. (2012) reviewed the microwave pasteurization of fluid food materials, which included apple juice and cider, coconut water, grapefruit juice, milk, sweet potato puree, peanut beverages, green pea and carrot puree, amongst others. Other examples of packaged foods treated by MWs include yoghurt (Neetoo et al. 2012), apple purée (Picouet et al. 2009), and pre-packaged apple juice (Cañumir et al. 2002). In the case of solid food materials, MW pasteurization has been applied to in-shell egg (Dev et al. 2008), yolk of shell eggs (Shenga et al. 2010), or pickled asparagus (Lau and Tang 2002). The application of a continuous MW treatment (2450 MHz, 20 kW, 50–60 seconds) showed great efficacy to control post-harvest brown rot caused by *Monilinia* spp. in peaches and nectarines when they were immersed in warm water (40 °C) (Sisquella et al. 2014). Furthermore, MW processing has been successfully applied on an industrial-scale for pasteurizing ready-to-eat meals, fresh pasta, bread, granola, and milk (Giese 1992; Neetoo et al. 2012). Several commercial systems of MW pasteurization for ready-to-eat meals exist (Tang et al. 2018).

The use of MW for drying food materials has been extensively studied (Chandrasekaran et al. 2013; Magdalena Zielinska et al. 2019). Compared to conventional drying (solar, convection, drum, spray, fluidized-bed drying, freeze drying, and osmotic drying), MW drying has the advantages of achieving fast drying rates and improving the quality of some food products (Chandrasekaran et al. 2013) and preventing shrinkage of food materials. However, MW drying alone has several limitations such as the non-uniformity of the electromagnetic field within the microwave cavity, and the high penetration depth within the dielectric materials. Additionally, fast mass transport of the microwave power may alter the quality of dried food (Zielinska et al. 2019) and the excessive temperature along the corner or edges of food products results in scorching and production of off-flavours, especially during final stages of drying. These have led to combining MW drying with other methods such as hot air convective drying (MWHACD), vacuum drying (MWVC), and vacuum freeze drying (MWVFD), amongst others (Zielinska et al. 2019). For example, Koné et al. (2013) studied the effect of MWHACD on the quality of tomato slices, which ensured better texture, and reduced shrinkage. The effect of hybrid and combined MW-assisted drying on the mechanical, optical, and nutritional values has been explored on whole blueberries and blueberry pomace (Zielinska and Michalska 2016, 2018), green peas (Zielinska et al. 2013), potato cubes (Bondaruk et al. 2007), apple cubes (Chong et al. 2014), and many other fruits and vegetables. Nowacka et al. (2019) found that the microwave–vacuum drying significantly reduced drying time of cranberry snacks in comparison to hot-air drying, resulting in a better retention of bioactive compounds.

Finally, MW have been also studied as an eco-innovative processing technology to assist traditional extraction procedures, called microwave-assisted extraction (MAE), such as high-added value compounds from plant materials (Putnik et al., 2018), proteins from plant origin for human consumption with preserved techno-functional properties (Pojić et al. 2018), and pectin from plant food wastes and by-products (Pojić et al. 2018). The advantages of MAE are the shortened time of processing, significant increase in yield, lower energy consumption, and the decreased usage of solvents, during MAE and environmentally friendly solution in processing (Chemat et al. 2012).

MW alone or in combination with other technologies have been proved to be a promising technology for several food applications due to some advantages in comparison to conventional techniques, including faster heating, shorter processing time, better quality retention in the processed product, suitability of application for thermosensitive, high viscosity, and multiphase fluids, safe handling with pasteurization after packaging, ease of operation, low energy consumption, and environmentally clean processing, amongst others (Martins et al. 2019). However, one of the main drawbacks of MW is the unpredictable and non-uniform energy distribution with the formation of overheated and under-heated spots, which are of importance for microbial safety reasons. Therefore, progress of MW at the industrial level has been relatively slow and more research is needed to overcome these problems.

9.4 Conclusions

In order to extend shelf life and obtain low-cost foods, the food industry has relied on chemical processing aids and intense thermal treatments. However, consumers are becoming more aware of the relationship between food, diet, and health and this has led to an increased interest in natural ingredients and to consumption of foods that are both tasty and nutritious. Negative perceptions about processed foods and processing technologies exist, but several consumer research studies have shown that if a novel technology allows the introduction of new products with health benefits, consumers are likely to accept it (Cullen et al. 2012). Overall, novel thermal technologies have proved to minimize the degradation of healthy compounds such as polyphenols and vitamins when compared to conventional thermal treatments. Moreover, with respect to safety, a large amount of scientific evidence suggests that novel technologies such as ohmic heating or thermo-sonication can lead to comparable, or even higher, microbial inactivation when compared to conventional thermal treatments. Furthermore, with respect to environmental impact, emerging thermal technologies can improve the overall energy efficiency of the process and/or reduce the consumption of non-renewable resources (Pereira and Vicente 2010). Novel thermal technologies are increasingly attracting the attention of food processors as they show potential for obtaining products with improved quality, added value, and an reduced environmental footprint.

Acknowledgements

This work was supported by the CERCA Programme of *Generalitat de Catalunya*. T. Lafarga, A. Vallverdú, and I. Aguiló-Aguayo are in receipt *Juan de la Cierva* (IJC2018-035287-I) and *Ramon y Cajal* RYC-2016-19355 and RYC-2016-19949 Grants, respectively, awarded by the Spanish Ministry of Economy, Industry, and Competitiveness.

Conflict of Interests

The authors declare no conflict of interests.

References

Achir, N., Dhuique-Mayer, C., Hadjal, T. et al. (2016). Pasteurization of citrus juices with ohmic heating to preserve the carotenoid profile. *Innovative Food Science & Emerging Technologies* 33: 397–404.

Agcam, E., Akyıldız, A., and Evrendilek, G.A. (2014). Comparison of phenolic compounds of orange juice processed by pulsed electric fields (PEF) and conventional thermal pasteurisation. *Food Chemistry* 143: 354–361.

Benlloch-Tinoco, M., Kaulmann, A., Corte-Real, J. et al. (2015). Chlorophylls and carotenoids of kiwifruit puree are affected similarly or less by microwave than by conventional heat processing and storage. *Food Chemistry* 187: 254–262.

Bondaruk, J., Markowski, M., and Błaszczak, W. (2007). Effect of drying conditions on the quality of vacuum-microwave dried potato cubes. *Journal of Food Engineering* 81 (2): 306–312.

Brochier, B., Mercali, G.D., and Marczak, L.D.F. (2016). Influence of moderate electric field on inactivation kinetics of peroxidase and polyphenol oxidase and on phenolic compounds of sugarcane juice treated by ohmic heating. *LWT - Food Science and Technology* 74: 396–403.

Brownmiller, C., Howard, L., and Prior, R. (2008). Processing and storage effects on monomeric anthocyanins, percent polymeric color, and antioxidant capacity of processed blueberry products. *Journal of Food Science* 73 (5): H72–H79.

Bugianesi, R., Salucci, M., Leonardi, C. et al. (2004). Effect of domestic cooking on human bioavailability of naringenin, chlorogenic acid, lycopene and β-carotene in cherry tomatoes. *European Journal of Nutrition* 43 (6): 360–366.

Cabrera, S. and Moon, K. (2015). Effects of processing treatments on the bioactive compounds of Campbell grape juice. *Asia Pacific Journal of Multidisciplinary Research* 3 (4): 1–4.

Cañumir, J.A., Celis, J.E., de Bruijn, J., and Vidal, L.V. (2002). Pasteurisation of apple juice by using microwaves. *LWT – Food Science and Technology* 35 (5): 389–392.

Cappato, L.P., Ferreira, M.V., Guimaraes, J.T. et al. (2017). Ohmic heating in dairy processing: relevant aspects for safety and quality. *Trends in Food Science & Technology* 62: 104–112.

Chandrasekaran, S., Ramanathan, S., and Basak, T. (2013). Microwave food processing—a review. *Food Research International* 52 (1): 243–261.

Chemat, F., Vian, M.A., and Cravotto, G. (2012). Green extraction of natural products: concept and principles. *International Journal of Molecular Sciences* 13 (7): 8615–8627.

Chong, C.H., Figiel, A., Law, C.L., and Wojdyło, A. (2014). Combined drying of apple cubes by using of heat pump, vacuum-microwave, and intermittent techniques. *Food and Bioprocess Technology* 7 (4): 975–989.

Chu, Y.H., Chang, C.L., and Hsu, H.F. (2000). Flavonoid content of several vegetables and their antioxidant activity. *Journal of the Science of Food and Agriculture* 80 (5): 561–566.

Colle, I., Lemmens, L., Van Buggenhout, S. et al. (2010a). Effect of thermal processing on the degradation, isomerization, and bioaccessibility of lycopene in tomato pulp. *Journal of Food Science* 75 (9): C753–C759.

Colle, I., Van Buggenhout, S., Van Loey, A., and Hendrickx, M. (2010b). High pressure homogenization followed by thermal processing of tomato pulp: influence on microstructure and lycopene *in vitro* bioaccessibility. *Food Research International* 43 (8): 2193–2200.

Colle, I.J., Lemmens, L., Tolesa, G.N. et al. (2010c). Lycopene degradation and isomerization kinetics during thermal processing of an olive oil/tomato emulsion. *Journal of Agricultural and Food Chemistry* 58 (24): 12784–12789.

Cullen, P., Tiwari, B.K., and Valdramidis, V.P. (2012). Status and trends of novel thermal and non-thermal technologies for fluid foods. In: *Novel Thermal and Non-thermal Technologies for Fluid Foods* (eds. P. Cullen, B.K. Tiwari and V.P. Valdramidis), 1–6. Academic Press.

Datta, N. and Anantheswaran, R.C. (2000). *Handbook of Microwave Technology for Food Applications*. New York;: Marcel Dekker Inc.

Davidov-Pardo, G., Arozarena, I., and Marín-Arroyo, M.R. (2011). Stability of polyphenolic extracts from grape seeds after thermal treatments. *European Food Research and Technology* 232 (2): 211–220.

Delgado, E., Sun, D.W., and Rubiolo, A.C. (2012). Thermal physical properties of foods. In: *Thermal Food Processing: New Technologies and Quality Issues*, vol. 2 (ed. D.W. Sun), 688. Boca Raton: CRC Press.

Dev, S.R.S., Raghavan, G.S.V., and Gariepy, Y. (2008). Dielectric properties of egg components and microwave heating for in-shell pasteurization of eggs. *Journal of Food Engineering* 86 (2): 207–214.

Dewanto, V., Wu, X., Adom, K.K., and Liu, R.H. (2002). Thermal processing enhances the nutritional value of tomatoes by increasing total antioxidant activity. *Journal of Agricultural and Food Chemistry* 50 (10): 3010–3014.

El Darra, N., Grimi, N., Vorobiev, E. et al. (2013). Extraction of polyphenols from red grape pomace assisted by pulsed ohmic heating. *Food and Bioprocess Technology* 6 (5): 1281–1289.

Engchuan, W., Jittanit, W., and Garnjanagoonchorn, W. (2014). The ohmic heating of meat ball: modeling and quality determination. *Innovative Food Science & Emerging Technologies* 23: 121–130.

Fellows, P.J. (2017). Overview of heat processing. In: *Food Processing Technology: Principles and Practice*, Woodhead Publishing Series in Food Science, Technology and Nutrition, 4e, 515–522. Elsevier Science.

Fernández-García, E., Rincon, F., and Pérez-Gálvez, A. (2008). Developing an emulsifier system to improve the bioaccessibility of carotenoids. *Journal of Agricultural and Food Chemistry* 56 (21): 10384–10390.

Fuleki, T. and Ricardo-da-Silva, J.M. (2003). Effects of cultivar and processing method on the contents of catechins and procyanidins in grape juice. *Journal of Agricultural and Food Chemistry* 51 (3): 640–646.

Gavahian, M., Tiwari, B.K., Chu, Y.-H. et al. (2019). Food texture as affected by ohmic heating: mechanisms involved, recent findings, benefits, and limitations. *Trends in Food Science & Technology* 86: 328–339.

Giese, J. (1992). Advances in microwave food procesing. *Food Technology* 46 (9): 118–123.

Goh, S.G., Adzahan, N.M., Leong, C.M. et al. (2012). Effect of thermal and ultraviolet treatments on the stability of antioxidant compounds in single strength pineapple juice throughout refrigerated storage. *International Food Research Journal* 19 (3): 1131–1136.

Goñi, I., Serrano, J., and Saura-Calixto, F. (2006). Bioaccessibility of β-carotene, lutein, and lycopene from fruits and vegetables. *Journal of Agricultural and Food Chemistry* 54 (15): 5382–5387.

Goula, A.M., Adamopoulos, K.G., Chatzitakis, P.C., and Nikas, V.A. (2006). Prediction of lycopene degradation during a drying process of tomato pulp. *Journal of Food Engineering* 74 (1): 37–46.

He, Z., Tao, Y., Zeng, M. et al. (2016). High pressure homogenization processing, thermal treatment and milk matrix affect in vitro bioaccessibility of phenolics in apple, grape and orange juice to different extents. *Food Chemistry* 200: 107–116.

Heras-Ramírez, M.E., Quintero-Ramos, A., Camacho-Dávila, A.A. et al. (2012). Effect of blanching and drying temperature on polyphenolic compound stability and antioxidant capacity of apple pomace. *Food and Bioprocess Technology* 5 (6): 2201–2210.

Howard, L.R., Castrodale, C., Brownmiller, C., and Mauromoustakos, A. (2010). Jam processing and storage effects on blueberry polyphenolics and antioxidant capacity. *Journal of Agricultural and Food Chemistry* 58 (7): 4022–4029.

Ito, R., Fukuoka, M., and Hamada-Sato, N. (2014). Innovative food processing technology using ohmic heating and aseptic packaging for meat. *Meat Science* 96 (2): 675–681.

Jaeschke, D.P., Marczak, L.D.F., and Mercali, G.D. (2016). Evaluation of non-thermal effects of electricity on ascorbic acid and carotenoid degradation in acerola pulp during ohmic heating. *Food Chemistry* 199: 128–134.

Jakób, A., Bryjak, J., Wójtowicz, H. et al. (2010). Inactivation kinetics of food enzymes during ohmic heating. *Food Chemistry* 123 (2): 369–376.

Jermann, C., Koutchma, T., Margas, E. et al. (2015). Mapping trends in novel and emerging food processing technologies around the world. *Innovative Food Science & Emerging Technologies* 31: 14–27.

Jiménez-Aguilar, D.M., Escobedo-Avellaneda, Z., Martín-Belloso, O. et al. (2015). Effect of high hydrostatic pressure on the content of phytochemical compounds and antioxidant activity of prickly pears (Opuntia ficus-indica) beverages. *Food Engineering Reviews* 7 (2): 198–208.

Karakaya, S. and Yılmaz, N. (2007). Lycopene content and antioxidant activity of fresh and processed tomatoes and in vitro bioavailability of lycopene. *Journal of the Science of Food and Agriculture* 87 (12): 2342–2347.

Kim, S.S. and Kang, D.H. (2015a). Comparison of pH effects on ohmic heating and conventional heating for inactivation of *Escherichia coli* O157: H7, salmonella enterica Serovar typhimurium and listeria monocytogenes in orange juice. *LWT - Food Science and Technology* 64 (2): 860–866.

Kim, S.S. and Kang, D.H. (2015b). Effect of milk fat content on the performance of ohmic heating for inactivation of Escherichia coli O157: H7, salmonella enterica Serovar typhimurium and listeria monocytogenes. *Journal of Applied Microbiology* 119 (2): 475–486.

Kim, S.S. and Kang, D.H. (2017). Combination treatment of ohmic heating with various essential oil components for inactivation of food-borne pathogens in buffered peptone water and salsa. *Food Control* 80: 29–36.

Kim, D.O. and Padilla-Zakour, O.I. (2004). Jam processing effect on phenolics and antioxidant capacity in anthocyanin-rich fruits: cherry, plum, and raspberry. *Journal of Food Science* 69 (9): S395–S400.

Kim, S.-S., Park, S.-H., and Kang, D.-H. (2018). Application of continuous-type pulsed ohmic heating system for inactivation of foodborne pathogens in buffered peptone water and tomato juice. *LWT - Food Science and Technology* 93: 316–322.

Knirsch, M.C., Dos Santos, C.A., de Oliveira Soares, A.A.M., and Penna, T.C.V. (2010). Ohmic heating–a review. *Trends in Food Science & Technology* 21 (9): 436–441.

Koné, K.Y., Druon, C., Gnimpieba, E.Z. et al. (2013). Power density control in microwave assisted air drying to improve quality of food. *Journal of Food Engineering* 119 (4): 750–757.

Kozempel, M.F., Annous, B.A., Cook, R.D. et al. (1998). Inactivation of microorganisms with microwaves at reduced Temperaturas. *Journal of Food Protection* 61 (5): 582–585.

Lafarga, T., Bobo, G., Viñas, I. et al. (2018). Effects of thermal and non-thermal processing of cruciferous vegetables on glucosinolates and its derived forms. *Journal of Food Science and Technology* 55 (6): 1973–1981.

Lafarga, T., Viñas, I., Bobo, G. et al. (2018). Effect of steaming and sous vide processing on the total phenolic content, vitamin C and antioxidant potential of the genus Brassica. *Innovative Food Science & Emerging Technologies* 47: 412–420.

Lascorz, D., Torella, E., Lyng, J.G., and Arroyo, C. (2016). The potential of ohmic heating as an alternative to steam for heat processing shrimps. *Innovative Food Science & Emerging Technologies* 37: 329–335.

Lau, M.H. and Tang, J. (2002). Pasteurization of pickled asparagus using 915 MHz microwaves. *Journal of Food Engineering* 51 (4): 283–290.

Lemmens, L., Colle, I.J., Van Buggenhout, S. et al. (2011). Quantifying the influence of thermal process parameters on in vitro β-carotene bioaccessibility: a case study on carrots. *Journal of Agricultural and Food Chemistry* 59 (7): 3162–3167.

Lešková, E., Kubíková, J., Kováčiková, E. et al. (2006). Vitamin losses: retention during heat treatment and continual changes expressed by mathematical models. *Journal of Food Composition and Analysis* 19 (4): 252–276.

Li, X., Ye, C., Tian, Y. et al. (2018). Effect of ohmic heating on fundamental properties of protein in soybean milk. *Journal of Food Process Engineering* 41 (3): e12660.

Loypimai, P., Moongngarm, A., Chottanom, P., and Moontree, T. (2015). Ohmic heating-assisted extraction of anthocyanins from black rice bran to prepare a natural food colourant. *Innovative Food Science & Emerging Technologies* 27: 102–110.

Makroo, H., Rastogi, N., and Srivastava, B. (2017). Enzyme inactivation of tomato juice by ohmic heating and its effects on physico-chemical characteristics of concentrated tomato paste. *Journal of Food Process Engineering* 40 (3): e12464.

Martins, C.P.C., Cavalcanti, R.N., Couto, S.M. et al. (2019). Microwave processing: current background and effects on the physicochemical and microbiological aspects of dairy products. *Comprehensive Reviews in Food Science and Food Safety* 18 (1): 67–83.

Medina-Meza, I.G. and Barbosa-Cánovas, G.V. (2015). Assisted extraction of bioactive compounds from plum and grape peels by ultrasonics and pulsed electric fields. *Journal of Food Engineering* 166: 268–275.

Mercali, G.D., Jaeschke, D.P., Tessaro, I.C., and Marczak, L.D.F. (2013). Degradation kinetics of anthocyanins in acerola pulp: comparison between ohmic and conventional heat treatment. *Food Chemistry* 136 (2): 853–857.

Mercali, G.D., Schwartz, S., Marczak, L.D.F. et al. (2014). Ascorbic acid degradation and color changes in acerola pulp during ohmic heating: effect of electric field frequency. *Journal of Food Engineering* 123: 1–7.

Morales-De La Peña, M., Salvia-Trujillo, L., Rojas-Graü, M., and Martín-Belloso, O. (2010). Isoflavone profile of a high intensity pulsed electric field or thermally treated fruit juice-soymilk beverage stored under refrigeration. *Innovative Food Science & Emerging Technologies* 11 (4): 604–610.

Morales-de La Peña, M., Salvia-Trujillo, L., Rojas-Graü, M., and Martín-Belloso, O. (2011). Changes on phenolic and carotenoid composition of high intensity pulsed electric field and thermally treated fruit juice–soymilk beverages during refrigerated storage. *Food Chemistry* 129 (3): 982–990.

Morales-de la Peña, M., Salvia-Trujillo, L., Rojas-Graü, M.A., and Martín-Belloso, O. (2017). Effects of high intensity pulsed electric fields or thermal pasteurization and refrigerated storage on antioxidant compounds of fruit juice-milk beverages. Part I: phenolic acids and flavonoids. *Journal of Food Processing and Preservation* 41 (3): e12912.

Nayak, B., Liu, R.H., and Tang, J. (2015). Effect of processing on phenolic antioxidants of fruits, vegetables, and grains—a review. *Critical Reviews in Food Science and Nutrition* 55 (7): 887–918.

Neetoo, H., Chen, H., and Hoover, D.G. (2012). Emerging methods for post-packaging microbial decontamination of food. In: *Woodhead Publishing Series in Food Science, Technology and Nutrition, Microbial Decontamination in the Food Industry* (eds. A. Demirci and M.O. Ngadi), 746–787. Cambridge: Woodhead Publishing Ltd.

Nowacka, M., Wiktor, A., Anuszewska, A. et al. (2019). The application of unconventional technologies as pulsed electric field, ultrasound and microwave-vacuum drying in the production of dried cranberry snacks. *Ultrasonics Sonochemistry* 56: 1–13.

Odriozola-Serrano, I., Soliva-Fortuny, R., and Martín-Belloso, O. (2008). Phenolic acids, flavonoids, vitamin C and antioxidant capacity of strawberry juices processed by high-intensity pulsed electric fields or heat treatments. *European Food Research and Technology* 228: 239–248.

Odriozola-Serrano, I., Soliva-Fortuny, R., Hernández-Jover, T., and Martín-Belloso, O. (2009). Carotenoid and phenolic profile of tomato juices processed by high intensity pulsed electric fields compared with conventional thermal treatments. *Food Chemistry* 112 (1): 258–266.

Ojha, P., Karki, T.B., and Sitaula, R. (2016). Physio-chemical and functional quality evaluation of mandarin peel powder. *Journal of Agricultural Science and Technology* 18: 575–582.

Park, I.-K. and Kang, D.-H. (2013). Effect of electropermeabilization by ohmic heating for inactivation of *Escherichia coli* O157: H7, salmonella enterica serovar typhimurium, and listeria monocytogenes in buffered peptone water and apple juice. *Applied and Environmental Microbiology* 79 (23): 7122–7129.

Parmar, P., Singh, A.K., Meena, G.S. et al. (2018). Application of ohmic heating for concentration of milk. *Journal of Food Science and Technology* 55 (12): 4956–4963.

Patras, A., Brunton, N.P., Da Pieve, S., and Butler, F. (2009). Impact of high pressure processing on total antioxidant activity, phenolic, ascorbic acid, anthocyanin content and colour of strawberry and blackberry purées. *Innovative Food Science & Emerging Technologies* 10 (3): 308–313.

Pereira, R. and Vicente, A. (2010). Environmental impact of novel thermal and non-thermal technologies in food processing. *Food Research International* 43 (7): 1936–1943.

Pereira, R.N., Rodrigues, R.M., Genisheva, Z. et al. (2016). Effects of ohmic heating on extraction of food-grade phytochemicals from colored potato. *LWT - Food Science and Technology* 74: 493–503.

Picouet, P.A., Landl, A., Abadias, M. et al. (2009). Minimal processing of a granny smith apple purée by microwave heating. *Innovative Food Science & Emerging Technologies* 10 (4): 545–550.

Pojić, M., Mišan, A., and Tiwari, B. (2018). Eco-innovative technologies for extraction of proteins for human consumption from renewable protein sources of plant origin. *Trends in Food Science & Technology* 75: 93–104.

Prajapati, V., Nema, P.K., and Rathore, S. (2011). Effect of pretreatment and drying methods on quality of value-added dried aonla (Emblica officinalis Gaertn) shreds. *Journal of Food Science and Technology* 48 (1): 45–52.

Putnik, P., Lorenzo, M.J., Barba, J.F. et al. (2018). Novel food processing and extraction technologies of high-added value compounds from plant materials. *Food* 7 (7): 106–121.

Rawson, A., Patras, A., Tiwari, B. et al. (2011). Effect of thermal and non thermal processing technologies on the bioactive content of exotic fruits and their products: review of recent advances. *Food Research International* 44 (7): 1875–1887.

Richardson, P. (2001). Introduction. In: *Thermal Technologies in Food Processing* (ed. P. Richardson), 1–3. Elsevier Science.

Roux, S., Courel, M., Birlouez-Aragon, I. et al. (2016). Comparative thermal impact of two UHT technologies, continuous ohmic heating and direct steam injection, on the nutritional properties of liquid infant formula. *Journal of Food Engineering* 179: 36–43.

Ryan, L., O'connell, O., O'Sullivan, L. et al. (2008). Micellarisation of carotenoids from raw and cooked vegetables. *Plant Foods for Human Nutrition* 63 (3): 127–133.

Sakr, M. and Liu, S. (2014). A comprehensive review on applications of ohmic heating (OH). *Renewable and Sustainable Energy Reviews* 39: 262–269.

Salazar-González, C., San Martín-González, M.F., López-Malo, A., and Sosa-Morales, M.E. (2012). Recent studies related to microwave processing of fluid foods. *Food and Bioprocess Technology* 5 (1): 31–46.

Sarkis, J.R., Jaeschke, D.P., Tessaro, I.C., and Marczak, L.D. (2013). Effects of ohmic and conventional heating on anthocyanin degradation during the processing of blueberry pulp. *LWT - Food Science and Technology* 51 (1): 79–85.

Shenga, E., Singh, R.P., and Yadav, A.S. (2010). Effect of pasteurization of shell egg on its quality characteristics under ambient storage. *Journal of Food Science and Technology* 47 (4): 420–425.

Sisquella, M., Picouet, P., Viñas, I. et al. (2014). Improvement of microwave treatment with immersion of fruit in water to control brown rot in stone fruit. *Innovative Food Science & Emerging Technologies* 26: 168–175.

Somavat, R., Mohamed, H.M., Chung, Y.-K. et al. (2012). Accelerated inactivation of Geobacillus stearothermophilus spores by ohmic heating. *Journal of Food Engineering* 108 (1): 69–76.

Somavat, R., Mohamed, H.M., and Sastry, S.K. (2013). Inactivation kinetics of Bacillus coagulans spores under ohmic and conventional heating. *LWT - Food Science and Technology* 54 (1): 194–198.

Tang, J., Hao, F., and Lau, M. (2002). Microwave heating in food processing. In: *Advances in Bioprocessing Engineering* (eds. X.H. Yang and J. Tang), 1–44. New York;: World Scientific.

Tang, J., Hong, Y.-K., Inanoglu, S., and Liu, F. (2018). Microwave pasteurization for ready-to-eat meals. *Current Opinion in Food Science* 23: 133–141.

Vadivambal, R. and Jayas, D.S. (2010). Non-uniform temperature distribution during microwave heating of food materials—a review. *Food and Bioprocess Technology* 3 (2): 161–171.

Vallverdú-Queralt, A., Medina-Remón, A., Casals-Ribes, I. et al. (2012). Effect of tomato industrial processing on phenolic profile and hydrophilic antioxidant capacity. *LWT - Food Science and Technology* 47 (1): 154–160.

Varghese, K.S., Pandey, M., Radhakrishna, K., and Bawa, A. (2014). Technology, applications and modelling of ohmic heating: a review. *Journal of Food Science and Technology* 51 (10): 2304–2317.

Velázquez-Estrada, R.M. (2011). Evaluation of the efficacy of ultra-high pressure homogenization technology to improve the safety and quality of liquid foods and especially of orange juice. PhD thesis. Universitat Autónoma de Barcelona.

Volden, J., Borge, G.I.A., Bengtsson, G.B. et al. (2008). Effect of thermal treatment on glucosinolates and antioxidant-related parameters in red cabbage (Brassica oleracea L. ssp. capitata f. rubra). *Food Chemistry* 109 (3): 595–605.

Wang, F., Du, B.-L., Cui, Z.-W. et al. (2017). Effects of high hydrostatic pressure and thermal processing on bioactive compounds, antioxidant activity, and volatile profile of mulberry juice. *Food Science and Technology International* 23 (2): 119–127.

Wolfe, K.L. and Liu, R.H. (2003). Apple peels as a value-added food ingredient. *Journal of Agricultural and Food Chemistry* 51 (6): 1676–1683.

Yildiz-Turp, G., Sengun, I., Kendirci, P., and Icier, F. (2013). Effect of ohmic treatment on quality characteristic of meat: a review. *Meat Science* 93 (3): 441–448.

Yu, X.-m., Zhu, P., Zhong, Q.-p. et al. (2015). Subcritical water extraction of antioxidant phenolic compounds from XiLan olive fruit dreg. *Journal of Food Science and Technology* 52 (8): 5012–5020.

Zell, M., Lyng, J.G., Cronin, D.A., and Morgan, D.J. (2010). Ohmic cooking of whole Turkey meat–effect of rapid ohmic heating on selected product parameters. *Food Chemistry* 120 (3): 724–729.

Zhu, J., Kuznetsov, A.V., and Sandeep, K.P. (2007). Mathematical modeling of continuous flow microwave heating of liquids (effects of dielectric properties and design parameters). *International Journal of Thermal Sciences* 46 (4): 328–341.

Zielinska, M. and Michalska, A. (2016). Microwave-assisted drying of blueberry (Vaccinium corymbosum L.) fruits: drying kinetics, polyphenols, anthocyanins, antioxidant capacity, colour and texture. *Food Chemistry* 212: 671–680.

Zielinska, M. and Michalska, A. (2018). The influence of convective, microwave vacuum and microwave-assisted drying on blueberry pomace physicochemical properties. *International Journal of Food Engineering* 14 (3): 20170332. https://doi.org/10.1515/ijfe-2017-0332.

Zielinska, M., Zapotoczny, P., Alves-Filho, O. et al. (2013). Microwave vacuum–assisted drying of green peas using heat pump and fluidized bed: a comparative study between atmospheric freeze drying and hot air convective drying. *Drying Technology* 31 (6): 633–642.

Zielinska, M., Ropelewska, E., Xiao, H.-W. et al. (2019). Review of recent applications and research progress in hybrid and combined microwave-assisted drying of food products: quality properties. *Critical Reviews in Food Science and Nutrition* 60 (13): 2212–2264.

10
Non-thermal Processing Technologies: High Pressure Processing and Others

Maneesha S. Mohan[1], Ingrid Aguiló-Aguayo[2], and Tomás Lafarga[2,3]

[1] *Department of Wine, Food and Molecular Biosciences, Faculty of Agriculture and Life Sciences, Lincoln University, Lincoln, New Zealand*
[2] *Institute of Agrifood Research and Technology (IRTA), Lleida, Spain*
[3] *Department of Chemical Engineering, University of Almería, Almería, Spain*

10.1 Introduction

The development of non-thermal emerging technologies in the past few decades include the advancement of high pressure processing (HPP), pulsed electric fields (PEFs), and ultrasound as alternatives to thermal processing. This chapter provides a comprehensive review of high pressure processing technologies and their effect on different food components. Further the chapter also includes an overview of other non-thermal processing technologies including PEF and ultrasound on the physico-chemical and nutritional properties of different food systems. These technologies are utilized for preservation of heat-sensitive nutritional compounds foods, inactivation of microorganisms, extraction of food constituents and disruption of particles in food processing.

10.2 Non-thermal Technologies

10.2.1 High Pressure Processing

HPP was first researched in the late nineteenth century for application in food processing (1880–1899). The commercial high pressure processed food products were manufactured only in the late twentieth century – pressure-treated jams and jellies (Meidi-Ya Food Co., Japan, 1990) and pressure-treated guacamole (Avomex., USA, 1997). The history of research and commercialization of HPP of foods has been reviewed by several researchers (Rivalain et al. 2010; Balasubramaniam et al. 2015;

Food Formulation: Novel Ingredients and Processing Techniques, First Edition.
Edited by Shivani Pathania and Brijesh K. Tiwari.

Sevenich and Mathys 2018). In the last two decades, HPP of foods have emerged as a promising processing technology and is forecast to rise from a \$30 billion market in 2016 to a \$42 billion market in 2026 (Hiperbaric 2018). The HPP technology is increasingly being accepted as an alternative to thermal processing, owing to its lower energy requirements compared to thermal processes. HPP provides anti-microbial effects with minimal loss of sensory and nutritive quality of foods. Pressure treatment is applied in food systems mainly by two methods: (i) high hydrostatic pressure (HHP) that applies hydrostatic pressure as high as 1000 MPa in a closed chamber; and (ii) high pressure homogenization (HPH), a continuous process with multiple forces (hydrostatic pressure, cavitation, shear stress, turbulence, and thermal energy) acting on the system, as shown in Figure 10.1 (Datta et al. 2005; Diels and Michiels 2006; Donsì et al. 2009). Maximum pressures so far tried for HPH is 400 MPa. Some of the characteristics of high pressure treatment and extraction are quick, safe, low energy consumption, high efficiency, environmentally friendly, and maintain the colour, flavour, and nutritional value of extracts and processed foods (Huang et al. 2013; Wang et al. 2016; Sevenich and Mathys 2018).

10.2.1.1 Principles of High Pressure Processing Pressure treatment of food affects its physico-chemical, functional, and technological properties, which reflect the changes occurring in the constitutive bio-molecules in the foods. The main principle involved in HPP is the isostatic principle, which states that the uniform application of pressure is independent of time and space and acts equally in all directions (Balasubramaniam et al. 2008). When exposed to pressure treatment, biomolecules follow the Le Chatelier-Braun principle, which states that when pressure is applied to a system in equilibrium, the system will counteract mainly by reducing the volume (Huppertz et al. 2002). The reduction in volume may be the result of closer packing of molecules or chemical reactions (Heremans 1982). However, the packing owing to high pressure affects parts of the molecules bound by weak bonds like hydrophobic, ionic, and hydrogen bonds (Federighi et al. 1995). These bonds are characterized by a negative change in volume, indicating that the volume of reactant is greater than that of the product on application of pressure (volume including intrinsic or van der Waals', solvation shell, and cavities volumes). In other words, hydrostatic pressure alters the

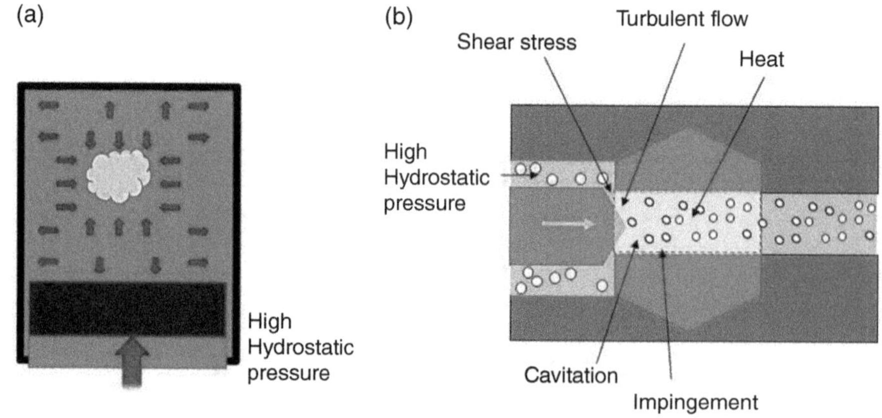

Figure 10.1 Forces acting on a (a) particle in a hydrostatic vessel and (b) in a liquid in a high pressure homogenization valve.

inter-atomic distances. In the case of electrostatic interactions, the force of the interactions is inversely related to the distance between the charged particles. Similarly, the strength of hydrogen bonding and van der Waals' forces are also dependent on the distance between particles, except covalent bonds. Hence, these distance-dependent bonds and forces will be affected by pressure changes (Mozhaev et al. 1994; Heremans 2001; Martínez-Monteagudo and Saldaña 2014). The relation of changes in pressure to Gibbs free energy and chemical reaction equilibrium constants have been elucidated by researchers (Brandts et al. 1970; Rivalain et al. 2010; Balasubramaniam et al. 2015). These changes in inter-atomic distances on application of high pressure have been related to changes in physical properties (melting point, solubility, density, and viscosity) and chemical processes (rates of reaction, dissociation, ionization, acid–base equilibria), as reviewed by Balasubramaniam et al. (2015).

Increase in pressure in a system is also associated with increase in temperature. Ardia et al. (2004) modelled a relationship for adiabatic heating at high pressure when compared with experimental values, and Balasubramaniam et al. (2008) have associated the thermal effects to the Arrhenius relationship of the chemical reaction rates involved in the high pressure process. A much more pronounced heating effect is observed in HPH induced by the dynamic pressure and other forces including cavitation, turbulence, shear stress, and thermal energy (Paquin 1999). A number of studies have investigated the effect of different types of pressure treatments including HHP up to 1000 MPa (HHP) and HPH up to 450 MPa.

Another recent application is high pressure shift freezing utilizing pressures up to 200 MPa. The principle involved is the depression of freezing point at high pressure. The high moisture food material is pressurized to about 200 MPa and allowed to cool to −20 °C; at this point the depression alleviates the freezing of the water molecules. Once the food and the water content attain −20 °C, the release of pressure will convert the unstable supercooled water rapidly to very small ice crystals distributed in the food matrix (Volkert et al. 2012; Wang et al. 2016). The growth and coarsening of ice particles in pressure shift freezing has been theoretically analyzed and modelled by Smith et al. (2013).

10.2.1.2 High Pressure Equipment The main components of an HHP system are the pressure vessel, high pressure pump and intensifier, pressure transmitting fluid, and product loading and unloading system. The common commercially utilized pressure transmitting liquid used is water and others are glycol–water solutions, silicone oil, sodium benzoate solutions, ethanol solutions, and castor oil (Balasubramaniam et al. 2008). Food products are typically vacuum packed and then loaded into the pressure chamber. The high pressure pump and intensifier work together to pump the pressure transmitting liquid at high pressure around the product in the pressure chamber. The pressure is maintained in the chamber for a prescribed amount of time and then released to unload the product. This system has been used commercially for batch processes as well as semi-continuous processes with automated product loading and unloading from the pressure chamber (Balasubramaniam et al. 2015). Other semi-continuous systems utilize two or more pressure chambers, where it is arranged in such a way that one chamber pressurizes when the other unloads, ensuring a continuous output (Ting 2010).

Most industrial pilot or lab-scale high pressure homogenizers have a plunger type pump and a valve made from hard abrasive materials. The commonly used types of valves are micro fluidics, ceramic needle and seat, ceramic ball and seat, and nozzle made from diamond, sapphire, or ruby. Another major component in an HPH system is the attenuator, which is a constant volume chamber that stabilizes fluctuations in pressure and other forces. So far, the systems developed to study food systems can achieve maximum pressures of up to 450 MPa. However, outside the foods and pharmaceutical industry, high pressure water is used for cutting very hard materials. The high pressure is generated by water jets and can achieve pressures as high as 600 MPa with the help of a nozzle made of diamond, sapphire, or ruby, with an orifice diameter of less than 0.35 mm. This system has a high pressure piston pump connected to an attenuator that provides the back pressure in the system to achieve very high pressures. Recent studies have utilized high pressure water jets to process milk (Mohan et al. 2016; Hettiarachchi et al. 2018; Tran et al. 2018). More research is warranted for improving the pressure limits, temperature control, and energy efficiency of HHP and HPH processing equipment.

10.2.1.3 Effect of High Pressure on Different Food Components Research has demonstrated a variety of applications of high pressure technology for food processing, including inactivation of microorganisms, extraction of food components, increase in digestibility, reduction of allergenicity, increased functional properties like emulsification, modification of structure of biopolymers, interactions of food biomolecules, and many more, as indicated in Table 10.1.

10.2.1.3.1 Effect of High Pressure Treatment on Proteins The effect of high pressure treatment on the proteins in foods has been of great interest. The structure of proteins in food is generally complex and consists of well-developed primary, secondary, tertiary, and quaternary structures. The primary structure of food proteins consists of covalent bonds that are not generally affected by high pressure treatments unless accompanied by very high thermal treatment (Mozhaev et al. 1994). The secondary structure of proteins is mainly stabilized by hydrogen bonds that have been shown to be affected by pressure treatment. At low pressures, hydrogen bonds are enhanced and are disrupted at very high pressures (Hendrickx et al. 1998). High pressure treatments destabilize electrostatic interactions between the charges in the vicinity of each ion caused by the decrease in volume of the water layers, associated with these charges or ions (Gross and Jaenicke 1994). The hydrophobic and ionic interactions contribute to the tertiary structure of proteins. With the reduction in volume owing to pressure, hydrophobic interactions are also affected, changing the proximity of hydrophobic groups (Heremans 1982; Mozhaev et al. 1994). Studies have shown that quaternary structures of proteins stabilized by non-covalent bonds are disrupted, even at pressures just above 150 MPa. Many multimeric proteins are broken down into monomeric proteins (Gross and Jaenicke 1994; Hendrickx et al. 1998; Balny 2002) and partially unfolded on exposure to high pressure treatments (Royer 2002). Dissociation of casein micelles stabilized mainly by the quaternary structure of proteins and hydrophobic interactions has been reported on high pressure treatments of above 300 MPa based on the total solid content of the solution (Anema 2008; Orlien et al. 2010). High pressure assisted protein refolding has been reported by researchers (Seefeldt et al. 2004, Schoner et al. 2005, Chura-Chambi et al. 2008) and used to improve the solubility and functionality of certain proteins, as reviewed by Aertsen et al. (2009). High pressure treatment and increase in total solid content affect the microstructure and strength of gels from milk proteins, as reviewed by Devi et al. (2013). Hydrogen bonds, electrostatic interactions, hydrophobic interactions, and

Table 10.1 Studies on the effect of pressure treatment on food components including proteins and lipids (since 2007).

Food components	Pressure Treatment	Effect on food component	Reference and related research
Proteins			
HHP			
Enzyme activity			
Whey proteins	Treatment with pepsin (at pH 2.5) and chymotrypsin (at pH 6.8) under 400 MPa on α-lactalbumin (α-La) and β-lactoglobulin (β-Lg)	Pressure treatment increased proteolysis of whey proteins depending on the enzyme and pH. Hydrolysates obtained were highly digestible.	(Chicón et al. 2009) High digestibility of pressure treated (up to 800 MPa with temperature not exceeding 40 °C) β-Lg, also reported by Zeece et al. (2008)
Carrot enzyme	PME activity in carrots on combining pressure from 0.1 to 600 MPa and temperatures of 30–55 °C	Increase in the enzyme catalytic activity when pressure increased to 600 MPa and temperature of 40 °C.	Sila et al. (2007)
Meat enzymes	HHP up to 600 MPa for 6 min for brined (5.5% NaCl and 3% dextrose in water) pork meat	Calpain inactivated above 200 MPa pressure treatment and cathepsin activity was more than doubled on treatment at 600 MPa pressure.	Grossi et al. (2012)
Disruption of structure			
Milk proteins	HHP up to 600 MPa on milk	Disruption of casein micelles at 300 MPa pressure for 10–15% total solids (TS) milk and aggregation at 300 MPa pressure for 20% TS milk.	(Anema 2008) Similar results of dissociation of casein micelles above 300 MPa pressure at pH 7 and 30 °C for reconstituted milk reported by (Orlien et al. 2010)

(Continued)

Table 10.1 (Continued)

Food components	Pressure Treatment	Effect on food component	Reference and related research
Gel formation			
Whey and fish proteins	HHP up to 600 MPa at 30 °C for 10 min	Pressure induced WPI becomes stronger at pH 6.8 and 8 as ionic strength increases to 0.2.	(He et al. 2010) Similar increase in gel strength observed in gels from frozen flying fish surimi subjected to up to 200 MPa pressure by Moreno et al. (2015). Another study reported that pressure induced whey protein gels can be formed with relatively low pressure of up to 250 MPa using longer pressurization time and higher protein concentration (He and Ruan 2009).
Denaturation and solubility			
Soy proteins	Soybean protein isolate solution (1% solution) subjected to HHP up to 600 MPa for 15 min at 25 °C.	Unfolding and partial denaturation of proteins observed during pressure treatment, but the proteins re-folded and formed secondary and tertiary structure on release of pressure.	Tang and Ma (2009)
Beef	Beef muscles subjected to HHP for 20 min at pressures up to 600 MPa for temperatures up to 30 °C.	Solubility of sarcoplasmic proteins ranged between 25 to 45% at 400 and 600 MPa pressures compared to 65 to 70% for 0 and 200 MPa treated samples at 10, 20 and 30 °C. The water holding capacity reduced for both 400 and 60 MPa treated beef muscle samples.	(Marcos et al. 2010) The solubility of the sarcoplasmic proteins is related to the degree of denaturation and formation of insoluble aggregates. Sarcoplasmic proteins, myofibrillar proteins, myosin and actin play an important role in the water-binding capacity of beef muscle. Lee et al. (2007) studied the effect of HHP up to 500 MPa on solubility of myofibrillar proteins. The effect of high pressure on meat proteins, enzymes and overall structure has been review by Buckow et al. (2013).

HPH			
Disruption of structure			
Milk proteins	High pressure jet processing up to 500 MPa	Disruption of casein micelles at lower pressure below 300 MPa pressure and aggregation leading to increase in particle size above 300 MPa pressure compared to raw milk.	(Mohan et al. 2016) Similar results of disruption of casein micelles at lower pressures (Zamora et al. 2007, Ciron et al. 2010) and aggregation related increase in size was observed on HPH at 330 MPa by Zamora et al. (2007).
Denaturation and emulsifying ability			
Milk and soy protein	HPH at 2200 and 300 MPa with inlet temperatures of 30 and 40 °C.	α-La and β-Lg denatures to different levels on HPH up to 300 MPa pressure.	(Pereda et al. 2009) Partial protein denaturation of soy proteins on HPH (Liu and Kuo 2016) has been associated with improved emulsifying ability by Fernández-Ávila et al. (2015), (Fernández-Ávila and Trujillo 2016).
Acid gelation			
Soy and milk protein	HPH at 200 and 300 MPa with inlet temperatures of 40 and 50 °C.	Soymilk subjected to HPH up to 300 MPa produces stronger acid gels compared to UHT soymilk.	(Cruz et al. 2009) However, skim milk subjected to HPH up to 300 MPa produced the weakest acid gels compare to thermally treated milk (Hernández and Harte 2008).
Enzyme activity			
Orange juice	HPH up to 300 MPa with inlet temperatures of 10 or 30 °C.	PME inactivated to above 96% for HPH treatment above 200 MPa similar to inactivation by pasteurization at 90 °C.	(Velázquez-Estrada et al. 2012) A study by Welti-Chanes et al. (2009) also reported inactivation of PME in orange juice but to a lower extend.

(*Continued*)

Table 10.1 (Continued)

Food components	Pressure Treatment	Effect on food component	Reference and related research
Lipids			
HHP			
Lipid oxidation			
Pork	Longissimus muscle subjected to HHP up to 500 MPa for 20 min at 20 °C.	Oxidation of lipids including triglycerides containing linoleic acid observed for samples treated to 500 MPa. About 33% of the PUFA in the phospholipids decreased on storage of 7 days after HHP at 500 MPa. These PUFA were hydrolysed and released as free fatty acids.	(He et al. 2012) Radicals initiate the lipid oxidation in meat products and the reaction order of radical formation is zero order up to 600 MPa and first order at 700 MPa at 25 °C (Bolumar et al. 2012).
Beef	Beef neck muscle samples subjected to pressure of 200 or 800 MPa for 20 min.	Lipid oxidation was not initiated after high pressure treatment of up to 800 MPa at 60 °C on day 0. However, the lipid oxidation (TBARS) increased during chilled storage for 6 days under vacuum above 200 MPa treatment.	(Tume et al. 2010) The addition of vitamin E in this study did not reduce lipid oxidation over storage, however addition of metal chelators like disodium EDTA reduces lipid oxidation. Hence, free metal ions produced during high pressure treatment have been associated to the increase in lipid oxidation in minced beef and chicken muscles. (Ma et al. 2007).
Pressure-induced crystallization			
Oil	Oil with 82% diacylglycerol and 18% triacylglycerol compressed to 450 MPa at different compressing rates between 0.02 to 1.23 MPa/s.	Compression rates higher than 0.18 MPa/s increased the induction time for crystallization as measure by the intensity of transmitted beam.	(Ko⊠ciesza et al. 2012) Fat crystals formed in corn germ oil and sunflower oil above 235 MPa at 20 °C.

HPH			
Disruption and formation of structure			
Soya oil – whey protein isolate emulsion	HPH at 50 to 200 MPa for first pass and 3 MPa for second pass	Decreasing the droplet size is one of the reasons for emulsion stability on increasing the HPH pressure. However, the HPH induced modification on the surfactant molecules like proteins also plays a significant role in emulsion stability	Lee et al. (2009)
Soymilk	HPH of 200 and 300 MPa at inlet temperature of 40 °C	The particle size distributions curve indicates a shoulder around 1.5–2.5 and 1.5–6 μm, respectively for 300 and 200 MPa pressure of HPH. This has been associated with formation of aggregates of coalescence of fat globules.	(Cruz et al. 2007) Similar results of agglomeration owing to decrease in availability of surfactant molecules in relation with increase in the surface area of fat globules has also been reported by Cortés-Muñoz et al. (2009) on increasing peanut oil content to 45% in an O/W emulsion prepared by singe stage HPH at pressure 100–300 MPa with 4.3% whey proteins.

intermolecular disulphide linkages are involved in the gelation of protein solutions (Clark 1991). pH and ionic strength also affect the strength of pressure-induced gels by controlling the cross-linking between proteins and phase separation occurring during gelation (He et al. 2010; Moreno et al. 2015). Another research group has associated gel strength of proteins on high pressure treatment to pressure release rate, with high pressure release rate leading to stronger gel formation of casein gels (Merel-Rausch et al. 2007). Proteins in eggs on exposure to HPH exhibit improved foaming capacity in egg whites owing to exposure of sulphhydryl groups (Van der Plancken et al. 2007; Yang et al. 2009), loss of solubility in the egg yolk (Yan et al. 2010), improved surface hydrophobicity of proteins (Acero-Lopez et al. 2012), and change in secondary structure of proteins at certain pressures (Acero-Lopez et al. 2012).

High pressure treatment leads to activation or inactivation of different enzyme isoforms at varying extent in different plant-based food products, as reviewed by Jolie et al. (2012) and Oey (2016). For instance, pectin methyl esterase (PME) exhibits an increase in catalytic activity up to pressures of 600 MPa at 30 °C in shredded carrots and a further increase in activity observed at elevated temperatures up to 50 °C in the same product (Sila et al. 2007). On the other hand, PME in orange juice was inactivated to reduce its activity to 20% at 30 °C after 20 minutes of high pressure treatment at 500 MPa (Torres et al. 2016). Polygalacturonase (PG) is pressure labile compared to PME and can be inactivated at room temperature in the pressure range of 450–500 MPa (Rodrigo et al. 2006; Houben et al. 2013). Inactivation of endogenous enzymes thereby prolong the shelf life of fruit and vegetable products. Increased lightness and change in the redness in colour attributes has been reported in meat on HHP owing to depolymerization of myofibrillar proteins, partial denaturation of nitrosylmyoglobin, and oxidation of ferrous or ferric myoglobin (Carlez et al. 1995; Andrés et al. 2004; Cava et al. 2009; Marcos et al. 2010; McArdle et al. 2010).

Most of the phenomena of pressure-induced unfolding of proteins and effect on bonds and interactions hold true mainly for HHP treatments. This is owing to the fact that HHP treatments are mainly associated with a single stress applied by hydrostatic pressure. On the other hand, HPH produces a much more complex phenomena due to the number of forces existing in it, but only for an exposure time in the order of milliseconds. The magnitude of these forces depends highly on the type of equipment used, the characteristics of the food system, temperature conditions, velocity, and many other factors.

Partial denaturation of proteins owing to HPH up to 300 MPa has been associated with better emulsifying and foaming properties for whey proteins and soy proteins (Pereda et al. 2009; Santana et al. 2011; Fernández-Ávila et al. 2015; Fernández-Ávila and Trujillo 2016; Liu and Kuo 2016). These effects are also related to the adiabatic heating during the HPH process. The amount of the denaturation varies based on the structural characteristics of the proteins (Pereda et al. 2009). In orange juice, the PME amount decreased to less than 10% after HPH above 200 MPa, which is comparable to pasteurization inactivation of PME (Velázquez-Estrada et al. 2012). Other researchers have only reported lower PME inactivation at similar pressures in orange juice, possibly owing to the variation in the homogenization equipment and temperatures during homogenization (Welti-Chanes et al. 2009). Pectate lyase in banana juice is extremely susceptible to HPH beyond 150 MPa pressures (Calligaris et al. 2012). A major factor affecting the varied susceptibility of different enzymes to HPH is the degree of denaturation or unfolding or aggregation, depending on the structure of these proteins.

For large protein structures like the casein micelles, the micelles were disrupted at HPH below 300 MPa and have been found to lead to further aggregation above 300 MPa HPH treatment (Zamora et al. 2007; Ciron et al. 2010). A similar effect of increase in aggregation of particles on HPH to 300 MPa was observed in almond milk (Briviba et al. 2016). Soy milk subjected to HPH to 300 MPa at inlet temperatures 40 and 50 °C produces stronger acid gels compared to UHT soy milk (Cruz et al. 2009). However, cow's milk, when subjected to HPH up to 300 MPa, produced the weakest acid gel compared to thermally treated milk (Hernández and Harte 2008; Ciron et al. 2010). Another study indicated that the firmness of acid gel from skimmed milk subjected to HPH of 100–400 MPa was higher than 500 MPa and pasteurized milk gels (Mohan 2014). The varied effect of HPH could be due to the difference in structure of the proteins contributing to the acid gel firmness and the degree of protein denaturation through HPH treatment. Overall high pressure treatment affects the quaternary, tertiary, and secondary structures of proteins, thus altering their binding, hydrophobicity, and solubility properties.

10.2.1.3.2 Effect of High Pressure Treatment on Lipids Le Chatelier-Braun's principle states that high pressure treatment of lipids in systems exhibiting the highest degree of decrease in activation volume are associated with the highest effect of pressure (Medina-Meza et al. 2014). Conversely, the application of pressure changes the reaction equilibrium towards a state achieving the smallest volume. Lipid assemblies are mostly governed by hydrophobic interactions which are weaker than covalent bonds, and hence highly susceptible to pressure-induced changes and reactions (Tauscher 1995). A decrease in volume of up to 2% has been reported for lipids in the pressure range of 350–1000 MPa (Wiśniewski et al. 1994; Guignon et al. 2009; Rostocki et al. 2013). Hence, the high pressure treatment favours reactions like disruption of structure, crystallization of fat, and lipid oxidation.

Oxidation of lipids and hydrolysis of fatty acids from triglycerides and phospholipids have been reported in pork, beef, and chicken muscle treated to above 500 MPa HHP pressure treatments at varying temperatures (Orlien et al. 2000; Wiggers et al. 2004; Ma et al. 2007; He et al. 2012). The disodium salt of ethylenediaminetetraacetic acid (EDTA) and citric acid, which are chelators, have been reported to decrease the lipid oxidation caused by HHP treatment, indicating that high pressure treatment releases pro-oxidant transition metal ions from insoluble complexes like the porphyrin core of myoglobin and also causes reduction of transition metal ions (Fe^{+3} to Fe^{+2}). These metal ions react with disrupted cell membranes rich in phospholipids and triglycerides, initiating lipid oxidation in meat and fish samples above 300 MPa HHP treatment (Cheah and Ledward 1997; Orlien et al. 2000; Ma et al. 2007; Yagiz et al. 2009; Tume et al. 2010; He et al. 2012). However, Orlien et al. (2000) concluded from their studies that the lipid oxidation caused by HHP above 500 MPa did not result from the free iron ions or conformational changes in the metmyoglobulin alone, but from membrane damage in the muscle. HHP treatment had little or no effect on lipid oxidation below a 300 MPa pressure of HHP. Moreover, free radical formation increased above the threshold of 400 MPa at 25 °C and 500 MPa at 5 °C for pressures of 400–800 MPa, temperatures of 5–40 °C, and time of 0–60 minutes (He et al. 2012). The radicals initiate lipid oxidation in meat products and the reaction order of radical formation is zero order up to 600 MPa and first order at 700 MPa at 25 °C (Bolumar et al. 2012). The effect of high pressure treatment on protein and lipid oxidation in meat has been comprehensively reviewed by Guyon et al. (2016) and

Medina-Meza et al. (2014). Therefore, HHP processing alters the flavour and colour of meat and meat products over their shelf life to some extent.

The effect on lipolytic activity is reflected in the free fatty acids (FFA) content in the HHP treated milk. The low increase in FFA content has been attributed to the partial inactivation of native lipoprotein lipase and structural rearrangement of milk fat globule membrane (MFGM), with adsorption of disintegrated casein micelles or denatured whey proteins to the MFGM surface (Felipe et al. 1997; Law et al. 1998). Temperature also had an effect, as indicated by the lower lipolysis at 4 and 50 °C, as compared to at 25 °C (Gervilla et al. 2001). Fat globule size reduction is desirable in soft cheeses and dairy gels due to the resultant high moisture content, producing softer, creamier, and smoother texture. The higher lipolytic sensitivity of smaller fat globules is also utilized in specialized products including blue cheeses (Chandrapala and Leong 2015). Trujillo et al. (2002) has reviewed the possibilities of utilization of the lipolytic and lipid oxidation effects caused by high pressure treatment (pressures of 100 to 600 MPa) for accelerated ripening of cheese (Yokoyama et al. 1992; Kolakowski et al. 1998; Saldo et al. 1999; O'Reilly et al. 2000; Saldo et al. 2002; Trujillo et al. 2016). Another factor affecting the lipolytic effect and FFA content of milk fat when subjected to HHP is the interaction between casein micelles, serum proteins, and MFGM. β-Lg, α-La, and κ-casein associate with each other and MFGM at varying levels as treatment pressure increases for the HHP process. β-Lg starts associating with MFGM at pressures above 100 MPa, depending on the pressure and treatment times, while α-La and κ-casein associate with MFGM at pressures greater than and equal to 700 and 500 MPa, respectively. Enzymes promoting lipid oxidation like xanthine oxidase and butyrophillin also bind with β-Lg, initiating the oxidation process of HHP (Ye et al. 2004). These mechanisms affecting the milk fat globule size and globule membrane constituents result in lipid oxidation and lipolysis on high pressure treatment of fat-rich dairy products. On the other hand, there is scope for utilizing high pressure treatment for accelerated ripening and flavour enhancement in dairy products like cheese.

High pressure induces lipid crystallization by elevation of melting temperature in a reversible manner by about 10 °C per 100 MPa pressure (Cheftel and Culioli 1997; Delgado et al. 2010; Ferstl et al. 2010). The kinetics of lipid crystallization under pressure is dependent on processing parameters like pressure, temperature, holding time, and compression rate, in addition to lipid composition (Zulkurnain et al. 2016). Formation of different polymorphic crystal forms occurs when the free energy of the solid phase is less than the liquid phase, governed by the melting temperature and enthalpy of the system. Studies on triolein have indicated that increase in pressure to 200 MPa indicated a linear relation of melting point with temperature, while above 200 MPa up to 450 MPa was found to be non-linear (Ferstl et al. 2010). The pressure–temperature phase diagram of lipids has been illustrated by Zulkurnain et al. (2016). The crystallization process involves nucleation and growth of crystals. High pressure treatment leads to an increase in rate of nucleation, with increase in pressure treatment in a multi-step pressure process. Studies have reported a 47–138 times increase in nucleation rate at different temperatures from about 200 to 450 MPa pressure treatments (Ferstl et al. 2010, 2011). Multi-step pressure treatment leads to a crystallization mechanism similar to supercooling (Ferstl et al. 2010; Kościesza et al. 2012). The initial crystal forms are the β' form and with time grow into stable β form crystals (Jaeger and Eggers 2005; Zulkurnain et al. 2016). The crystal growth rate increases with increase in pressure and decreases with increase in temperature (Ferstl et al. 2011). Pressure treatment as a single rapid compression step produces instantaneous onset of crystallization as less stable crystal polymorphs like the α form (Tefelski et al. 2008; Kościesza et al. 2012). Several studies have been conducted on the effect of pressure on crystallization in lipid-based food systems like vegetable oils, cocoa butter, and dairy creams, as detailed by Zulkurnain et al. (2016) and Mohan et al. (2020). Most of the studies of the food system exhibit an increase in crystallization on application of pressure treatment.

Pressure treatments have also been shown to impact the disruption and aggregation of oil or fat droplets, globules, and emulsions. Studies of milk fat globules have reported some degree of disrupting the effect of HHP on the size distribution of the globules dependent on the temperature (Kanno et al. 1998; Gervilla et al. 2001; Ye et al. 2004). The decrease in the particle sizes has been shown to improve emulsion stability. The effect of HPH, which involved more cavitation and turbulence forces compared to HHP, has been shown to cause disruption or aggregation of emulsion droplets, depending on the homogenization pressure, number of passes, inlet temperature, and effect of pressure on the surfactant molecules (Thiebaud et al. 2003; Lee et al. 2009). Increase in pressure from 0 to 300 MPa HPH treatment decreased the mean diameter significantly at 4 and 24 °C. The size distribution of the milk fat globules at 200, 250, and 300 MPa were relatively similar. At 200 MPa, two passes of HPH pressure significantly decreased the mean diameter of the milk fat globules at 4, 14, and 24 °C (Thiebaud et al. 2003). Utilization of a second stage of homogenization at 10–20% pressure of the first stage has shown a reduction in milk fat globule size. The observation of shoulders in the particle size distribution of soy milk subjected to single-stage HPH at 200 and 300 MPa, has been associated with aggregation of fat globules (Cruz et al. 2007). Similarly, the mean diameter of the fat globule increased significantly from 200 to 250 MPa on HPH of raw whole bovine milk with an inlet temperature of 45 °C. This has been attributed to the partial agglomeration of small globules with surfaces not covered by surfactant molecules, like caseins, denatured whey proteins, and phospholipids. The agglomeration possibly increased the mean diameter of the fat globule at 250 MPa HPH pressure, similar to that observed for conventional homogenized and pasteurized milk samples (Hayes et al. 2005). Cortés-Muñoz et al. (2009) has associated limited availability of surfactant molecules (4.3% of whey proteins) with increasing surface area of peanut oil droplets in an o/w emulsion to aggregation of emulsion droplets and increase in the scan of particle size distribution.

10.2.2 Pulsed Electric Field

PEF is an emerging non-thermal processing technology gaining consideration in the food industry due to its wide range of food applications. PEF has extensively proved to be a potential alternative to thermal pasteurization methods for extending shelf life of a high range of liquid products, including beverages and dairy products (Aguiló-Aguayo et al. 2015; Datta and Tomasula 2015; Sampedro and Rodrigo 2015; McAuley et al. 2016; Domínguez Avila et al. 2018; Gabrić et al. 2018; Sharma et al. 2018; Tiwari 2018; Mahendran et al. 2019). The induction of irreversible electroporation induced by PEF and low processing temperatures achieved during the treatment (normally less than 40 °C) not only led to the inactivation of different microorganisms but also the preservation of heat-sensitive bioactive compounds with no alteration of food sensory properties (Barbosa-Cánovas and Altunakar 2006; Toepfl et al. 2006). This represents an important advantage in the production of functional fruit- or vegetable-based beverages, since PEF-processed products could assure the daily intake of health-promoted compounds (Domínguez Avila et al. 2018). The application of PEF in a wider range of food materials in solid state has also been demonstrated to preserve or promote their nutrient content. Exposing solid vegetable material to PEF treatment has also been proved as an effective pre-treatment method to promote production of health-related compounds such as production of glucosinolates in PEF-treated broccoli florets and sprouts (Aguiló-Aguayo et al. 2015; Subedi et al. 2019). In onion bulbs, the increment of the electric field strength enhanced fructan leakage (Liu et al. 2019). In whole PEF-treated apples, an increment in the initial flavonol content led to apples with enhanced functional value (Ribas-Agustí et al. 2019).

The electroporation PEF principle is based on the application of a critical field strength (between 10 and 80 kV/cm), helping to reach the dielectric breakdown of cell membranes, resulting in changes in the membrane structure and permeability (Figure 10.2) (Gabrić et al. 2018). Based on this principle, PEF is a food technology that could improve other food applications including extraction in plant-based food. During extraction, the increment of the mass transfer rate by disturbing the membrane structure of the processed cell walls could potentially extract nutritionally valuable compounds (Toepfl et al. 2006). There are many reports regarding the feasibility of PEF as a pre-treatment method in the enhancement of extraction processes from plant materials (Barba et al. 2015; Parniakov et al. 2016; Liu et al. 2018; Gagneten et al. 2019; Jaeschke et al. 2019), biorecovery from wastes and biomasses (Barbosa-Pereira et al. 2018; Martínez et al. 2018; Redondo et al. 2018; Nadia et al. 2019; Peiró et al. 2019), or increasing juice yield as an alternative of mechanical processing operations (Pataro et al. 2017; El Kantar et al. 2018).

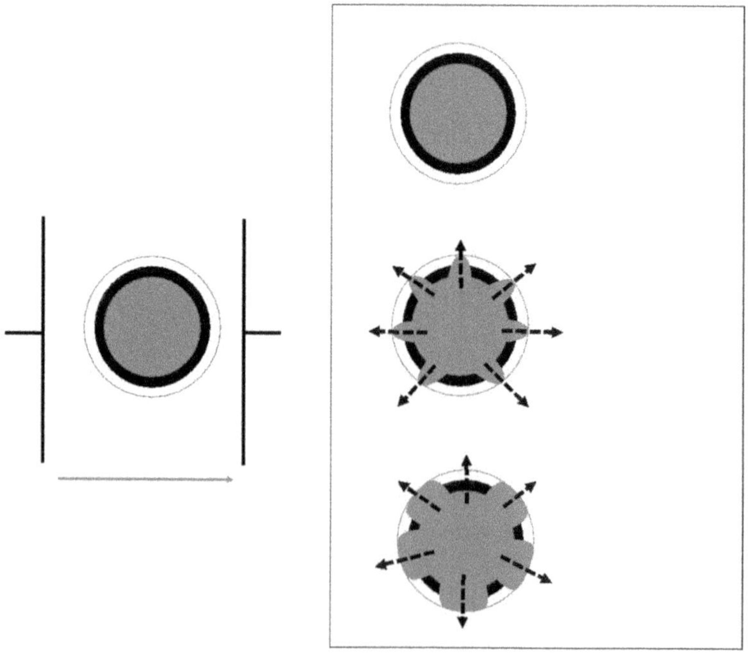

Figure 10.2 Mechanism of membrane permeabilization depending on the application of an external electrical field. Legend – E: external electrical field and Ecrit: critical external electrical field.

Some PEF extraction studies are exploring the potential applications in microalgae (Parniakov et al. 2015; Zocher et al. 2016; Scherer et al. 2019), which are one of the hot trends now present in the food industry (Lafarga 2019). Application of PEF treatment conditions at 80 V/cm for 10 minutes in *Chlorella vulgaris*, *Scenedesmus* sp. and *Chloroccom* sp. resulted in diffusion out from cells of protein (Samsu et al. 2018). PEF have also been shown to promote selective extraction of carbohydrates (36%, w/w, of total carbohydrates) and low molecular weight proteins (5.2%, w/w, of total proteins) when processing microalgae *C. vulgaris* suspensions (1.2%, w/w) (Carullo et al. 2018). Moreover, the application of PEF as an oil extraction process has been reported as an effective pretreatment in the oil industry (Puértolas et al. 2016; Shorstkii et al. 2017; Sukardi et al. 2017). Oil extraction from cannabis seeds has been optimized by applying a PEF intensity of up to 3 kV/cm (Haji-Moradkhani et al. 2018). The intensification of the extraction by PEF of essential oils has been reported as an alternative to the current process of hydro-distillation generally employed in the extraction process (Miloudi et al. 2018; Yajun et al. 2017). In addition, PEF has been reported to be a fast and energy-efficient way to separate oil and water, known as the demulsification process in crude oil emulsions, showing a 50% time and energy saving compared to chemical demulsification (Yajun et al. 2017).

The potential application of PEF in the meat and fish processing industries has also been reported to improve the extraction of high-added value compounds, but also to enhance processes such as preservation, tenderization, and ageing (Gómez et al. 2019). Other potential food applications to consider in food formulation, especially for the development of new food products, is the effect of PEF on the enhancement of water loss during drying (Janositz et al. 2011; Onwude et al. 2017; Huang et al. 2019). PEF-pretreatment applied in freeze-dried apples led to better preservation of shape, crunchiness, and polyphenol content (Lammerskitten et al. 2019). For instance, the application of combined PEF and ultrasound (US) technologies led to reduced sugar content in dried cranberry snacks (Nowacka et al. 2019). The combination of US- and PEF-assisted extraction has been shown as a promising technique in the extraction process of bioactive compounds from agro-industrial by-products, as an alternative to the current extraction solvent procedures (Kumari et al. 2018) or value ingredients such as sunflower oil (Moradi and Rahimi 2019) by reducing time and cost.

Another important point to take into account in the application of PEF in food formulation is the bioavailability of nutrients during the digestive process of PEF-treated products. The *in vitro* digestibility test showed an increase in the portion of rapidly digestible starch of PEF-treated solid food systems including wheat, potato, and pea starches (Li et al. 2019). Structural and functional changes in food proteins have been related to PEF, inducing an alteration in the hydrolysis of meat proteins by gastrointestinal proteases during the digestion of meat products (Bhat et al. 2019). The release of pro-inflammatory cytokines and chemokines confirming immunomodulatory activity was observed in PEF-treated brewer's spent grains, showing significantly increased yields of carbohydrate, protein, and starch (Kumari et al. 2019).

10.2.3 Ultrasound Technologies

Other thermal processing technologies include combinations of non-thermal technologies such as sonication and HPP with mild temperatures. The former is known as thermosonication. Several authors highlighted the antimicrobial effects of this technology

(Marx et al. 2011; Arroyo et al. 2012; Bermúdez-Aguirre and Barbosa-Cánovas 2012; Kiang et al. 2013; Ferrario et al. 2015; Sánchez-Rubio et al. 2016; Garud et al. 2017; Lafarga et al. 2019a). Microbial reductions reported during the last decade vary within the range of 1–9 log cycles (Anaya-Esparza et al. 2017). The observed differences depend on several factors such as juice type, microorganism studied, initial microbial content, and treatment conditions. Several hypotheses have been suggested on the mechanisms that cause microbial inactivation during thermosonication, the cavitation phenomenon being the most widely accepted. Previous studies also suggest that thermosonication could lead to a lower degradation of healthy compounds and to an improved retention of the organoleptic and overall quality of food products, especially of liquid foods such as fruit juices (Aguilar et al. 2017). For example, Tomas Lafarga et al. (2019b) reported that thermosonication allowed an increased retention of colour attributes as well as polyphenol, lycopene, and anthocyanin content of a tomato juice enriched with strawberry-derived anthocyanins previously. In that study, the authors also reported increased amounts of bio-accessible polyphenols after thermosonication, when compared to a conventional heat treatment. Similar results were reported in a different study, where a matrix-dependant increase in bio-accessibility was observed after sonication of vegetables (Lafarga et al. 2019c). Thermosonication can also inactivates enzymes responsible for food spoilage. For example, Sulaiman et al. (2015) reported that thermosonication for 10 minutes at 32 °C can inactivate polyphenoloxidases in pear, strawberry, and apple purees, allowing the use of a much lower temperature when compared to thermal processing alone. Similar results were reported by other research groups (Cruz et al. 2006; Siwach and Kumar 2012; Aadil et al. 2015; Jabbar et al. 2015; Koshani et al. 2015). The combination of thermosonication with HPP, known as manothermosonication, has been suggested as a strategy to promote cavitation's disruptive effect. External pressures applied can theoretically counter pressure on cavitation from ultrasounds and lead to a more vigorous bubble implosion. Different authors reported that pressurized sonication and heat cooperate as a synergistic effect. For example, Chantapakul et al. (2019) recently reported reductions in the population of *Escherichia coli* and *Staphylococcus aureus* of 6.25 and 4.6-log cfu/ml after five minutes of processing at 50 °C and 400 kPa. Similar results were published by Kahraman et al. (2017), who obtained a 5-log cfu/ml reduction of *E. coli* O157:H7 after processing of an apple-carrot blended juice for 30 seconds at 60 °C and 200 kPa.

10.3 Conclusions and Future Studies

Non-thermal technologies are being used commercially to process fruits and vegetable products, including purees, juices, sauces, dairy, and meat products. This chapter provides the first comprehensive review of the effect of high pressure processes, PEF, and ultrasound on the physico-chemical and functional properties of different food systems.

High pressure treatment affects the quaternary, tertiary, and secondary structures of proteins and enzymes, modifying their binding, hydrophobicity, and solubility properties. Chemical and structural modifications of lipid assemblies subjected to high pressure treatments include reduction in particle sizes, lipid oxidation, and emulsification in food matrices such as meat and dairy products. There are numerous studies on the effect of HPP on the functional properties of food products; however, there are very few studies on the nutritional and health aspects of the modified food components. For example, there is limited research on the possible application of HPP for producing human milk fat substitutes (HMFS). HMFS are being tailored from bovine milk fat, vegetable oils, and phospholipids to improve the bioavailability and nutrition content in infant and toddler milks and formulae. Improvement and re-designing of HPP technologies is needed to provide better energy efficiency, temperature control, and wider pressure ranges for HPH and HHP equipment. On the other hand, PEF technology is widely used in the pasteurization of liquid products by extending their shelf life and as a resulting and alternative technology to thermal treatments. The irreversible electroporation induced by PEF and the preservation of heat-sensitive bioactive compounds in food matrices represents an important advantage in processing functional products and assuring non-alteration of the nutritional content, including health-related phytochemicals with anti-inflammatory or antioxidant properties or even to improve digestibility of PEF-treated products. This technology has also been reported as a potential method to extract valuable food industry compounds such as carbohydrates, proteins, or oils.

Further investigation is required to combine high pressure treatment with other novel processing technologies like ultrasound and PEF. Potential microbial reductions in juices have been reported after thermosonication, due to the cavitation process induced by the ultrasound or its combination with HPH.

Some diversification has been attempted through technologies like high pressure shift freezing, high pressure extraction processes, and utilization of high pressure jets for HPH. Moreover, a greater understanding of the interaction effects between different food components within the food matrix undergoing non-thermal treatments is needed. Research is also required to establish novel applications of these technologies in the form of ingredients and food products, with commercial viability, modified functionality, and added value.

Acknowledgements

T. Lafarga, and I. Aguiló-Aguayo are in receipt *Juan de la Cierva* (IJC2018-035287-I) and *Ramon y Cajal* (RYC-2016-19949) Grants, respectively, awarded by the Spanish Ministry of Economy, Industry, and Competitiveness.

References

Aadil, R.M., Zeng, X.A., Zhang, Z.H. et al. (2015). Thermosonication: a potential technique that influences the quality of grapefruit juice. *International Journal of Food Science & Technology* 50 (5): 1275–1282.

Acero-Lopez, A., Ullah, A., Offengenden, M. et al. (2012). Effect of high pressure treatment on ovotransferrin. *Food Chemistry* 135 (4): 2245–2252.

Aertsen, A., Meersman, F., Hendrickx, M.E.G. et al. (2009). Biotechnology under high pressure: applications and implications. *Trends in Biotechnology* 27 (7): 434–441.

Aguilar, K., Garvín, A., Ibarz, A., and Augusto, P.E. (2017). Ascorbic acid stability in fruit juices during thermosonication. *Ultrasonics Sonochemistry* 37: 375–381.

Aguiló-Aguayo, I., Suarez, M., Plaza, L. et al. (2015). Optimization of pulsed electric field pre-treatments to enhance health-promoting glucosinolates in broccoli flowers and stalk. *Journal of the Science of Food and Agriculture* 95 (9): 1868–1875.

Anaya-Esparza, L.M., Velázquez-Estrada, R.M., Roig, A.X. et al. (2017). Thermosonication: an alternative processing for fruit and vegetable juices. *Trends in Food Science & Technology* 61: 26–37.

Andrés, A.I., Møller, J.K.S., Adamsen, C.E., and Skibsted, L.H. (2004). High pressure treatment of dry-cured Iberian ham. Effect on radical formation, lipid oxidation and colour. *European Food Research and Technology* 219 (3): 205–210.

Anema, S.G. (2008). Effect of milk solids concentration on whey protein denaturation, particle size changes and solubilization of casein in high-pressure-treated skim milk. *International Dairy Journal* 18 (3): 228–235.

Ardia, A., Knorr, D., and Heinz, V. (2004). Adiabatic heat modelling for pressure build-up during high-pressure treatment in liquid-food processing. *Food and Bioproducts Processing* 82 (1): 89–95.

Arroyo, C., Cebrián, G., Pagán, R., and Condón, S. (2012). Synergistic combination of heat and ultrasonic waves under pressure for Cronobacter sakazakii inactivation in apple juice. *Food Control* 25 (1): 342–348.

Balasubramaniam, V., Farkas, D., and Turek, E.J. (2008). Preserving foods through high-pressure processing. *Food Technology* 62 (11): 32–38.

Balasubramaniam, V.M., Martínez-Monteagudo, S.I., and Gupta, R. (2015). Principles and application of high pressure–based technologies in the food industry. *Annual Review of Food Science and Technology* 6 (1): 435–462.

Balny, C. (2002). High pressure and protein oligomeric dissociation. *High Pressure Research* 22 (3–4): 737–741.

Barba, F.J., Parniakov, O., Pereira, S.A. et al. (2015). Current applications and new opportunities for the use of pulsed electric fields in food science and industry. *Food Research International* 77: 773–798.

Barbosa-Cánovas, G.V. and Altunakar, B. (2006). Pulsed electric fields processing of foods: an overview. In: *Pulsed Electric Fields Technology for the Food Industry* (eds. J. Raso and V. Heinz), 3–26. MA, USA: Springer.

Barbosa-Pereira, L., Guglielmetti, A., and Zeppa, G. (2018). Pulsed electric field assisted extraction of bioactive compounds from cocoa bean shell and coffee silverskin. *Food and Bioprocess Technology* 11 (4): 818–835.

Bermúdez-Aguirre, D. and Barbosa-Cánovas, G.V. (2012). Inactivation of Saccharomyces cerevisiae in pineapple, grape and cranberry juices under pulsed and continuous thermo-sonication treatments. *Journal of Food Engineering* 108 (3): 383–392.

Bhat, Z.F., Morton, J.D., Mason, S.L. et al. (2019). Pulsed electric field: a new way to improve digestibility of cooked beef. *Meat Science* 155: 79–84.

Bolumar, T., Skibsted, L.H., and Orlien, V. (2012). Kinetics of the formation of radicals in meat during high pressure processing. *Food Chemistry* 134 (4): 2114–2120.

Brandts, J.F., Oliveira, R.J., and Westort, C. (1970). Thermodynamics of protein denaturation. Effect of pressure on the denaturation on ribonuclease A. *Biochemistry* 9 (4): 1038–1047.

Briviba, K., Gräf, V., Walz, E. et al. (2016). Ultra high pressure homogenization of almond milk: Physico-chemical and physiological effects. *Food Chemistry* 192: 82–89.

Buckow, R., Sikes, A., and Tume, R. (2013). Effect of high pressure on physicochemical properties of meat. *Critical Reviews in Food Science and Nutrition* 53 (7): 770–786.

Calligaris, S., Foschia, M., Bartolomeoli, I. et al. (2012). Study on the applicability of high-pressure homogenization for the production of banana juices. *LWT – Food Science and Technology* 45 (1): 117–121.

Carlez, A., Veciana-Nogues, T., and Cheftel, J.-C. (1995). Changes in colour and myoglobin of minced beef meat due to high pressure processing. *LWT – Food Science and Technology* 28 (5): 528–538.

Carullo, D., Abera, B.D., Casazza, A.A. et al. (2018). Effect of pulsed electric fields and high pressure homogenization on the aqueous extraction of intracellular compounds from the microalgae Chlorella vulgaris. *Algal Research* 31: 60–69.

Cava, R., Ladero, L., González, S. et al. (2009). Effect of pressure and holding time on colour, protein and lipid oxidation of sliced dry-cured Iberian ham and loin during refrigerated storage. *Innovative Food Science & Emerging Technologies* 10 (1): 76–81.

Chandrapala, J. and Leong, T. (2015). Ultrasonic processing for dairy applications: recent advances. *Food Engineering Reviews* 7 (2): 143–158.

Chantapakul, T., Lv, R., Wang, W. et al. (2019). Manothermosonication: inactivation of Escherichia coli and Staphylococcus aureus. *Journal of Food Engineering* 246: 16–24.

Cheah, P.B. and Ledward, D.A. (1997). Catalytic mechanism of lipid oxidation following high pressure treatment in pork fat and meat. *Journal of Food Science* 62 (6): 1135–1139.

Cheftel, J.C. and Culioli, J. (1997). Effects of high pressure on meat: a review. *Meat Science* 46 (3): 211–236.

Chicón, R., Belloque, J., Alonso, E., and López-Fandiño, R. (2009). Antibody binding and functional properties of whey protein hydrolysates obtained under high pressure. *Food Hydrocolloids* 23 (3): 593–599.

Chura-Chambi, R.M., Genova, L.A., Affonso, R., and Morganti, L. (2008). Refolding of endostatin from inclusion bodies using high hydrostatic pressure. *Analytical Biochemistry* 379 (1): 32–39.

Ciron, C.I.E., Gee, V.L., Kelly, A.L., and Auty, M.A.E. (2010). Comparison of the effects of high-pressure microfluidization and conventional homogenization of milk on particle size, water retention and texture of non-fat and low-fat yoghurts. *International Dairy Journal* 20 (5): 314–320.

Clark, A.H. (1991). Structural and mechanical properties of biopolymer gels. In: *Food Polymers, Gels and Colloids* (ed. E. Dickinson), 322–338. Cambridge: The Royal Society of Chemistry.

Cortés-Muñoz, M., Chevalier-Lucia, D., and Dumay, E. (2009). Characteristics of submicron emulsions prepared by ultra-high pressure homogenisation: effect of chilled or frozen storage. *Food Hydrocolloids* 23 (3): 640–654.

Cruz, R.M., Vieira, M.C., and Silva, C.L. (2006). Effect of heat and thermosonication treatments on peroxidase inactivation kinetics in watercress (Nasturtium officinale). *Journal of Food Engineering* 72 (1): 8–15.

Cruz, N., Capellas, M., Hernández, M. et al. (2007). Ultra high pressure homogenization of soymilk: microbiological, physicochemical and microstructural characteristics. *Food Research International* 40 (6): 725–732.

Cruz, N.S., Capellas, M., Jaramillo, D.P. et al. (2009). Soymilk treated by ultra high-pressure homogenization: acid coagulation properties and characteristics of a soy-yogurt product. *Food Hydrocolloids* 23 (2): 490–496.

Datta, N. and Tomasula, P.M. (2015). *Emerging Dairy Processing Technologies: Opportunities for the Dairy Industry*. Wiley Blackwell.

Datta, N., Hayes, M.G., Deeth, H.C., and Kelly, A.L. (2005). Significance of frictional heating for effects of high pressure homogenisation on milk. *Journal of Dairy Research* 72 (4): 393–399.

Delgado, A., Kulisiewicz, L., Rauh, C., and Benning, R. (2010). Basic aspects of phase changes under high pressure. *Annals of the New York; Academy of Sciences* 1189 (1): 16–23.

Devi, A.F., Buckow, R., Hemar, Y., and Kasapis, S. (2013). Structuring dairy systems through high pressure processing. *Journal of Food Engineering* 114 (1): 106–122.

Diels, A.M. and Michiels, C.W. (2006). High-pressure homogenization as a non-thermal technique for the inactivation of microorganisms. *Critical Reviews in Microbiology* 32 (4): 201–216.

Domínguez Avila, J.A., Wall Medrano, A., Ruiz Pardo, C.A. et al. (2018). Use of nonthermal technologies in the production of functional beverages from vegetable ingredients to preserve heat-labile phytochemicals. *Journal of Food Processing and Preservation* 42 (2): e13506.

Donsì, F., Ferrari, G., Lenza, E., and Maresca, P. (2009). Main factors regulating microbial inactivation by high-pressure homogenization: operating parameters and scale of operation. *Chemical Engineering Science* 64 (3): 520–532.

El Kantar, S., Boussetta, N., Lebovka, N. et al. (2018). Pulsed electric field treatment of citrus fruits: improvement of juice and polyphenols extraction. *Innovative Food Science & Emerging Technologies* 46: 153–161.

Federighi, M., Vidon, M., Mescle, J., and Pilet, M. (1995). Traitement hautes pressions et denrées alimentaires: Revue bibliographique: 2ème partie: Effets des hautes pressions sur les microorganismes et applications aux denrées alimentaires. *Microbiologie, Aliments, Nutrition* 13 (3): 225–239.

Felipe, X., Capellas, M., and Law, A.J. (1997). Comparison of the effects of high-pressure treatments and heat pasteurization on the whey proteins in goat's milk. *Journal of Agricultural and Food Chemistry* 45 (3): 627–631.

Fernández-Ávila, C. and Trujillo, A.J. (2016). Ultra-high pressure homogenization improves oxidative stability and interfacial properties of soy protein isolate-stabilized emulsions. *Food Chemistry* 209: 104–113.

Fernández-Ávila, C., Escriu, R., and Trujillo, A.J. (2015). Ultra-high pressure homogenization enhances physicochemical properties of soy protein isolate-stabilized emulsions. *Food Research International* 75: 357–366.

Ferrario, M., Alzamora, S.M., and Guerrero, S. (2015). Study of the inactivation of spoilage microorganisms in apple juice by pulsed light and ultrasound. *Food Microbiology* 46: 635–642.

Ferstl, P., Gillig, S., Kaufmann, C. et al. (2010). Pressure-induced phase transitions in triacylglycerides. *Annals of the New York; Academy of Sciences* 1189 (1): 62–67.

Ferstl, P., Eder, C., Ruß, W., and Wierschem, A. (2011). Pressure-induced crystallization of triacylglycerides. *High Pressure Research* 31 (2): 339–349.

Gabrić, D., Barba, F., Roohinejad, S. et al. (2018). Pulsed electric fields as an alternative to thermal processing for preservation of nutritive and physicochemical properties of beverages: a review. *Journal of Food Process Engineering* 41 (1): e12638.

Gagneten, M., Leiva, G., Salvatori, D. et al. (2019). Optimization of pulsed electric field treatment for the extraction of bioactive compounds from blackcurrant. *Food and Bioprocess Technology* https://doi.org/10.1007/s11947-11019-02283-11941.

Garud, S.R., Priyanka, B.S., Negi, P.S., and Rastogi, N.K. (2017). Effect of thermosonication on bacterial count in artificially inoculated model system and natural microflora of sugarcane juice. *Journal of Food Processing and Preservation* 41 (2): e12813.

Gervilla, R., Ferragut, V., and Guamis, B. (2001). High hydrostatic pressure effects on color and milk-fat globule of Ewe's milk. *Journal of Food Science* 66 (6): 880–885.

Gómez, B., Munekata, P.E., Gavahian, M. et al. (2019). Application of pulsed electric fields in meat and fish processing industries: an overview. *Food Research International* 123: 90–105.

Gross, M. and Jaenicke, R. (1994). Proteins under pressure. *European Journal of Biochemistry* 221 (2): 617–630.

Grossi, A., Gkarane, V., Otte, J.A. et al. (2012). High pressure treatment of brine enhanced pork affects endopeptidase activity, protein solubility, and peptide formation. *Food Chemistry* 134 (3): 1556–1563.

Guignon, B., Aparicio, C., and Sanz, P.D. (2009). Volumetric properties of sunflower and olive oils at temperatures between 15 and 55 °C under pressures up to 350 MPa. *High Pressure Research* 29 (1): 38–45.

Guyon, C., Meynier, A., and de Lamballerie, M. (2016). Protein and lipid oxidation in meat: a review with emphasis on high-pressure treatments. *Trends in Food Science & Technology* 50: 131–143.

Haji-Moradkhani, A., Rezaei, R., and Moghimi, M. (2018). Optimization of pulsed electric field-assisted oil extraction from cannabis seeds. *Journal of Food Process Engineering* 2019: e13028.

Hayes, M.G., Fox, P.F., and Kelly, A.L. (2005). Potential applications of high pressure homogenisation in processing of liquid milk. *Journal of Dairy Research* 72 (01): 25–33.

He, J.-S. and Ruan, K. (2009). Kinetics of phase separation during pressure-induced gelation of a whey protein isolate. *Food Hydrocolloids* 23 (7): 1729–1733.

He, J.-S., Azuma, N., and Yang, H. (2010). Effects of pH and ionic strength on the rheology and microstructure of a pressure-induced whey protein gel. *International Dairy Journal* 20 (2): 89–95.

He, Z., Huang, Y., Li, H. et al. (2012). Effect of high-pressure treatment on the fatty acid composition of intramuscular lipid in pork. *Meat Science* 90 (1): 170–175.

Hendrickx, M., Ludikhuyze, L., Van den Broeck, I., and Weemaes, C. (1998). Effects of high pressure on enzymes related to food quality. *Trends in Food Science & Technology* 9 (5): 197–203.

Heremans, K. (1982). High pressure effects on proteins and other biomolecules. *Annual Review of Biophysics and Bioengineering* 11 (1): 1–21.

Heremans, K. (2001). The effects of high pressure on biomaterials. In: *Ultra High Pressure Treatments of Foods*, Food Engineering Series (eds. M.E.G. Hendrickx, D. Knorr, L. Ludikhuyze, et al.), 23–51. New York;: Springer.

Hernández, A. and Harte, F.M. (2008). Manufacture of acid gels from skim milk using high-pressure homogenization. *Journal of Dairy Science* 91 (10): 3761–3767.

Hettiarachchi, C.A., Corzo-Martínez, M., Mohan, M.S., and Harte, F.M. (2018). Enhanced foaming and emulsifying properties of high-pressure-jet-processed skim milk. *International Dairy Journal* 87: 60–66.

Hiperbaric. (2018). Beverage products consolidates as the fastest growing HPP sector in the world. https://www.hiperbaric.com/en/beverage-products-consolidates-as-the-fastest-growing-hpp-sector-in-the-world (accessed 2 February 2020).

Houben, K., Jamsazzadeh Kermani, Z., Van Buggenhout, S. et al. (2013). Thermal and high-pressure stability of pectinmethylesterase, polygalacturonase, β-Galactosidase and α-Arabinofuranosidase in a tomato matrix: towards the creation of specific endogenous enzyme populations through processing. *Food and Bioprocess Technology* 6 (12): 3368–3380.

Huang, H.-W., Hsu, C.-P., Yang, B.B., and Wang, C.-Y. (2013). Advances in the extraction of natural ingredients by high pressure extraction technology. *Trends in Food Science & Technology* 33 (1): 54–62.

Huang, W., Feng, Z., Aila, R. et al. (2019). Effect of pulsed electric fields (PEF) on physicochemical properties, β-carotene and antioxidant activity of air-dried apricots. *Food Chemistry* 291: 253–262.

Huppertz, T., Kelly, A.L., and Fox, P.F. (2002). Effects of high pressure on constituents and properties of milk. *International Dairy Journal* 12 (7): 561–572.

Jabbar, S., Abid, M., Hu, B. et al. (2015). Exploring the potential of thermosonication in carrot juice processing. *Journal of Food Science and Technology* 52 (11): 7002–7013.

Jaeger, P. and Eggers, R. (2005). Liquid–liquid interphases at high pressures in presence of compressible fluids. *Thermochimica Acta* 438 (1): 16–21.

Jaeschke, D.P., Mercali, G.D., Marczak, L.D.F. et al. (2019). Extraction of valuable compounds from Arthrospira platensis using pulsed electric field treatment. *Bioresource Technology* 283: 207–212.

Janositz, A., Noack, A.-K., and Knorr, D. (2011). Pulsed electric fields and their impact on the diffusion characteristics of potato slices. *LWT – Food Science and Technology* 44 (9): 1939–1945.

Jolie, R.P., Christiaens, S., De Roeck, A. et al. (2012). Pectin conversions under high pressure: implications for the structure-related quality characteristics of plant-based foods. *Trends in Food Science & Technology* 24 (2): 103–118.

Kahraman, O., Lee, H., Zhang, W., and Feng, H. (2017). Manothermosonication (MTS) treatment of apple-carrot juice blend for inactivation of Escherichia coli 0157: H7. *Ultrasonics Sonochemistry* 38: 820–828.

Kanno, C., Uchimura, T., Hagiwara, T. et al. (1998). Effect of hydrostatic pressure on the physicochemical properties of bovine milk fat globules and the milk fat globule membrane. In: *High Pressure Food Science, Bioscience and Chemistry* (ed. N. Isaacs), 182–192. Woodhead Publishing.

Kiang, W.S., Bhat, R., Rosma, A., and Cheng, L.H. (2013). Effects of thermosonication on the fate of E scherichia coli O157: H7 and Salmonella E nteritidis in mango juice. *Letters in Applied Microbiology* 56 (4): 251–257.

Kolakowski, P., Reps, A., and Babuchowski, A. (1998). Characteristics of pressurized ripened cheeses. *Polish Journal of Food and Nutrition Sciences* 7 (3): 473–483.

Ko⊠ciesza, R., Tefelski, D.B., Ptasznik, S. et al. (2012). A study of the high pressure phase transition of diacylglycerol oil by means of light transmission and scattering. *High Pressure Research* 32 (2): 323–329.

Koshani, R., Ziaee, E., Niakousari, M., and Golmakani, M.T. (2015). Optimization of thermal and thermosonication treatments on pectin methyl esterase inactivation of sour orange juice (Citrus aurantium). *Journal of Food Processing and Preservation* 39 (6): 567–573.

Kumari, B., Tiwari, B.K., Hossain, M.B. et al. (2018). Recent advances on application of ultrasound and pulsed electric field technologies in the extraction of bioactives from agro-industrial by-products. *Food and Bioprocess Technology* 11 (2): 223–241.

Kumari, B., Tiwari, B.K., Walsh, D. et al. (2019). Impact of pulsed electric field pre-treatment on nutritional and polyphenolic contents and bioactivities of light and dark brewer's spent grains. *Innovative Food Science & Emerging Technologies* 54: 200–210.

Lafarga, T. (2019). Effect of microalgal biomass incorporation into foods: nutritional and sensorial attributes of the end products. *Algal Research* 41: 101566.

Lafarga, T., Colás-Medà, P., Abadias, M. et al. (2019a). Strategies to reduce microbial risk and improve quality of fresh and processed strawberries: a review. *Innovative Food Science & Emerging Technologies* 52: 197–212.

Lafarga, T., Rodríguez-Roque, M.J., Bobo, G. et al. (2019b). Effect of ultrasound processing on the bioaccessibility of phenolic compounds and antioxidant capacity of selected vegetables. *Food Science and Biotechnology* 28: 1713–1721.

Lafarga, T., Ruiz-Aguirre, I., Abadias, M. et al. (2019c). Effect of thermosonication on the bioaccessibility of antioxidant compounds and the microbiological, physicochemical, and nutritional quality of an anthocyanin-enriched tomato juice. *Food and Bioprocess Technology* 12 (1): 147–157.

Lammerskitten, A., Wiktor, A., Siemer, C. et al. (2019). The effects of pulsed electric fields on the quality parameters of freeze-dried apples. *Journal of Food Engineering* 252: 36–43.

Law, A.J., Leaver, J., Felipe, X. et al. (1998). Comparison of the effects of high pressure and thermal treatments on the casein micelles in goat's milk. *Journal of Agricultural and Food Chemistry* 46 (7): 2523–2530.

Lee, E.-J., Kim, Y.-J., Lee, N.-H. et al. (2007). Differences in properties of myofibrillar proteins from bovine semitendinosus muscle after hydrostatic pressure or heat treatment. *Journal of the Science of Food and Agriculture* 87 (1): 40–46.

Lee, S.-H., Lefèvre, T., Subirade, M., and Paquin, P. (2009). Effects of ultra-high pressure homogenization on the properties and structure of interfacial protein layer in whey protein-stabilized emulsion. *Food Chemistry* 113 (1): 191–195.

Li, Q., Wu, Q.-Y., Jiang, W. et al. (2019). Effect of pulsed electric field on structural properties and digestibility of starches with different crystalline type in solid state. *Carbohydrate Polymers* 207: 362–370.

Liu, H.-H. and Kuo, M.-I. (2016). Ultra high pressure homogenization effect on the proteins in soy flour. *Food Hydrocolloids* 52: 741–748.

Liu, Z.W., Zeng, X.A., and Ngadi, M. (2018). Enhanced extraction of phenolic compounds from onion by pulsed electric field (PEF). *Journal of Food Processing and Preservation* 42 (9): e13755.

Liu, T., Burritt, D.J., and Oey, I. (2019). Understanding the effect of pulsed electric fields on multilayered solid plant foods: bunching onions (Allium fistulosum) as a model system. *Food Research International* 120: 560–567.

Ma, H.J., Ledward, D.A., Zamri, A.I. et al. (2007). Effects of high pressure/thermal treatment on lipid oxidation in beef and chicken muscle. *Food Chemistry* 104 (4): 1575–1579.

Mahendran, R., Ramanan, K.R., Barba, F.J. et al. (2019). Recent advances in the application of pulsed light processing for improving food safety and increasing shelf life. *Trends in Food Science & Technology* 88: 67–79.

Marcos, B., Kerry, J.P., and Mullen, A.M. (2010). High pressure induced changes on sarcoplasmic protein fraction and quality indicators. *Meat Science* 85 (1): 115–120.

Martínez, J.M., Delso, C., Angulo, J. et al. (2018). Pulsed electric field-assisted extraction of carotenoids from fresh biomass of Rhodotorula glutinis. *Innovative Food Science & Emerging Technologies* 47: 421–427.

Martínez-Monteagudo, S.I. and Saldaña, M.D.A. (2014). Modeling the retention kinetics of conjugated linoleic acid during high-pressure sterilization of milk. *Food Research International* 62: 169–176.

Marx, G., Moody, A., and Bermúdez-Aguirre, D. (2011). A comparative study on the structure of Saccharomyces cerevisiae under nonthermal technologies: high hydrostatic pressure, pulsed electric fields and thermo-sonication. *International Journal of Food Microbiology* 151 (3): 327–337.

McArdle, R., Marcos, B., Kerry, J.P., and Mullen, A. (2010). Monitoring the effects of high pressure processing and temperature on selected beef quality attributes. *Meat Science* 86 (3): 629–634.

McAuley, C.M., Singh, T.K., Haro-Maza, J.F. et al. (2016). Microbiological and physicochemical stability of raw, pasteurised or pulsed electric field-treated milk. *Innovative Food Science & Emerging Technologies* 38: 365–373.

Medina-Meza, I.G., Barnaba, C., and Barbosa-Cánovas, G.V. (2014). Effects of high pressure processing on lipid oxidation: a review. *Innovative Food Science & Emerging Technologies* 22: 1–10.

Merel-Rausch, E., Kulozik, U., and Hinrichs, J. (2007). Influence of pressure release rate and protein concentration on the formation of pressure-induced casein structures. *Journal of Dairy Research* 74 (3): 283–289.

Miloudi, K., Hamimed, A., Benmimoun, Y. et al. (2018). Intensification of essential oil extraction of the Marrubium vulgare using pulsed electric field. *Journal of Essential Oil-Bearing Plants* 21 (3): 811–824.

Mohan, M.S. (2014). Casein Micelles and their properties: polydispersity, association with vitamin A and effect of ultra-high pressure homogenization. PhD dissertation. University of Tennessee, Knoxville, Tennessee, USA.

Mohan, M., Ye, R., and Harte, F. (2016). Initial study on high pressure jet processing using a modified waterjet on physicochemical and rennet coagulation properties of pasteurized skim milk. *International Dairy Journal* 55: 52–58.

Mohan, M.S., O'Callaghan, T.F., Kelly, P., and Hogan, S.A. (2020). Milk fat: opportunities, challenges and innovation. *Critical Reviews in Food Science and Nutrition* 1–33.

Moradi, N. and Rahimi, M. (2019). Effect of ultrasound-and pulsed electric field-assisted enzymatic treatment on the recovery and quality of sunflower oil. *Separation Science and Technology* 54 (6): 1043–1054.

Moreno, H.M., Bargiela, V., Tovar, C.A. et al. (2015). High pressure applied to frozen flying fish (Parexocoetus brachyterus) surimi: effect on physicochemical and rheological properties of gels. *Food Hydrocolloids* 48: 127–134.

Mozhaev, V.V., Heremans, K., Frank, J. et al. (1994). Exploiting the effects of high hydrostatic pressure in biotechnological applications. *Trends in Biotechnology* 12 (12): 493–501.

Nadia, J., Ismail, M., Shahbaz, K., and Farid, M. (2019). The potential use of pulsed electric field to assist in polygodial extraction from horopito (Pseudowintera colorata) leaves. *Korean Journal of Chemical Engineering* 36 (2): 272–280.

Nowacka, M., Wiktor, A., Anuszewska, A. et al. (2019). The application of unconventional technologies as pulsed electric field, ultrasound and microwave-vacuum drying in the production of dried cranberry snacks. *Ultrasonics Sonochemistry* 56: 1–13.

Oey, I. (2016). Effects of high pressure on enzymes. In: *High Pressure Processing of Food: Principles, Technology and Applications* (eds. V.M. Balasubramaniam, G.V. Barbosa-Cánovas and H.L.M. Lelieveld), 391–431. New York;, NY: Springer.

Onwude, D.I., Hashim, N., Janius, R. et al. (2017). Non-thermal hybrid drying of fruits and vegetables: a review of current technologies. *Innovative Food Science & Emerging Technologies* 43: 223–238.

O'Reilly, C.E., O'Connor, P.M., Murphy, P.M. et al. (2000). The effect of exposure to pressure of 50 MPa on Cheddar cheese ripening. *Innovative Food Science & Emerging Technologies* 1 (2): 109–117.

Orlien, V., Hansen, E., and Skibsted, L.H. (2000). Lipid oxidation in high-pressure processed chicken breast muscle during chill storage: critical working pressure in relation to oxidation mechanism. *European Food Research and Technology* 211 (2): 99–104.

Orlien, V., Boserup, L., and Olsen, K. (2010). Casein micelle dissociation in skim milk during high-pressure treatment: effects of pressure, pH, and temperature. *Journal of Dairy Science* 93 (1): 12–18.

Paquin, P. (1999). Technological properties of high pressure homogenizers: the effect of fat globules, milk proteins, and polysaccharides. *International Dairy Journal* 9 (3–6): 329–335.

Parniakov, O., Barba, F.J., Grimi, N. et al. (2015). Pulsed electric field and pH assisted selective extraction of intracellular components from microalgae Nannochloropsis. *Algal Research* 8: 128–134.

Parniakov, O., Barba, F.J., Grimi, N. et al. (2016). Extraction assisted by pulsed electric energy as a potential tool for green and sustainable recovery of nutritionally valuable compounds from mango peels. *Food Chemistry* 192: 842–848.

Pataro, G., Carullo, D., Bobinaite, R. et al. (2017). Improving the extraction yield of juice and bioactive compounds from sweet cherries and their by-products by pulsed electric fields. *Chemical Engineering Transactions* 57: 1717–1722.

Peiró, S., Luengo, E., Segovia, F. et al. (2019). Improving polyphenol extraction from lemon residues by pulsed electric fields. *Waste and Biomass Valorization* 10 (4): 889–897.

Pereda, J., Ferragut, V., Quevedo, J.M. et al. (2009). Heat damage evaluation in ultra-high pressure homogenized milk. *Food Hydrocolloids* 23 (7): 1974–1979.

Puértolas, E., Koubaa, M., and Barba, F.J. (2016). An overview of the impact of electrotechnologies for the recovery of oil and high-value compounds from vegetable oil industry: energy and economic cost implications. *Food Research International* 80: 19–26.

Redondo, D., Venturini, M.E., Luengo, E. et al. (2018). Pulsed electric fields as a green technology for the extraction of bioactive compounds from thinned peach by-products. *Innovative Food Science & Emerging Technologies* 45: 335–343.

Ribas-Agustí, A., Martín-Belloso, O., Soliva-Fortuny, R., and Elez-Martínez, P. (2019). Enhancing hydroxycinnamic acids and flavan-3-ol contents by pulsed electric fields without affecting quality attributes of apple. *Food Research International* 121: 433–440.

Rivalain, N., Roquain, J., and Demazeau, G. (2010). Development of high hydrostatic pressure in biosciences: pressure effect on biological structures and potential applications in biotechnologies. *Biotechnology Advances* 28 (6): 659–672.

Rodrigo, D., Cortés, C., Clynen, E. et al. (2006). Thermal and high-pressure stability of purified polygalacturonase and pectinmethylesterase from four different tomato processing varieties. *Food Research International* 39 (4): 440–448.

Rostocki, A.J., Tarakowski, R., Kiełczyński, P. et al. (2013). The ultrasonic investigation of phase transition in olive oil up to 0.7 GPa. *Journal of the American Oil Chemists' Society* 90 (6): 813–818.

Royer, C.A. (2002). Revisiting volume changes in pressure-induced protein unfolding. *Biochimica et Biophysica Acta (BBA) – Protein Structure and Molecular Enzymology* 1595 (1): 201–209.

Saldo, J., Sendra, E., and Guamis, B. (1999). Evaluation of a method to accelerate ripening of goat cheeses using high hydrostatic pressure. *Alimentaria* 36 (308): 75–78.

Saldo, J., McSweeney, P., Sendra, E. et al. (2002). Proteolysis in caprine milk cheese treated by high pressure to accelerate cheese ripening. *International Dairy Journal* 12 (1): 35–44.

Sampedro, F. and Rodrigo, D. (2015). Pulsed electric fields (PEF) processing of milk and dairy products. In: *Emerging Dairy Processing Technologies: Opportunities for the Dairy Industry* (eds. N. Datta and P.M. Tomasula), 115–148. Sussex, UK: Wiley-Blackwell.

Samsu, S., Yassin, F.M., and Dayou, J. (2018). Extraction of protein from microalgae using low pulsed electric field. *ASM Science Journal* 11 (2): 106–111.

Sánchez-Rubio, M., Taboada-Rodríguez, A., Cava-Roda, R. et al. (2016). Combined use of thermo-ultrasound and cinnamon leaf essential oil to inactivate Saccharomyces cerevisiae in natural orange and pomegranate juices. *LWT – Food Science and Technology* 73: 140–146.

Santana, R.C., Perrechil, F.A., Sato, A.C.K., and Cunha, R.L. (2011). Emulsifying properties of collagen fibers: effect of pH, protein concentration and homogenization pressure. *Food Hydrocolloids* 25 (4): 604–612.

Scherer, D., Krust, D., Frey, W. et al. (2019). Pulsed electric field (PEF)-assisted protein recovery from Chlorella vulgaris is mediated by an enzymatic process after cell death. *Algal Research* 41: 101536.

Schoner, B.E., Bramlett, K.S., Guo, H., and Burris, T.P. (2005). Reconstitution of functional nuclear receptor proteins using high pressure refolding. *Molecular Genetics and Metabolism* 85 (4): 318–322.

Seefeldt, M.B., Ouyang, J., Froland, W.A. et al. (2004). High-pressure refolding of bikunin: efficacy and thermodynamics. *Protein Science* 13 (10): 2639–2650.

Sevenich, R. and Mathys, A. (2018). Continuous versus discontinuous ultra-high-pressure Systems for Food Sterilization with focus on ultra-high-pressure homogenization and high-pressure thermal sterilization: a review. *Comprehensive Reviews in Food Science and Food Safety* 17 (3): 646–662.

Sharma, P., Oey, I., Bremer, P., and Everett, D.W. (2018). Microbiological and enzymatic activity of bovine whole milk treated by pulsed electric fields. *International Journal of Dairy Technology* 71 (1): 10–19.

Shorstkii, I., Mirshekarloo, M.S., and Koshevoi, E. (2017). Application of pulsed electric field for oil extraction from sunflower seeds: electrical parameter effects on oil yield. *Journal of Food Process Engineering* 40 (1): e12281.

Sila, D.N., Smout, C., Satara, Y. et al. (2007). Combined thermal and high pressure effect on carrot pectinmethylesterase stability and catalytic activity. *Journal of Food Engineering* 78 (3): 755–764.

Siwach, R. and Kumar, M. (2012). Comparative study of thermosonication and thermal treatments on pectin methylesterase inactivation in mosambi juice. *Journal of Dairying Foods & Home Sciences* 31 (4): 290–296.

Smith, N., Burlakov, V., and Ramos, A.M. (2013). Mathematical modeling of the growth and coarsening of ice particles in the context of high pressure shift freezing processes. *The Journal of Physical Chemistry B* 117 (29): 8887–8895.

Subedi, L., Cho, K., Park, Y.U. et al. (2019). Sulforaphane-enriched broccoli sprouts Pretreated by pulsed electric fields reduces Neuroinflammation and ameliorates scopolamine-induced amnesia in mouse brain through its antioxidant ability via Nrf2-HO-1 activation. *Oxidative Medicine and Cellular Longevity* 2019: ID3549274.

Sukardi, S., Soeparman, S., Argo, B.D., and Irawan, Y.S. (2017). Optimization of patchouli oil (Pogostemon cablin, Benth) with steam distillation assisted by pulsed electric field via response surface methodology. *Journal of Engineering Science and Technology* 12 (8): 2106–2119.

Sulaiman, A., Soo, M.J., Farid, M., and Silva, F.V. (2015). Thermosonication for polyphenoloxidase inactivation in fruits: modeling the ultrasound and thermal kinetics in pear, apple and strawberry purees at different temperatures. *Journal of Food Engineering* 165: 133–140.

Tang, C.-H. and Ma, C.-Y. (2009). Effect of high pressure treatment on aggregation and structural properties of soy protein isolate. *LWT - Food Science and Technology* 42 (2): 606–611.

Tauscher, B. (1995). Pasteurization of food by hydrostatic high pressure: chemical aspects. *Zeitschrift für Lebensmittel-Untersuchung und -Forschung* 200 (1): 3–13.

Tefelski, D.B., Siegoczyński, R.M., Rostocki, A.J. et al. (2008). The investigation of the dynamics of the phase transformation in triolein and oleic acid under pressure. *Journal of Physics: Conference Series* 121 (14): 142004.

Thiebaud, M., Dumay, E., Picart, L. et al. (2003). High-pressure homogenisation of raw bovine milk. Effects on fat globule size distribution and microbial inactivation. *International Dairy Journal* 13 (6): 427–439.

Ting, E. (2010). High-pressure processing equipment fundamentals. In: *Nonthermal Processing Technologies for Food* (eds. H.Q. Zhang, G.V. Barbosa-Cánovas, V.M. Balasubramaniam, et al.), 20–27. Chichester, UK: Wiley Blackwell.

Tiwari, U. (2018). Production of fruit-based smoothies. In: *Fruit Juices: Extraction, Composition, Quality and Analysis* (eds. G. Rajauria and B.K. Tiwari), 261–278. MS, USA: Elsevier.

Toepfl, S., Heinz, V., and Knorr, D. (2006). Applications of pulsed electric fields technology for the food industry. In: *Pulsed Electric Fields Technology for the Food Industry* (eds. J. Raso and V. Heinz), 197–221. MA, USA: Springer.

Torres, E., González-M, G., Klotz, B., and Rodrigo, D. (2016). Effects of high hydrostatic pressure and temperature increase on Escherichia coli spp. and pectin methyl esterase inactivation in orange juice. *Food Science and Technology International* 22 (2): 173–180.

Tran, M., Roberts, R., Felix, T.L., and Harte, F.M. (2018). Effect of high-pressure-jet processing on the viscosity and foaming properties of pasteurized whole milk. *Journal of Dairy Science* 101 (5): 3887–3899.

Trujillo, A.J., Capellas, M., Saldo, J. et al. (2002). Applications of high-hydrostatic pressure on milk and dairy products: a review. *Innovative Food Science & Emerging Technologies* 3 (4): 295–307.

Trujillo, A.J., Ferragut, V., Juan, B. et al. (2016). Processing of dairy products utilizing high pressure. In: *High Pressure Processing of Food: Principles, Technology and Applications* (eds. V.M. Balasubramaniam, G.V. Barbosa-Cánovas and H.L.M. Lelieveld), 553–590. New York;, NY: Springer.

Tume, R.K., Sikes, A.L., and Smith, S.B. (2010). Enriching M. sternomandibularis with α-tocopherol by dietary means does not protect against the lipid oxidation caused by high-pressure processing. *Meat Science* 84 (1): 66–70.

Van der Plancken, I., Van Loey, A., and Hendrickx, M.E. (2007). Foaming properties of egg white proteins affected by heat or high pressure treatment. *Journal of Food Engineering* 78 (4): 1410–1426.

Velázquez-Estrada, R., Hernández-Herrero, M., Guamis-López, B., and Roig-Sagués, A. (2012). Impact of ultra high pressure homogenization on pectin methylesterase activity and microbial characteristics of orange juice: a comparative study against conventional heat pasteurization. *Innovative Food Science & Emerging Technologies* 13: 100–106.

Volkert, M., Puaud, M., Wille, H.-J., and Knorr, D. (2012). Effects of high pressure–low temperature treatment on freezing behavior, sensorial properties and air cell distribution in sugar rich dairy based frozen food foam and emulsions. *Innovative Food Science & Emerging Technologies* 13: 75–85.

Wang, C.-Y., Huang, H.-W., Hsu, C.-P., and Yang, B.B. (2016). Recent advances in food processing using high hydrostatic pressure technology. *Critical Reviews in Food Science and Nutrition* 56 (4): 527–540.

Welti-Chanes, J., Ochoa-Velasco, C., and Guerrero-Beltrán, J. (2009). High-pressure homogenization of orange juice to inactivate pectinmethylesterase. *Innovative Food Science & Emerging Technologies* 10 (4): 457–462.

Wiggers, S.B., Kröger-Ohlsen, M.V., and Skibsted, L.H. (2004). Lipid oxidation in high-pressure processed chicken breast during chill storage and subsequent heat treatment: effect of working pressure, packaging atmosphere and storage time. *European Food Research and Technology* 219 (2): 167–170.

Wiśniewski, R., Jerzejewski, J., Siegoczyński, R.M., and Tkacz, A. (1994). Volume changes of castor oil during its transformation to the high pressure phase. *High Pressure Research* 11 (6): 385–391.

Yagiz, Y., Kristinsson, H.G., Balaban, M.O. et al. (2009). Effect of high pressure processing and cooking treatment on the quality of Atlantic salmon. *Food Chemistry* 116 (4): 828–835.

Yajun, Z., Changmei, X., Susu, Z. et al. (2017). Effects of high intensity pulsed electric fields on yield and chemical composition of rose essential oil. *International Journal of Agricultural and Biological Engineering* 10 (3): 295–301.

Yan, W., Qiao, L., Gu, X. et al. (2010). Effect of high pressure treatment on the physico-chemical and functional properties of egg yolk. *European Food Research and Technology* 231 (3): 371–377.

Yang, R.-X., Li, W.-Z., Zhu, C.-Q., and Zhang, Q. (2009). Effects of ultra-high hydrostatic pressure on foaming and physical-chemistry properties of egg white. *Journal of Biomedical Science and Engineering* 2 (08): 617.

Ye, A., Anema, S.G., and Singh, H. (2004). High-pressure induced interactions between milk fat globule membrane proteins and skim milk proteins in whole milk. *Journal of Dairy Science* 87 (12): 4013–4022.

Yokoyama, H., Sawamura, N., and Motobayashi, N. (1992). Method for accelerating cheese ripening. European Patent Office application EP 0469857B1, filed 30 July 1991 and published 26 January 1994.

Zamora, A., Ferragut, V., Jaramillo, P.D. et al. (2007). Effects of ultra-high pressure homogenization on the cheese-making properties of milk. *Journal of Dairy Science* 90 (1): 13–23.

Zeece, M., Huppertz, T., and Kelly, A. (2008). Effect of high-pressure treatment on in-vitro digestibility of β-lactoglobulin. *Innovative Food Science & Emerging Technologies* 9 (1): 62–69.

Zocher, K., Banaschik, R., Schulze, C. et al. (2016). Comparison of extraction of valuable compounds from microalgae by atmospheric pressure plasmas and pulsed electric fields. *Plasma Medicine* 6: 273–302.

Zulkurnain, M., Maleky, F., and Balasubramaniam, V.M. (2016). High pressure processing effects on lipids Thermophysical properties and crystallization kinetics. *Food Engineering Reviews* 8 (4): 393–413.

11
Formulation for Food Intolerance

Iuliana Vintilă
Dunarea de Jos University, Galati, Romania

Summary

Food intolerances are food-related diseases largely spread throughout modern society. As general practice, the management of food intolerances involves elimination diets, total abstinance and elimination of the intolerance vector-food. The alteration of the local immune system existing in the small intestine has a positive correlation with celiac disease (CD) development. Food scientists' response to the increasing needs for gluten-free products is to create genetic-modified wheat varieties with gluten-free characteristics; to biotechnologically modify gluten by complete proteolysis during dough fermentation, in order to create gluten/gliadin-free products; to technologically modify food formula by gluten cereals substitution with pseudo-cereals (buckwheat, rice, tapioca, sago, pea, sorghum), and/or legumes (peas, potatoes, beans, lentils, lupines); to produce functional gluten-free foods based on gluten/gliadin removal and replacement with dietary powders (starch, polenta, flours of sorghum, soya, buckwheat, etc.); and the processing of pre-fermented or cooked paste (rice, buckwheat, peas), corn flakes, expended cereals, sauces/milks (soya), tofu, oilseeds (line, sunflower, pumpkin).

Lactose intolerance is a food-related disease caused by the incapacity of the human body to metabolize and assimilate one food component, lactose, because the enterocytes from the intestinal epithelium has a lactase (β-galactosidase) production/secretion deficiency. The processing solutions for manufacturing reduced/lactose-free milk are generally microfiltration, ultrafiltration, nanofiltration, and reverse osmosis systems. The foods tailored for specific intolerances should be compliant with the international regulations regarding the maximum content of the intolerance triggering-compound (food additive, a chemical component, or a contaminant).

Food Formulation: Novel Ingredients and Processing Techniques, First Edition.
Edited by Shivani Pathania and Brijesh K. Tiwari.

11.1 Introduction

Food intolerances are food-related diseases largely affecting modern society, manifested as structural-linked malabsorptions, with non-immune mediated reactions (pseudo-allergy) at ingestion of one food ingredient (lactose, gluten, caffeine, salicylates, histamines, fructose and other short-chain carbohydrates, sulphites, etc.) or functional intolerances, with immunological or non-immunological aggression triggers (Figure 11.1).

In 2009, the world population with one diagnosis of food intolerance was estimated as up to 20%, with a prevalence of 2–5% immunological-mediated hypersensitivity in the adult population and 5–10% in the child population (Zopf et al. 2009). Food intolerance is a hypersensitivity of the human body (digestive system, skin, eyes, etc.) of one or multiple food components, as a result of an immune system defence reaction, several hours after the vector-food is ingestion (two to eight hours).

The cause of food intolerance type I is a non-tolerated component, non-recognized as valuable/necessary by the defence system of the human body, which initiates and

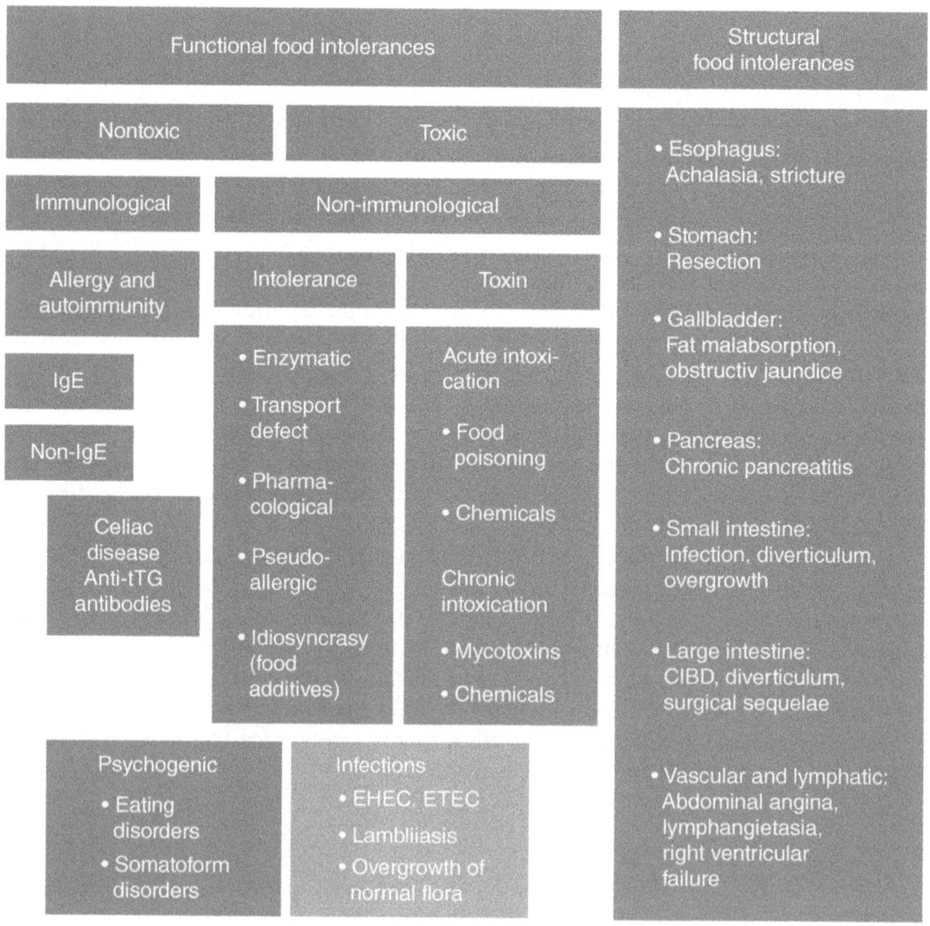

Figure 11.1 Types of food intolerances. *Source*: Zopf et al. 2009. © Deutscher Arzte-Verlag GmbH; permission conveyed through Copyright Clearance Center.

leads an immunological reaction of protection and rejection (inflammation, irritation, vasodilatation, diarrhoea).

The pseudo-allergies are type II food intolerances named food idiosyncrasies caused by food additives (flavours, sulphites, benzoate, vitamins, or pro-vitamin A, etc.), not associated with an immunologically cause but by triggering protection reactions such as atopic dermatitis herpetiformis (DH), chronic urticarial, eye inflammation, etc.

The causes of food intolerance type III are functionally- (enzymopathies) or structurally-induced malabsorption (anatomical and morphological alteration of the gastrointestinal tract) of one or multiple food components (usually lactose, gluten, or histamines), which create physiological discomfort (flatulence, abdominal pain, and diarrhoea) after the vector-food ingestion (one to three hours).

The food intolerance diagnosis, confirmation, and management involve complex medical investigation, including:

- serum analysis (serology): transglutaminase, anti-enterocyte antibodies, etc. used in cases of immune mediated reactions due to functional deficiencies (deficiency in lactase secretion, antibody deficiencies in IgA/IgE) or pre-existing clinical status (infection, lymphoma, mastocytosis, or tumours, etc.);
- imaging medical procedures (endoscopy, histopathology-intestinal/duodenal biopsy, radiology, and stool examinations); used in cases of structural intolerances manifested with gallstones, reflux esophagitis, or pancreatic insufficiency;
- monitoring of patient behaviour and well-being status after the dietary free/restricted diets.

11.2 Celiac Disease

Celiac disease (CD), also named gluten-induced enteropathy, is a food-related chronic disease of the atrophied jejune-duodenal intestine caused by the autoimmune-mediated local response of the human body defence system to dietary gluten.

The 'gluten-related disorders' named, generally CD, include the following types of medical diagnosis:

- DH;
- gluten ataxia (Al-Toma et al. 2019; Sapone et al. 2012);
- non-celiac gluten/wheat sensitivity (NCGS);
- CD (gluten-induced enteropathy);
- gluten/wheat allergy;
- inflammatory bowel diseases (IBD) (Casella et al. 2010).

The medical protocol approach of CD is usually diagnosis confirmation, by biopsy and serology tests, and dietary protocol in cases of positive results, under a strict medical dietary routine, as presented in Figure 11.2.

The prevalence of CD is estimated as approximately 1% of the worldwide population, mostly developed at a paediatric age (Lionetti et al. 2015).

The alteration of the local immune system existing in the small intestine by the artificial mode of childbirth, inadequate food formulation in childhood, viral and bacterial infections, and use of antibiotics, all show a positive correlation with CD development (Pagliari et al. 2015).

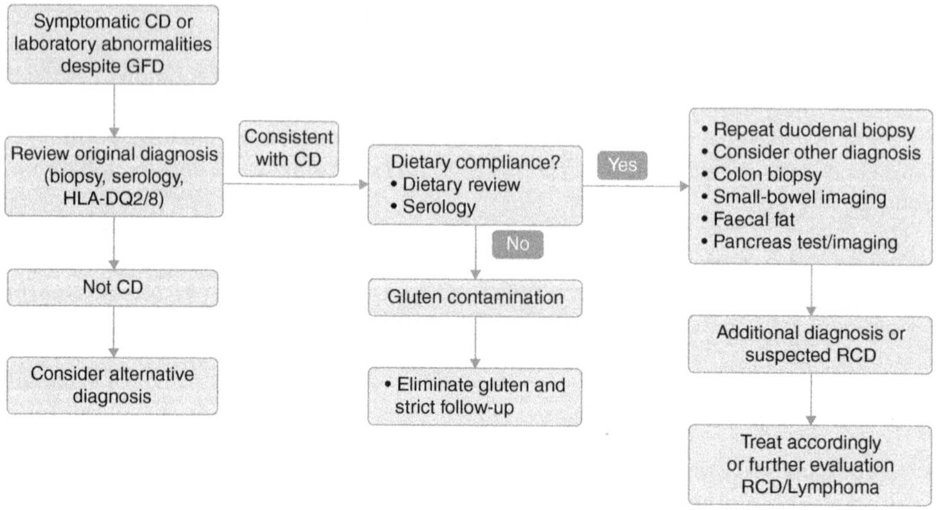

Figure 11.2 Medical management of CD. *Source*: Al-Toma et al. 2019. Reproduced with permission of SAGE Publications, Inc.

11.3 Celiac Disease Food Formulations, Technology, and Quality

Gluten is the main cereal protein fraction that gives dough its elasticity and represents approximatively 85% wheat proteins (Banu et al. 2010). The α-gliadin fraction from the cereal's gluten is involved in the CD, because of a small intestinal peptidase deficiency, with genetic determinants. The partial proteolysis of cereals gliadin generates CD toxic gliadin oligopeptides (α-9, α-6, α-2 gliadin), which have the potential to destroy the intestinal pylorus, migrate into the small intestine lumen and generate, in time, the enteropathy. The enteropathy is caused by the antigen role acted by the CD toxic gliadin oligopeptides, which activates the T-helper cells (lymphocytes) and induces lumen inflammation, and pylorus destruction with sclerotic tissue creation and, finally, necrosis.

The continual increase in gluten-related disorders/diseases creates a necessity for functional food products, tailored-made for this particular food intolerance (Gobbetti et al. 2018). From 2013 to 2015, the global gluten-free industry grew by 136% (Mintel 2015; Reilly 2016), reaching $11 billion in 2015 (Foschia et al. 2016) (Figure 11.3).

Gluten-free products are functional foods containing up to 20 mg/kg dietary gluten, according to the Codex Alimentarius Commission (2008) of the World Health Organization (WHO).

In the European Union, the Commission Implementing Regulation (EU) No. 828/2014 (2014/2016) set out the conditions under which foods claimed as 'gluten-free' or 'very-low gluten' have been effectively applied since 20 July 2016 (EU 2014/2016). The labels associating the EU with 'free-from' foods are presented in Figure 11.4.

The food science response to the increasing needs of gluten-free products was to create the following scientific and technological solutions:

- Genetic-modified wheat varieties with gluten-free characteristics;
- Biotechnologically modified gluten by complete proteolysis during dough fermentation, in order to create gluten-/gliadin-free products;

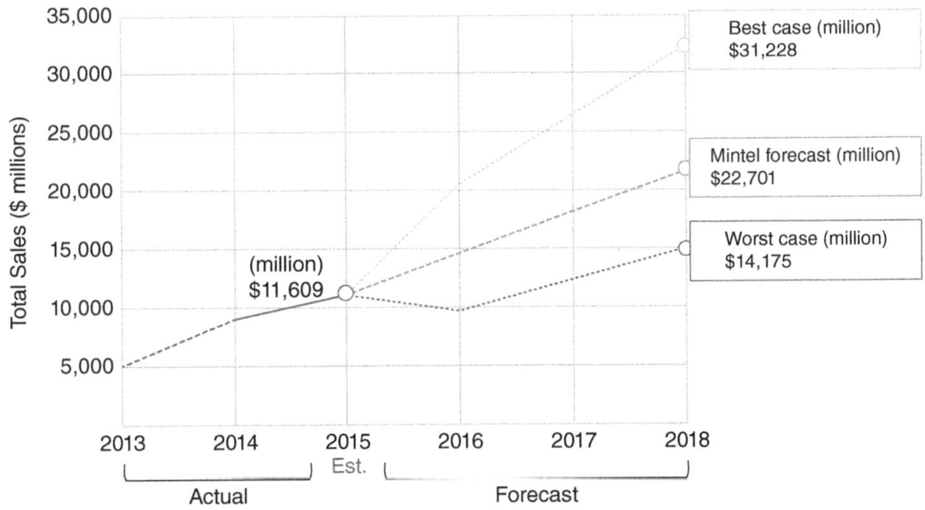

Figure 11.3 US Sales of gluten-free foods 2013–2018 (Mintel 2015; Foschia et al. 2016; SPINS, Information Resources, Inc. 2019).

Figure 11.4 The 'free-from' EU labels (EUFIC 2020).

- Technologically-modified food formula by gluten cereals substitution with pseudo-cereals (buckwheat, rice, tapioca, sago, pea, sorghum) and/or legumes (peas, potatoes, beans, lentils, lupines);
- Functional gluten-free foods based on gluten/gliadin removal and replacement with dietary powders (starch, polenta, flours of sorghum, soya, buckwheat, etc.), pre-fermented or cooked paste (rice, buckwheat, peas), corn flakes, expended cereals, sauces/milks (soya), tofu, oilseeds (line, sunflower, pumpkin);

Also, the gluten-free diet is correlated with medical/therapeutically treatment to compensate for the dietary gluten substitution, such as administration of supplements with glutathione, glucosamine, probiotics, and gliadin proteolysis enzymes (Figure 11.5).

The food science and technology act in the following engineering directions for gluten-free products development:

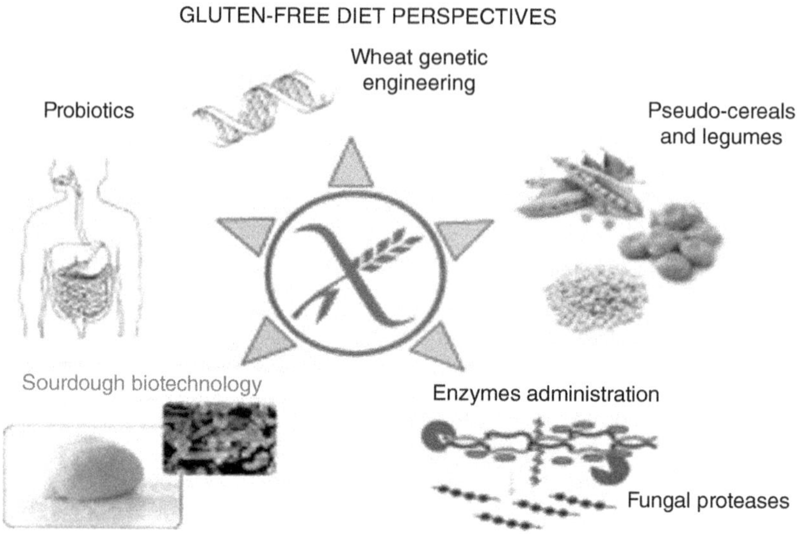

Figure 11.5 Gluten free diets. *Source*: Gobbetti et al. 2018. Reproduced with permission of Elsevier.

11.3.1 Food Formulation: Gluten Replacement and Fortification with Functional Ingredients

Legumes and pseudo-cereals are the principle protein replacement ingredients in gluten-free foods, used in their natural state (pseudo-cereals, seeds, legumes, and nuts) or technologically transformed (germinated, fermented, malting, gelatinized, granulated, expanded powders, sauces, pasta, etc.).

The formulations of gluten-free foods include the following groups of ingredients (Table 11.1):

- Cereals (corn, rice);
- Pseudo-cereals (amaranth, quinoa, millet, sorghum, chickpeas, buckwheat, peas, tapioca, sago);
- Oilseeds (soybean, line, flaxseed, sunflower, pumpkin);
- Legumes (peas, potatoes, beans, lentils, lupines);
- Combination of cereals with pseudo-cereals (soy, rice and corn) and/or oilseeds and legumes;
- Additives for improving water and gas-binding during dough formation (hydrocolloids such as hydroxypropyl methylcellulose cellulose, xanthan, guar gum), enzymes (transglutaminase, proteases, glucose oxidase, laccase);
- Fortification agents: animal proteins: milk (caseinate) or egg proteins (albumin), supplements of dietary fibres (*Psyllium* fibre, β-glucan, locust bean gum).
- Gluten-free bread production by substituting rice flour with carob flour and resistant starch offer fibre-rich formulations (Tsatsaragkou et al. 2014; Wronkowska et al. 2013).

Buckwheat flour improves the textural properties (Demirkesen et al. 2010) and chestnut flour improves the functional and technological properties of dough as well as the colour and flavour properties of breads (Figures 11.6 and 11.7).

Table 11.1 The ingredients commonly used in commercial gluten-free bread formulations.

Bread Commodities	Ingredient	Amount (%)[a]
Flours	Rice	59.3
	Maize	40.7
	Buckwheat	22.2
	Whole grain maize	18.5
	Tapioca	11.1
	Potato	7.4
	Millet	7.4
	Quinoa	3.7
Starches	Maize	88.9
	Potato	70.4
	Rice	59.3
	Tapioca	59.3
	Whole grain maize	18.5
	Wheat	3.0
Proteins	Egg white	63.0
	Pea	25.9
	Soya	18.5
	Whey	7.4
	Dried skim milk	7.4
	Milk	3.7
Hydrocolloids	HPMC	70.4
	Cellulose	40.7
	Xanthan	29.6
	Guar gum	25.9
	SCMC	11.1
	Agar agar	7.4
Fibres	Psyllium	74.1
	Rice bran extract	18.5
	Millet flakes	11.1
	Rice bran	7.4
	Soya bran	3.7
	Apple fibre	3.7
	Flaxseeds	3.7
Sourdoughs	Rice flour	22.2
	Maize flour	11.1
	Fermented quinoa	7.4

Source: Foschia et al. (2016). Reproduced with permission of Elsevier.
[a] Based on 27 bread samples from leading companies producing gluten-free foods.

Over 400 different ingredients with gluten-replacing abilities were included in the Italian biscuit commercial formulations, the most usual including maize starch, rice flour, maize flour, potato starch, guar gum, eggs, sugar, sunflower oil, and butter.

do Nascimento et al. (2013) found that rice flour, cassava fecula, natural maize, soy flour, and rice were the most frequent ingredients with gluten-replacing abilities in Brazilian gluten-free formulations such as baked products, granola, and pasta.

Figure 11.6 Gluten-free breads. *Source*: Mert et al. 2015. Reproduced with permission of Elsevier.

Maize starch and rice flours are present in 80.5 and 78.0%, respectively, of biscuits, whereas more nutritious flours, like buckwheat, teff, and millet, are present in less than 5% of biscuit formulations.

Legume flours are used for the purpose of increasing protein content. The content of minerals was generally higher in gluten-free formulations from buckwheat, millet, chickpea, oats, amaranth, teff, and quinoa than in products made with widely-used raw materials such rice, maize, and potato (Rybicka and Gliszczyńska-Świgło 2017).

In order to improve the nutritional quality (i.e. increase the phenol content with higher antioxidant activity) of sorghum, pearl millet. and so-based biscuits (Omoba et al. 2015), adding sourdough to the formulation is a good technological solution. Also, better sensory characteristics were obtained, with an increase in sour taste and rich fermented flavour.

Otherwise, the non-wheat flours may lead to a decay of sensory properties of biscuits which became harder and darker, and dry and with a sandy mouthfeel, also associated with unpleasant appearance and taste (Schober et al. 2003).

The refined flours are usually preferred in gluten-free biscuit formulation, because the addition of bran leads to a sandy product, darker and difficult to chew (Duta and Culetu 2015).

The phenol compounds and the higher ash content in non-refined flours may be responsible for the darker colour and the anti-nutritional compounds may give a bitter aftertaste (Di Cairano et al. 2018).

The presence of revani flour in gluten-free formulations offers optimal technologically results (Table 11.2, Figure 11.8). Also, the buckwheat flour improves the biscuit's flavour (Torbica et al. 2011).

Ten percent β-conglycinin concentrate (β-CC) was able to create a gluten-like, similar structure, acting as a structuring agent and protein quality promoter (Table 11.3) in gluten-free formulations (Espinosa-Ramírez et al. 2018).

11.3.2 Novel Food Technology Application

A gluten-free formulation shows particular differences in nutritional composition, sensory characteristics, and technological behaviour in comparison to natural gluten foods, in principal as follows:

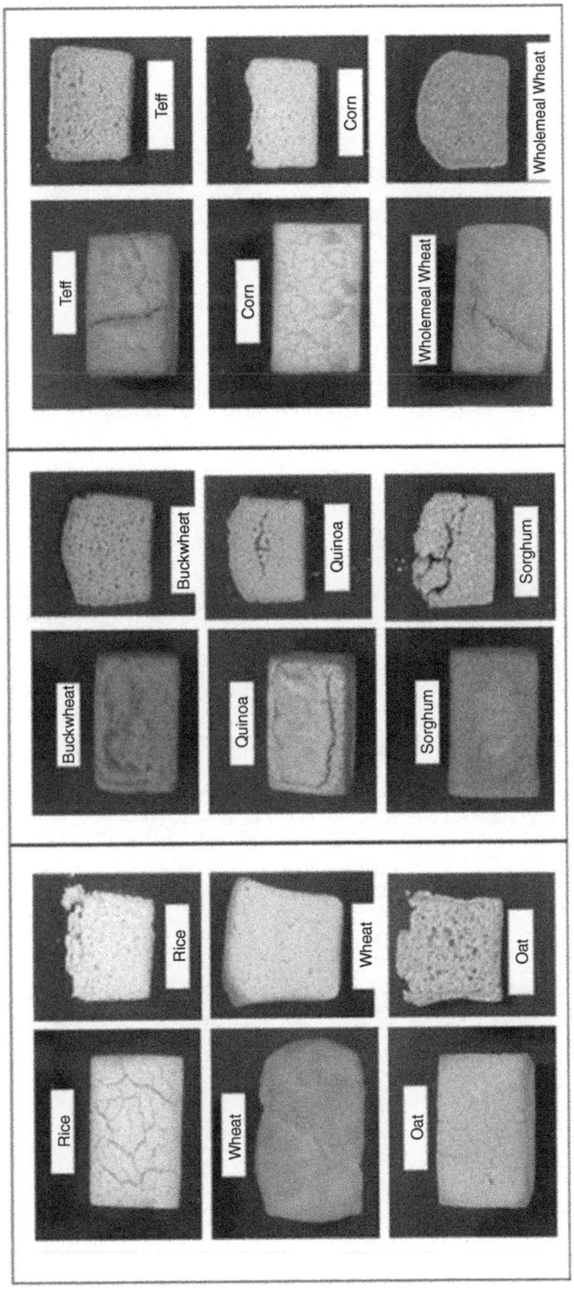

Figure 11.7 Gluten-free dough and crumb structure compared with conventional wheat flour bread. *Source:* Foschia et al. 2016. Reproduced with permission of Elsevier.

Table 11.2 Gluten-free revani formulation with four flour combinations.

	Flour A	Flour B	Flour C	Flour D
Corn flour (CF)	62.5		37.5	37.5
Rice flour (RF)	37.5	62.5		
Potato flour (PF)			31.25	31.25
Corn starch (CS)		37.5		31.25
Tapioca starch (TS)			31.25	
Flour Type	Protein Source			
	Soy (SP)	Pea (PP)	Transglutaminase (TG)	
Flour A1, B1, C1, D1				
Flour A2, B2, C2, D2	5			
Flour A3, B3, C3, D3		5		
Flour A4, B4, C4, D4	5		0.2 IU	
Flour A5, B5, C5, D5		5	0.2 IU	

Source: Yildirim et al. 2018. Reproduced with permission of Elsevier.

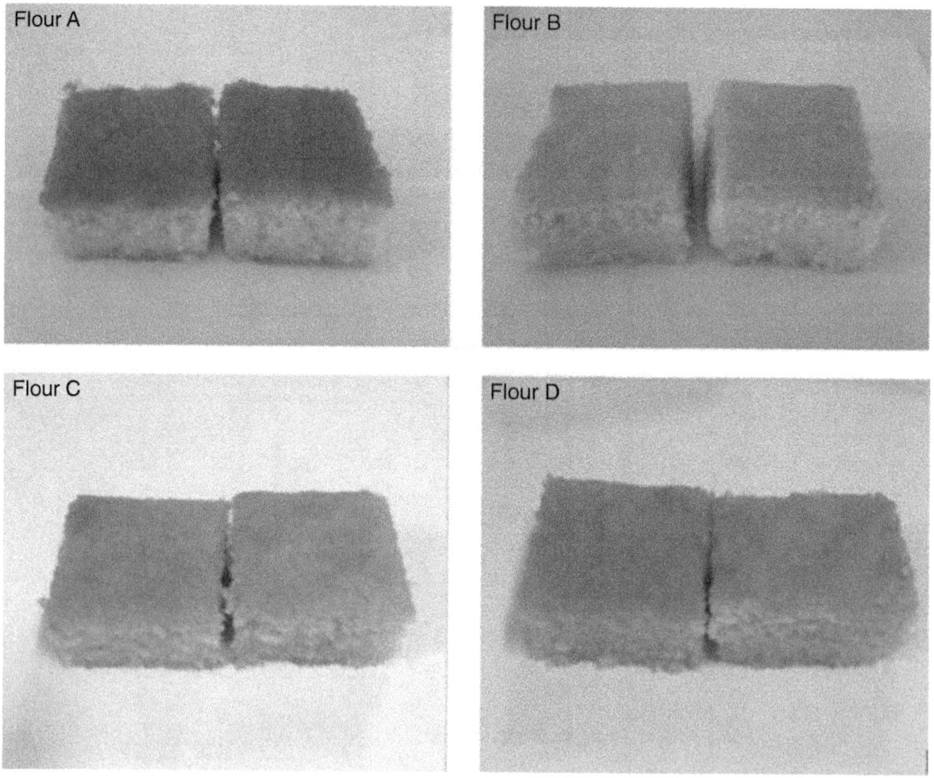

Figure 11.8 Gluten-free revani with four flour combinations. *Source*: Yildirim et al. 2018. Reproduced with permission of Elsevier.

Table 11.3 Protein nutritional quality of gluten-free formulations.

Composition	**Formulation**			**WHO/ FAO**[d]
	100% Rice flour	**90% Rice flour +10% Gluten**	**90% Rice flour+10% βCC**	
Protein (%, db)	7.81 ± 0.1[c]	1531 ± 0.1[b]	16 ± 0.1[a]	
	Amino acid content (g/100 g protein)			
Alanine	5.78	4.45	4.66	
Arginine	8.06	5.70	8.40	
Aspartic + Asparagine	9.14	6.01	11.78	
Glutamic+ Glutamine	17.88	26.83	20.36	
Glycine	4.70	3.97	3.79	
Histidine	2.42	2.26	2.42	1.8
Isoleucine	4.44	4.15	4.77	3.1
Leucine	8.60	7.73	8.24	6.3
Lysine	4.30	2.99	5.91	5.2
Methionine + Cysteine	2.82 + 2.42	2.15 + 2.27	1.72 + 1.52	2.6
Phenylalanine + Tyrosine	5.65 + 3.23	5.51 + 3.41	5.85 + 3.31	4.6
Proline	4.30	8.57	4.74	
Serine	4.84	4.25	4.36	
Threonine	3.63	3.05	2.96	2.7
Tryptophan	1.08	1.06	0.85	0.74
Valine	6.32	5.19	5.10	4.2
Limiting amino acid score	0.83 (lys)	0.57 (lys)	1.1[d]	
PDCAAS	0.65	0.47	0.78	

Source: Espinosa-Ramírez et al. (2018). Reproduced with permission of Elsevier.
[a] βCC = β-conglycinin concentrate; PDCAAS= Protein digestibility corrected amino acid score.
[b] Values are means and standard deviations of three replicates; means with different letters within the same parameter differ significantly ($P<0.05$).
[c] Suggested requirements for 1–2 years infants. Values are expressed in g/100 g protein.
[d] Value was truncated to 1.0 for the calculation of PDCAAS as recommended by WHO/FAO/UNU (2007).

- higher starch content, generally proportional with the removed gliadin/gluten;
- lower protein content, if there is no partial/total gluten replacement with other gluten-free protein by fortification;
- higher water content (Hager et al. 2012) due to the replacement of gluten protein with a more water-binding and holding compound, starch, in general;
- dough with less elasticity, specific volume, and regular alveolar structure, because the gluten fraction is the principal component involved in dough/bread structure and volume;

- dough less palatable in terms of flavour, taste, texture, and with decreased potential of flavour and crust colour formation under Maillard reactions;
- nutritional value decreased because the gluten replacement eliminates an important source of vegetable protein with balanced dietary amino acids.

Consequently, the food technologies need to change not only the food formulation but also the industrial processing routine in order to create valuable commercial gluten-free products (Table 11.4).

The non-gluten foods are created with the following modification of the industrial processing routes:

- use of non-glutenic genetically-modified grains, with active immunologically gliadin-free characteristics, sequences not identified in the grain genome (Mulder et al. 2001);
- germinated gluten-free grains, cereals, pseudo-cereals;
- use of sourdough endogenous microbiota (15–20% inoculum): lactic acid bacteria: specific and non-specific species of lactic acid bacteria – *Lactobacillus fermentum*, *Lactobacillus plantarum*, *Lactobacillus gallinarum*, *Lactobacillus graminis*, *Lactobacillus sakei*, *Pediococcus pentosaceus* (De Vuyst et al. 2014; Ruehmkorf et al. 2012; Wolter et al. 2014d) with the following positive results:
  - decreasing the glycaemia index (Capriles and Arêas 2016), probably because of the release of satiety hormones (Lundin et al. 2008) and the chemical changes that occur during sourdough fermentation (Östman 2003);
  - decreasing the concentration of anti-nutritional factors (acidification during fermentation due to the pH decreasing and increasing of β-glycosidase activity).
  - increasing the protein bio-availability (Yousif and El Tinay 2000) and the intake of essential amino acids (Gobbetti et al. 2007) such lysine in oats, rice, millet, and maize (Lee et al. 1999), lysine, methionine, and tryptophan (Nanson and Fields 1984).
  - increasing the bio-availability of iron, zinc, calcium, and magnesium by phytate enzyme degradation.
  - vitamins fortification: folate from yeast cells (Keagy et al. 1975), vitamin B_{12} produced by *Lactobacillus reuteri* (Taranto et al. 2003).
  - use of biotechnologically fermented sourdough (Table 11.5) with fungal proteases, in the first stage, and bacteria proline-specific peptidases for complete proteolysis (Rizzello et al. 2009) and transglutaminase for texture improvement (Babiker 2000; Alp and Bilgiçli 2008; Gaspar and de Góes-Favoni 2015).
    - ☐ pre-fermentation and freezing of dough;
    - ☐ pre-fermentation with glutenase (Table 11.6);
    - ☐ fermentation with microbial glutenase (*Flavobacterium meningosepticum*, *Lactobacillus helveticus*, *Myxococcus xanthus*, *Rothia mucilaginosa*; *Pseudomonas aeruginosa*), *Aspergillus niger* (prolyl endoprotease, AN-PEP) or glutenase produced by *Rhizopertha dominica* (Scherf et al. 2016);
    - ☐ dough pre-baking and freezing.

Table 11.4 Food technology solutions for improving the shelf life and rheological characteristics of gluten-free foods.

Food technology solution	Benefits	References
Pre-baking and frozen storage	Decrease the staling process and the crumb firmness	Sciarini et al. (2011)
Milk Protein fortification (caseinate)	Increase the dough stability	Houben et al. (2010) and Foschia et al. (2016)
Egg Protein fortification (albumins)	Increase the dough stability and porosity	Houben et al. (2010) and Foschia et al. (2016)
Whey Protein fortification (α-lactalbumin, β-lactoglobulin)	Increase the dough stability and bread shelf life	Houben et al. (2010) and Foschia et al. (2016)
Soy flour supplementation	Increase the dough stability, firmness, and specific volume; decreased staling rate	Sciarini et al. (2010) and Foschia et al. (2016)
Quinoa flour supplementation	Increased the bread volume; crumb featured homogeneous and finely distributed gas bubbles	Foschia et al. (2016)
Hydrocolloid additivities (carboxymethylcellulose)	Improved the bread-specific volume and crumb structure; reduced crumb hardness	Sciarini et al. (2011)
Xanthan gum additivities	Improved strengthened, firmness, elasticity and lightness.	Lazaridou et al. (2007), Foschia et al. (2016)
Hydrocolloid additives (β-glucans)	Increase the dough volume, porosity and firmness	Lazaridou et al. (2007) and Foschia et al. (2016)
Hydrocolloid additives (hydroxypropyl methylcellulose and carboxymethylcellulose)	Improved the dough viscoelastic properties	Foschia et al. (2016)
Zeine starch supplementation	Increase the dough consistency	Foschia et al. (2016)
Cassava starch supplementation	Improved crumb properties	Foschia et al. (2016)
Tapioca starch supplementation	Spongy crumb, high volume and good sensory acceptance	Milde et al. (2012) and Foschia et al. (2016)
Rice starch supplementation	Improved crumb properties.	Foschia et al. (2016)
Dietary fibres supplementation (*Psyllium*, oat, bamboo, potato, pea, and maize fibre, β-glucan, and locust bean gumcan)	Decrease the dough consistency, improve the specific volumes; improve the nutritional value.	Foschia et al. (2016)

Table 11.5 Sourdough biotechnology formulations.

Gluten-free Formulation	Sourdough Properties	Effect on batter/dough/ bread properties	References
Sorghum flour (*L. reuteri* Y2, *L. reuteri* VIP and *W. cibaria* MG1)	Formation of EPS (0.6–8.0 g/kg) and oligosaccharides	Decreased dough strength and elasticity	Galle et al. (2012)
Amaranth (*Amaranthus hypochondriacus*) *Lactobacillus plantarum* and *Lactobacillus paralimentarius*	*L. plantarum* dough with best textural characteristics	–	Houben et al. (2010)
Buckwheat flour (*Lactobacillus plantarum, Lactobacillus brevis, Lactobacillus paralimentarius* and *Weissella cibaria*)	Decreased elasticity	Significant increase in the gluten-like network strength	Moroni et al. (2011)
Buckwheat, quinoa, sorghum and teff flour (*Weissella cibaria, Lactobacillus plantarum*)	Resistant starch significantly decreased	Predicted glycemic index was reduced in sorghum and teff sourdough breads	Wolter et al. (2014a, b, c)
Buckwheat, quinoa, sorghum, teff flour (*Weissella cibaria*)	Quinoa sourdough showed the highest acidification level	Decreased dough strength Reduced crumb hardness and staling rate; increased crumb porosity	Wolter et al. (2014c)
Buckwheat, oat, quinoa, sorghum, teff flour (*Lactobacillus plantarum*)	Changes in protein profiles	Softer dough. Crumb porosity significantly increased.	Wolter et al. (2014a)
Refined maize meal (*L. plantarum*)	Starch granule modification	Maize dough softer and less elastic, bread volume and texture improved.	Falade et al. (2014)
Buckwheat flour, Buckwheat core, quinoa, rice (*L. animalis, L. reuteri* and *L. curvatus*)	Increased EPS	*In situ* produced EPS can replace added hydrocolloids	Ruehmkorf et al. (2012)
Teff and buckwheat flour (*Lactobacillus, Pediococcus, Leuconostoc, Kazachstania* and *Candida*)	*Acidification*	pH increased	Moroni et al. (2011)
Oat batter (*Leuconostoc argentinum, Pediococcus pentosaceus* and *Weissella cibaria, Lactobacillus coryniformis*)	Hydrolysis of proteins	Decreased starch viscosity of oat batter	Hüttner et al. (2009)
Fanio dough	Changes in starch characteristics	Reduction in starch gel firmness, increased pasting viscosity	Edema et al. (2013)
Sorghum batter (*L. plantarum*)	Acidification	Formation of strong starch gel	Schober et al. (2007)

Source: Deora et al. (2014), Foschia et al. (2016).

Table 11.6 Enzymes able to hydrolyse the peptides responsible for the immune response in celiac disease.

Enzyme[a] – source[b]	Effect	References
PEP – *Sp. Capsulate*	Proline-specific cleaving	Collado et al. (2009), Lähdeaho et al. (2014), Lebwohl et al. (2014), M'hir et al. (2012)
PEP – *Fl. Meningosepticum* PEP – *My. Xanthus* PEPO2 – *L. Helveticus*	Hydrolyzing effect on 33-mer peptide	M'hir et al. (2012)
PEP – *A. Niger*	Gluten degradation	Scherf et al. (2016), Lindfors et al. (2008), Tennyson et al. (2009)
Prolyl-specific peptidase – *Rh. Dominica*	Hydrolyzing effect on wheat and barley toxic peptides	Mika et al. (2015)
Cell-associated enzymes – *R. Mucilaginosa*	Hydrolyzing effect on 33-mer or 26-mer peptides	Tian et al. (2014)
Pseudolysin (lasB) – *Ps. Aeruginosa*	Gliadin degradation	Wei et al. (2015)
Kuma030 – re-designed	Gliadin degradation	Wolf et al. (2015)
Cysteine-endoprotease – Barley	Glutamine-specific cleaving	Collado et al. (2009), Lebwohl et al. (2014), M'hir et al. (2012)
Enzymes – germinated wheat	Rye secalin and gliadin degradation	Stenman et al. (2010)
Proteases – germinated wheat	Degradation of celiac toxic peptides	Hartmann et al. (2006), Stenman et al. (2009)
Cysteine protease triticain-α – wheat	Multiple cleavage sites within the sequences of 33-merpeptides	Savvateeva et al. (2015)

[a] PEP, prolyl-endopeptidase PepN, general aminopeptidase type N; PepX, prolyl dipeptidyl aminopeptidase; PepO, endopeptidase.
[b] *A., Aspergillus; Fl., Flavobacterium; L., Lactobacillus; My., Myxococcus; R., Rothia; Rh., Rhizopertha;* sp., *Sphingomas; Ps., Pseudomonas.*

The gluten-free certification organization (GFCO 2020) is a guarantee that a food product contains maximum admitted levels of 20-ppm, as measured by using the testing methods agreed by the Association of Official Agricultural Chemists (AOAC).

11.3.3 The Medical Treatments

The probiotic ameliorates the CD dyspepsia by the mechanism of gut repopulation with gastrointestinal microbiota involved in the gliadin proteolysis and responsible for local immune system enhancement.

11.4 Lactose Intolerance

Lactose intolerance is a food-related disease caused by the incapacity of the human body to metabolize and assimilate one food component, lactose, because the enterocytes from the intestinal epithelium has a lactase (β-galactosidase) production/secretion deficiency.

The intestinal non-metabolized lactose is subjected to colon fermentation after one to three hours of lactose-food ingestion, leading to abdominal pain, meteorism, flatulence, diarrhoea, etc. (Figure 11.9).

Galactosaemia is caused by a three-enzyme deficiency involved in the galactose assimilation process. According to the opinion of the European Food Safety Authority (EFSA) (EFSA Panel on Dietetic Products, Nutrition and Allergies 2010), lactose intolerance, as lactose maldigestion syndrome with genetically determinism, was displayed in up to 6–12 g of ingested lactose, depending on individuals. The main dietetic principle involves the elimination of all galactose sources, including human milk.

The dietetic management is started in childhood with lactose-free infant and follow-on formulae with low lactose content of up to 10 mg/100 kcal. In older infants, children, and adults, foods containing milk, milk derivate, or lactose supplements as an ingredient must be avoided. The overall daily lactose intake should be about 25 mg/100 kcal (EFSA Panel on Dietetic Products, Nutrition and Allergies 2010).

The medical therapy adopted in cases of lactose intolerance is orientated in two directions:

- substitution therapy, in which all lactose sources from foods, drugs, etc. are totally replaced with lactose-free products (lactose-free diet) and supplements of lactase enzymes;
- palliative therapy based on intestinal microflora boosting using functional foods (probiotics) such kefir, yoghurt, fermented milk, and supplements (glutamine, quercetin, chitinase, polyphenols extracts, etc.) (D'Angelo et al. 2009; Lee et al. 2014; Li et al. 2016; Wang et al. 2015).

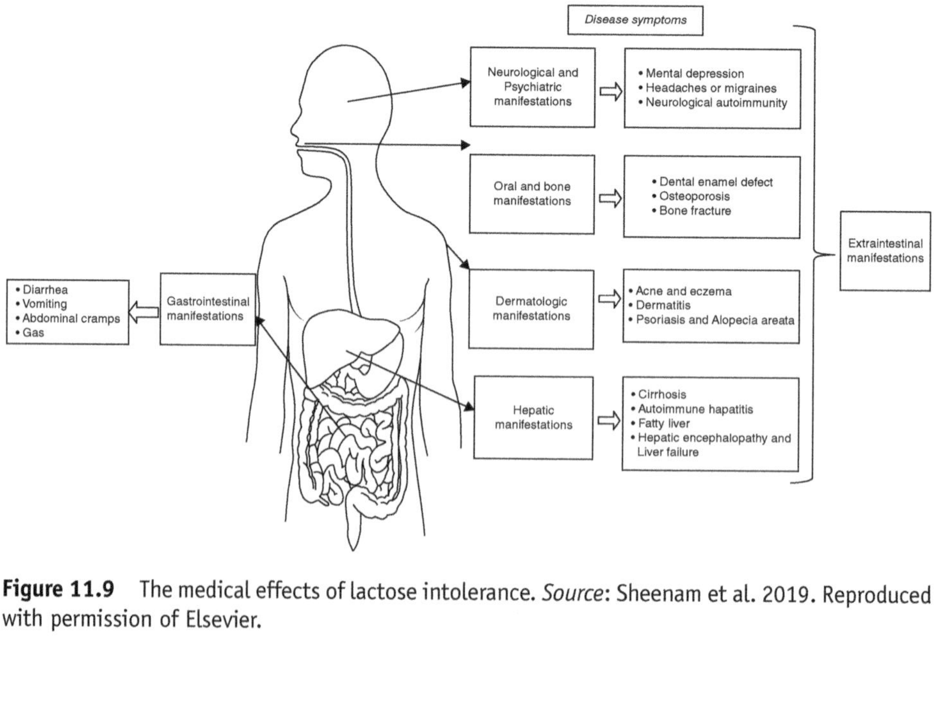

Figure 11.9 The medical effects of lactose intolerance. *Source*: Sheenam et al. 2019. Reproduced with permission of Elsevier.

11.5 Lactose-Free Food Formulations, Technology, and Quality

The lactose-free foods should not contain lactose as a formulation ingredient which means, for analytical reasons, a standard amount of up to 10mg lactose per 100g product.

The FSSAI Regulations nominate the ingredients admitted as lactose substitutes in 'lactose-reduced' or 'lactose-free' formulations, as follows:

- Soy protein-based, lactose-free formula shall have soy-protein and carbohydrate as glucose, dextrose, dextrin/maltodextrin, maltose, and/or sucrose;
- Lactose-free milk-based formulas shall have carbohydrate as glucose, dextrose, dextrin/maltodextrin, maltose, and sucrose (Tables 11.7–11.9).

Table 11.7 Sources of calcium in food intolerances.

Calcium Source	Carbohydrates (%)	Proteins (%)	Fat (%)	Moisture (%)	Calcium (ppm)
Soy Milk	1.5	3	1.5	94	3.90
Rice Milk	45–57.3	15.5	0.79–2.5	18.7	1.5–1.6
Coconut Milk	5.5–8.3	2.8–4.4	32.2–40.0	64.1–80.4	9.40
Oat Milk	12	0.7–1.0	1.3–1.52	–	–
Lactose Hydrolyzed Milk	4.8	3.2–3.37	1.5–4.3	87.1–89	112.7–120

Source: Sheenam et al. 2019. Reproduced with permission of Elsevier.

Table 11.8 Lactase enzymes and probiotics along with their effect on the lactose intolerance.

Key-compound/formulation	Benefits
Kluyveromyces lactis β-galactosidase	• intolerance symptoms suppressed in 60–77% cases
β-galactosidase medical formulation: lactogest, lactaid and dairyease	• improving lactose tolerance
Tilactase/*Lactobacillus reuteri*	• intolerance symptoms suppressed, especially in case of tilase
B. *Bassiana* β-galactosidase	• alternative source of β-galactosidase. •
Bifidobacterium longum	• alleviates malabsorbtion symptoms
Bifidobacterium animalis and *Lactobacillus plantarum*	• alleviates some gastrointestinal symptoms, such diarrhoea.
Lactobacillus bulgaricus and *Streptococcus thermophiles*	• ameliorated the maldigestion symptoms
Lactobacillus reuteri	• reducing nausea, flatulence and diarrhoea
Lactobacillus acidophilus	• avoid diarrhoea, cramping, and vomiting during an acute episode
Lactobacillus rhamnosus	• reducing inflammation

Source: Sheenam et al. (2019). Reproduced with permission of Elsevier.

Table 11.9 Lactose-free dairy foods.

Food	Raw materials	Benefits
Milk	Soybean	Anti-cholesterol effect Prevents heart disease, postmenopausal syndromes, cancers, osteoporosis
	Coconut	Boost immunity Prevents dyslipidaemia
	Rice	Prevents osteoporosis
	Oat	Prevents dyslipidaemia
Cheese	Soy milk	Anti-cholesterol effect Prevents heart diseases, postmenopausal syndromes, cancers, and osteoporosis.
	Coconut milk	Increases the protein intake
Creamed cottage cheese	Lactose Hydrolysed cream	Improves the textural characteristics
Milk powders	Coconut milk	More flavourful products
	Soy milk	Improves the protein and essential nutrients intake
Kefir	Soy milk	Good source of isoflavone genistein that suppresses the growth of cancer cells.
	Oat milk	Increases the beta-glucan content and promote the gut microbiota
Functional beverage	Peanut-Soy milk	Good source of essential amino acids
Yoghurt	Soy milk	Good source of protein (3%) Heart-healthy source of plant-based fat (7%).
Ice-cream	Rice milk Lactose hydrolyzed whey	High nutritive value
Chocolate bar	Soy milk Soy milk powder	Reduced beany flavour, improved texture

Source: Sheenam et al. (2019). Reproduced with permission of Elsevier.

The infant lactose- and galactose-free formulas are soy protein based, with standard quality criteria described in Directive 2006/141/EC4. The residual lactose permitted is 1.4 mg/dl bound galactose, with a daily galactose intake of maximum 10–50 mg (EFSA Panel on Dietetic Products, Nutrition and Allergies 2010).

The Lactose-free certification label (Figure 11.10), named Lfree, certifies a lactose content of under 0.01% (i.e. 0.01 g/100 g of total product). The 'Naturally lactose-free' label could be used for the naturally lactose-free dairy products with lactose removal under the technological process of up to 0.01%. On 13 July 2017, the EU Commission adopted a Notice on the provision of information on substances or products causing allergies or intolerances with new requirements of Regulation (EU) No 1169/2011 (2011) related to the indication of the presence of certain substances or products causing intolerances.

The processing solutions for manufacturing reduced/lactose-free milk are generally microfiltration, ultrafiltration, nanofiltration, and reverse osmosis systems (Figure 11.11).

Figure 11.10 The Lfree registered trademark in European Union (EU), USA, China, Japan, Norway, Australia, San Marino, and Switzerland (Elle Free 2019).

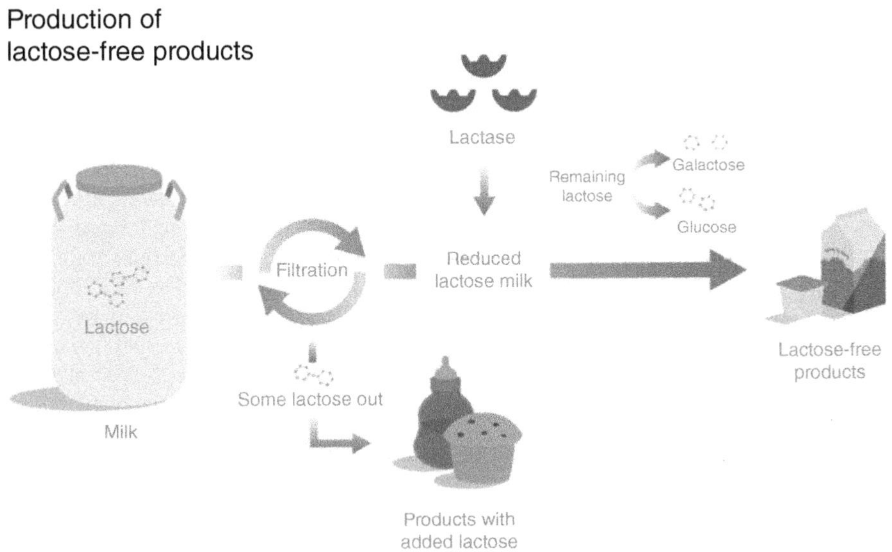

Figure 11.11 Smart technologically solutions for lactose-free products (GEA 2020).

11.6 Formulations for Other Food Intolerances

As general practice, the management of other types of food intolerances (Table 11.10) involve elimination diets, totally abstinance, and avoiding the intolerance vector food.

The food intolerances triggers which must be eliminate from the food formulation are generally a food additive, a chemical component, or a contaminant, such as:

- Monosodium glutamate (MSG, E 620 and E 621) from Camembert cheese, Parmesan cheese, tomatoes, soy sauce, and mushrooms;
- Caffeine;
- Alcohol;
- Artificial sweeteners;
- Histamine (mushrooms, pickled and cured foods, and alcoholic drinks);
- Toxins, viruses, bacteria, or parasites that have contaminated food; and
- Artificial food colours, preservatives, or flavour enhancers.

Table 11.10 Food intolerances due to enzyme deficiency and transport disorder.

Enzymatic structure involved in food intolerance	Target structure	Primary deficiency	Secondary deficiency/ disorder
Carbohydrates			
Combined disaccharide malabsorption syndrome	Lactose, sucrose, and other disaccharides	Autosomal recessive deficiency	Bowel inflammation (infections, celiac disease, CIBD[a])
Isolated disaccharide intolerances	Disaccharide		Bowel inflammation (infections, celiac disease, CIBD)
GLUT 5 transport defect	Fructose		Bowel inflammation (infections, celiac disease, CIBD)
Lactase (β-galactosidase)	Lactose	Congenital Autosomal recessive deficiency (very rare)	Bowel inflammation (infections, celiac disease, CIBD)
Saccharase (sucrase isomaltase)	Sucrose	Autosomal recessive sucrase-isomaltase deficiency	Bowel inflammation (infections, celiac disease, CIBD)
Maltase (alpha-glucosidase)	Maltose	Autosomal recessive deficiency	Medication with acarbose, miglitol
Trehalase	Trehalose	Autosomal recessive deficiency	Bowel inflammation (infections, celiac disease, CIBD)
Galactase	Galactose	Autosomal recessive deficiency	Bowel inflammation (infections, celiac disease, CIBD)
Biogenic amines			
e.g. Diamine oxidase	Histamine, etc.	Autosomal recessive deficiency	Bowel inflammation (infections, celiac disease, CIBD)
Fructose intolerance aldolase B	Fructose	Autosomal recessive deficiency	
Glucose-6-phosphate dehydrogenase	Fava beans	X-chromosomally inherited enzyme defect	Medication with sulphonamides
Alcohol dehydrogenase	Acetaldehyde		Medication with metronidazole

Source: Zopf et al. (2009).

[a]CIBD, chronic inflammatory bowel diseases.

Also, natural food substances could cause food intolerance symptoms and food technologies should eliminate them from the special food formulation tailored for food intolerances. The main intolerance-triggering compounds are as follows:

- Vasoactive amines such as tyramine, serotonin, and histamine naturally occurring in pineapples, bananas, baked meat, red wine, wood-matured wine, avocados, dark chocolate, citrus fruits, and mature cheese;
- Salicylates, natural aspirin-like compounds from herbs, spices, fruits, and vegetables. Can trigger urticaria by acting directly on skin mast cells;
- Endogenous toxins from fish, shellfish, etc. which contain histidine transformed into histamine by the endogenous gut microbiota activated in incorrect storage conditions;
- Caffeine and curry powders could act as gut irritants.

Other adverse reactions, associated or not with the food intolerances, are as follows:

- Feeling unwell, associated with heartburn, arytmia, sweating, etc. after a fatty or spicy meal, a large quantity of red wine, chocolate, coffee, green tea, etc.
- Food aversion, an emotional state characterized by food dislikes, impossibility to eat in public places, etc. associated with medical symptoms such eczema, urticaria, etc.
- Food anxiety associated with rapid breathing, dizziness, tight chest, blurred vision (ASCIA 2014).

11.7 Conclusions

Food intolerances could include some food allergies symptoms but the immune-mediated reaction makes the medical differences between the two different diagnosis. The immunological status, given by the natural way of birth, healthy way of life, success stress control, using drugs (especially antibiotics) under strict medical control, could avoid the food intolerance developing and, for the individuals with this diagnosis, to make them irreversible or only occur sporadically within a long medical cycle. The consistent medical diagnosis and the strictly avoidance diet followed by all the patients during their lifetime, associated with palliative supplements to alleviate the negative nutritional effects generated by the restricted diet, are the basic principles in food intolerance management. The foods tailored for specific intolerances should be compliant with the international regulations regarding the maximum content of the intolerance triggering compound. The global nutritional regulation and the food certification process need to be harmonized to benefit consumers and food industry global trading success.

References

Alp, H. and Bilgiçli, N. (2008). Effect of transglutaminase on some properties of cake enriched with various protein sources. *Journal of Food Science* 73: 209–214.

Al-Toma, A., Volta, U., Auricchio, R. et al. (2019). European Society for the Study of Coeliac Disease (ESsCD) guideline for coeliac disease and other gluten-related disorders. *United European Gastroenterology Journal* 7: 583–613. doi: 10.1177/2050640619844125.

Australian Society of Clinical Immunology and Allergy (ASCIA) (2014). Food intolerances. *25th Annual Conference*, Melbourne, Australia (10–13 September 2014).

Babiker, E.E. (2000). Effect of transglutaminase treatment on the functional properties of native and chymotrypsin-digested soy protein. *Food Chemistry* 70: 139–145.

Banu, C. et al. (2010). *Alimentatia in Bolile Digestive*. Bucuresti: ASAB.

Capriles, V.D. and Arêas, J.A.G. (2016). Approaches to reduce the glycemic response of gluten-free products: in vivo and in vitro studies. *Food & Function* 7: 1266–1272. https://doi.org/10.1039/c5fo01264c.

Casella, G., D'Incà, R., Oliva, L. et al. (2010). Prevalence of celiac disease in inflammatory bowel diseases: an IG-IBD multicentre study. *Digestive and Liver Disease* 42: 175–178.

Codex Alimentarius Commission (2008). http://www.fao.org/fao-who-codexalimentarius/en (accessed 8 December 2010).

Collado, M.C., Donat, E., Ribes-Koninckx, C. et al. (2009). Specific duodenal and faecal bacterial groups associated with paediatric coeliac disease. *Journal of Clinical Pathology* 62: 264–269.

D'Angelo, S., Morana, A., Salvatore, A. et al. (2009). Protective effect of polyphenols from Glycyrrhiza glabra against oxidative stress in Caco-2 cells. *Journal of Medicinal Food* 12 (6): 1326–1333. https://doi.org/10.1089/jmf.2008.0285.

De Vuyst, L., Van Kerrebroeck, S., Harth, H. et al. (2014). Microbial ecology of sourdough fermentations: diverse or uniform? *Food Microbiology* 37: 11–29.

Demirkesen, I., Mert, B., Sumnu, G., and Sahin, S. (2010). Utilization of chestnut flour in gluten-free bread formulations. *Journal of Food Engineering* 101: 329–336.

Deora, N., Deswal, A., and Mishra, H. (2014). Alternative approaches towards gluten-free dough development recent trends. *Food Engineering Reviews* 6: 89–104.

Di Cairano, M., Galgano, F., Tolve, R. et al. (2018). Focus on gluten free biscuits: ingredients and issues. *Trends in Food Science & Technology* 81 (11): 203–212.

do Nascimento, A.B., Fiates, G.M.R., Anjos, A.D., and Teixeira, E. (2013). Analysis of ingredient lists of commercially available gluten-free and gluten-containing food products using the text mining technique. *International Journal of Food Sciences and Nutrition* 64 (2): 217–222.

Duta, D.E. and Culetu, A. (2015). Evaluation of rheological, physicochemical, thermal, mechanical and sensory properties of oat-based gluten free cookies. *Journal of Food Engineering* 162: 1–8. doi: 10.1016/j.jfoodeng.2015.04.002.

Edema, M.O., Emmambux, M.N., and Taylor, J.R.N. (2013). Improvement of fonio dough properties through starch modification by sourdough fermentation. *Starch* 65 (9–10): 730–737. https://doi.org/10.1002/star.201200248.

EFSA Panel on Dietetic Products, Nutrition and Allergies (NDA) (2010). Scientific opinion on lactose thresholds in lactose intolerance and galactosaemia. *EFSA Journal* 8 (9): 1777. https://doi.org/10.2903/j.efsa.2010.1777.

Elle Free (2019). http://www.ellefree.com/eng (accessed 8 December 2020).

Espinosa-Ramírez, J., Garzon, R., Serna-Saldivar, S.O. et al. (2018). Functional and nutritional replacement of gluten in gluten-free yeast-leavened breads by using β-conglycinin concentrate extracted from soybean flour. *Food Hydrocolloids* 84 (11): 353–360.

EU Commission Implementing Regulation (EU) No. 828/2014 (2014/2016). https://ec.europa.eu/food/safety/labelling_nutrition/special_groups_food (accessed 8 December 2020)

EU Regulation No. 1169/2011 (2011). Related to the indication of the presence of certain substances or products causing intolerances. https://eur-lex.europa.eu (accessed 8 December 2020).

European Food Information Council (EUFIC) (2020). https://www.eufic.org (accessed 9 December 2020).

Falade, A.T., Emmambux, M.N., Buys, E.M., and Taylor, J.R.N. (2014). Improvement of maize bread quality through modification of dough rheological properties by lactic acid bacteria fermentation. *Journal of Cereal Science* 60: 471–476.

Foschia, M., Horstmann, S., Arendt, E.K. et al. (2016). Nutritional therapy – facing the gap between coeliac disease and gluten-free food. *International Journal of Food Microbiology* 239 (12): 113–124.

Galle, S., Schwab, C., Dal Bello, F. et al. (2012). Influence of in-situ synthesized exopolysaccharides on the quality of gluten-free sorghum sourdough bread. *International Journal of Food Microbiology* 155: 105–112.

Gaspar, A.L.C. and de Góes-Favoni, S.P. (2015). Action of microbial transglutaminase (MTGase) in the modification of food proteins: a review. *Food Chemistry* 171: 315–322.

GEA Group (2020). www.gea.com (accessed 9 December 2020).

Gobbetti, M., Rizzello, C.G., Di Cagno, R., and De Angelis, M. (2007). Sourdough lactobacilli and celiac disease. *Food Microbiology* 24: 187–196.

Gobbetti, M., Pontonio, E., Filannino, P. et al. (2018). How to improve the gluten-free diet: the state of the art from a food science perspective. *Food Research International* 110 (8): 22–32. https://doi.org/10.1016/j.foodres.2017.04.010.

Hager, A.S., Wolter, A., Jacob, F. et al. (2012). Nutritional properties and ultra-structure of commercial gluten free flours from different botanical sources compared to wheat flours. *Journal of Cereal Science* 56: 239–247.

Hartmann, G., Koehler, P., and Wieser, H. (2006). Rapid degradation of gliadin peptides toxic for coeliac disease patients by proteases from germinating cereals. *Journal of Cereal Science* 44: 368–371.

Houben, A., Götz, H., Mitzscherling, M., and Becker, T. (2010). Modification of the rheological behaviour of amaranth (*Amaranthus hypochondriacus*) dough. *Jounal of Cereal Science* 51: 350–356.

Hüttner, E.K., Dal Bello, F., Poutanen, K., and Arendt, E.K. (2009). Fundamental evaluation of the impact of high hydrostatic pressure on oat batters. *Jounal of Cereal Science* 49: 363–370.

Keagy, P.M., Stokstad, E.L.R., and Fellers, D.A. (1975). Folacin stability during bread processing and family flour storage. *Cereal Chemistry* 52: 348–356.

Lähdeaho, M.L., Kaukinen, K., Laurila, K. et al. (2014). Glutenase ALV003 attenuates gluten-induced mucosal injury in patients with celiac disease. *Gastroenterology* 146: 1649–1658.

Lazaridou, A., Duta, D., Papageorgiou, M. et al. (2007). Effects of hydrocolloids on dough rheology and bread quality parameters in gluten-free formulations. *Journal of Food Engineering* 79: 1033–1047.

Lebwohl, B., Ludvigsson, J.F., and Green, P.H. (2014). The unfolding story of celiac disease risk factors. *Clinical Gastroenterology and Hepatology* 12: 632–635.

Lee, J.H., Lee, S.K., Park, K.H. et al. (1999). Fermentation of rice using amylolytic *Bifidobacterium*. *International Journal of Food Microbiology* 50: 155–161.

Lee, I.A., Kamba, A., Low, D., and Mizoguchi, E. (2014). Novel methylxanthine derivative-mediated anti-inflammatory effects in inflammatory bowel disease. *World Journal of Gastroenterology* 20 (5): 1127–1138. https://doi.org/10.3748/wjg.v20.i5.1127.

Li, Y., Yao, J., Han, C. et al. (2016). Quercetin, inflammation and immunity. *Nutrients* 8 (3): 167. https://doi.org/10.3390/nu8030167.

Lindfors, K., Blomqvist, T., Juuti-Uusitalo, K. et al. (2008). Live probiotic *Bifidobacterium lactis* bacteria inhibit the toxic effects induced by wheat gliadin in epithelial cell culture. *Clinical and Experimental Immunology* 152: 552–558.

Lionetti, E., Gatti, S., Pulvirenti, A., and Catassi, C. (2015). Celiac disease from a global perspective. *Best Practice & Research Clinical Gastroenterology* 29: 365–379.

Lundin, L., Golding, M., and Wooster, T.J. (2008). Understanding food structure and function in developing foods for appetite control. *Nutrition and Dietetics* 65: S79–S85.

Mert, S., Sahin, S., and Sumnu, G. (2015). Development of gluten-free wafer sheet formulations. *LWT – Food Science and Technology* 63 (2): 1121–1127.

M'hir, S., Ziadi, M., Chammem, N., and Hamd, M. (2012). Gluten proteolysis as alternative therapy for celiac patients: a mini-review. *African Journal of Biotechnology* 11: 7323–7330.

Mika, N., Gorshkov, V., Spengler, B. et al. (2015). Characterization of novel insect associated peptidases for hydrolysis of food proteins. *European Food Research and Technology* 240: 431–439.

Milde, L.B., Ramallo, L.A., and Puppo, M.C. (2012). Gluten-free bread based on tapioca starch: texture and sensory studies. *Food and Bioprocess Technology* 5: 888–896.

Mintel (2015). Gluten-free foods – US, 2015: consumer market research report. www.mintel.com (accessed 8 December 2020).

Moroni, A.V., Arendt, E.K., and Dal Bello, F. (2011). Biodiversity of lactic acid bacteria and yeasts in spontaneously-fermented buckwheat and teff sourdoughs. *Food Microbiology* 28: 497–502.

Mulder, C.J., Wahab, P.J., Meijer, J.W., and Metselaar, E. (2001). A pilot study of recombinant human interleukin-10 in adults with refractory coeliac disease. *European Journal of Gastroenterology & Hepatology* 13: 1183–1188.

Nanson, N.J. and Fields, M.L. (1984). Influence of temperature of fermentation on the nutritive value of lactic acid fermented cornmeal. *Journal of Food Science* 49: 958–959.

Reilly, N.R. (2016). The gluten-free diet: recognizing fact, fiction, and fad. *The Journal of Pediatrics* 175: 206–210. https://doi.org/10.1016/j.jpeds.2016.04.014.

Omoba, O.S., Taylor, J.R.N., and de Kock, H.L. (2015). Sensory and nutritive profiles of biscuits from whole grain sorghum and pearl millet plus soya flour with and without sourdough fermentation. *International Journal of Food Science and Technology* 50 (12): 2554–2561. https://doi.org/10.1111/ijfs.12923.

Östman, E. (2003). Fermentation as a means of optimizing the glycaemic index – food mechanisms and metabolic merits with emphasis on lactic acid in cereal products. PhD thesis, Lund University, Department of Applied Nutrition and Food Chemistry.

Pagliari, D., Urgesi, R., Frosali, S. et al. (2015). Interaction among microbiota, immunity, and genetic and dietary factors is the condicio sine qua non celiac disease can develop. *Journal of Immunology Research* 2015: 123653. https://doi.org/10.1155/2015/123653.

Rizzello, G.C., Coda, R., De Angelis, M. et al. (2009). Long-term fungal inhibitory activity of watersoluble extract from *Amaranthus* spp.seeds during storage of gluten-free and wheat flour breads. *International Journal of Food Microbiology* 131: 189–196.

Ruehmkorf, C., Jungkunz, S., Wagner, M., and Vogel, R.F. (2012). Optimization of homoexopolysaccharide formation by lactobacilli in gluten-free sourdoughs. *Food Microbiology* 32 (2): 286–294.

Rybicka, A. and Gliszczyńska-Świgło, A. (2017). Minerals in grain gluten-free products. The content of calcium, potassium, magnesium, sodium, copper, iron, manganese, and zinc. *Journal of Food Composition and Analysis* 59: 61–67. https://doi.org/10.1016/j.jfca.2017.02.006.

Sapone, A., Bai, J.C., Ciacci, C. et al. (2012). Spectrum of gluten-related disorders: consensus on new nomenclature and classification. *BMC Medicine* 10: 13. https://doi.org/10.1186/1741-7015-10-13.

Savvateeva, L.V., Gorokhovets, N.V., Makarov, V.A. et al. (2015). Glutenase and collagenase activities of wheat cysteine protease triticain-α: feasibility for enzymatic therapy assays. *The International Journal of Biochemistry & Cell Biology* 62: 115–124.

Scherf, K.A., Wieser, H., and Koehler, P. (2016). Novel approaches for enzymatic gluten degradation to create high-quality gluten-free products. *Food Research International* 110: 62–72. https://doi.org/10.1016/j.foodres.2016.11.021.

Schober, T.J., O'Brien, C.M., McCarthy, D. et al. (2003). Influence of gluten-free flour mixes and fat powders on the quality of gluten-free biscuits. *European Food Research and Technology* 216 (5): 369–376. https://doi.org/10.1007/s00217-003-0694-3.

Schober, T., Bean, S.R., and Boyle, D.L. (2007). Gluten-free *Sorghum* bread improved by sourdough fermentation: biochemical, rheological, and microstructural background. *Journal of Agricultural and Food Chemistry* 55 (13): 5137–5146. https://doi.org/10.1021/jf0704155.

Sciarini, L.S., Ribotta, P.D., Leon, A.E., and Perez, G.T. (2010). Effect of hydrocolloids on gluten-free batter properties and bread quality. *International Journal of Food Science and Technology* 45: 2306–2312.

Sciarini, L.S., Ribotta, P.D., Leon, A.E., and Perez, G.T. (2011). Influence of gluten-free flours and their mixtures on batter properties and bread quality. *Food and Bioprocess Technology* 3: 577–585.

Sheenam, S., Kumar, V., Prasad, R. et al. (2019). Considerations for development of lactose-free food. *Journal of Nutrition & Interemediary Metabolism* 15: 24–37. https://doi.org/10.1016/j.jnim.2018.11.003.

SPINS, Information Resources, Inc. (2019). https://www.spins.com (accessed 8 December 2020).

Stenman, S.M., Venäläinen, J.I., Lindfors, K. et al. (2009). Enzymatic detoxification of gluten by germinating wheat proteases: implications for new treatment of celiac disease. *Annals of Medicine* 41: 390–400.

Stenman, S.M., Lindfors, K., Venäläinen, J.I. et al. (2010). Degradation of coeliac disease-inducing rye secalin by germinating cereal enzymes: diminishing toxic effects in intestinal epithelial cells. *Clinical and Experimental Immunology* 161: 242–249.

Taranto, M.P., Vera, J.L., Hugenholtz, J. et al. (2003). *Lactobacillus reuteri* CRL1098 produces cobalamin. *Journal of Bacteriology* 185: 5643–5647.

Tennyson, A., Lewis, S.K., and Green, P.H.R. (2009). New and developing therapies for celiac disease. *Therapeutic Advances in Gastroenterology* 2: 303–309.

The Gluten-Free Certification Organization (2020). https://gfco.org/get-certified/faq (accessed 8 December 2020).

Tian, N., Wei, G., Schuppan, D., and Helmerhorst, E.J. (2014). Effect of *Rothia mucilaginosa* enzymes on gliadin (gluten) structure, deamidation, and immunogenic epitopes relevant to celiac disease. *American Journal of Physiology* 307: G769–G776.

Torbica, A., Hadnađev, M., Dapčević, H., and Dokić, T.P. (2011). Functional gluten alternatives. In: *Gluten: Properties, Modifications and Dietary Intolerance* (ed. D.S. Fellstone), 50–73. Nova Science Publishers Inc.

Tsatsaragkou, K., Gounaropoulos, G., and Mandala, I. (2014). Development of gluten free bread containing carob flour and resistant starch. *LWT – Food Science and Technology* 58: 124–129.

Wang, B., Wu, G., Zhou, Z. et al. (2015). Glutamine and intestinal barrier function. *Amino Acids* 47 (10): 2143–2154. https://doi.org/10.1007/s00726-014-1773-4.

Wei, G., Tian, N., Valery, A.C. et al. (2015). Identification of pseudolysin (lasB) as an aciduric gluten-degrading enzyme with high therapeutic potential for celiac disease. *The American Journal of Gastroenterology* 110: 899–908.

WHO/FAO/UNU (2007). PDCAAS recommendation.

Wolf, C., Siegel, J.B., Tinberg, C. et al. (2015). Engineering of Kuma030: a gliadin peptidase that rapidly degrades immunogenic gliadin peptides in gastric conditions. *Journal of the American Chemical Society* 137: 13106–13113.

Wolter, A., Hager, A.-S., Zannini, E. et al. (2014a). Impact of sourdough fermented with *Lactobacillus plantarum* FST 1.7 on baking and sensory properties of gluten-free breads. *European Food Research and Technology* 239: 1–12.

Wolter, A., Hager, A.S., Zannini, E. et al. (2014b). Evaluation of exopolysaccharide producing *Weissella cibaria* MG1 strain for the production of sourdough from various flours. *Food Microbiology* 37: 44–50.

Wolter, A., Hager, A.-S., Zannini, E. et al. (2014c). Influence of dextran-producing *Weissella cibaria*on baking properties and sensory profile of gluten-free and wheat breads. *International Journal of Food Microbiology* 172: 83–91.

Wolter, A., Hager, A.S., Zannini, E. et al. (2014d). Impact of sourdough fermented with *Lactobacillus plantarum* FST 1.7 on baking and sensory properties of gluten-free breads. *European Food Research and Technology* 239: 1–12.

Wronkowska, M., Haros, M., and Soral-Smietana, M. (2013). Effect of starch substitution by buckwheat flour on gluten-free bread quality. *Food and Bioprocess Technology* 6: 1820–1827.

Yildirim, R.M., Gumus, T., Arici, M. et al. (2018). Optimization of a gluten free formulation of the Turkish dessert revani using different types of flours, protein sources and transglutaminase. *LWT – Food Science and Technology* 95 (9): 72–77.

Yousif, N.E. and El Tinay, A.H. (2000). Effect of fermentation on protein fractions and in vitro protein digestibility of maize. *Food Chemistry* 70: 181–184.

Zopf, Y., Baenkler, H.W., Silbermann, A. et al. (2009). The differential diagnosis of food intolerance. *Detsch Arztebl International* 106 (21): 359–369. https://doi.org/10.3238/arztebl.2009.0359.

12 Prebiotic and Probiotic Food Formulations

Pradip Behare[1,2], Shaik Abdul Hussain[1,3], Desirée Román Naranjo[2], Prateek Sharma[3,5], and Olivia McAuliffe[2,4]

[1] *Dairy Microbiology Division, ICAR-National Dairy Research Institute, Karnal, Haryana, India*
[2] *Department of Food Biosciences, Teagasc Food Research Centre, Moorepark, Cork, Ireland*
[3] *Department of Food Chemistry and Technology, Teagasc Food Research Centre, Moorepark, Cork, Ireland*
[4] *VistaMilk SFI Research Centre, Cork, Ireland*
[5] *Department of Nutrition, Dietetics, and Food Sciences, Utah State University, USA*

Summary

Probiotic species can beneficially influence human health by modulating and maintaining immune function in the gut. While prebiotics promote the growth of probiotics, their combinations with probiotics in synergistic way impart multifarious advantages to the host. Owing to their physiological benefits, demand for dairy and non-dairy food products supplemented with probiotics and prebiotics is ever increasing. However, their incorporation in food products not only alters techno-functional properties of food but also affects viability and bioactivity of these molecules. Hence, a careful selection of probiotics, prebiotics and food matrix besides optimizing food processing conditions is paramount for successful production of probiotic food.

12.1 Introduction

Studies performed over the past number of years have clearly demonstrated the relationship between food, microbes, and human health. These studies have revealed that the type of food consumed impacts on the microbial ecology in the gut, thus influencing human health (Conlon and Bird 2015). The gastrointestinal tract harbours $10^{13} - 10^{14}$ of microorganisms, comprising various genera (Backhed 2005). Some of these organisms forming the natural flora of the host are commensal, while non-commensal forms enter through the diet (Conlon and Bird 2015). These microbes undergo a symbiotic co-evolution and play numerous important functions within the host (Markowiak and Slizewska 2017). The fascinating work of Eli Metchnikoff, the Russian Nobel prize winner, attributed the consumption of fermented milk containing beneficial bacteria

Food Formulation: Novel Ingredients and Processing Techniques, First Edition.
Edited by Shivani Pathania and Brijesh K. Tiwari.

on prolonged youth and healthy ageing in Bulgarian peasants, giving researchers the impetus to further investigate the role of these organisms. Besides, certain non-digestible carbohydrates have been known to be selectively fermented by beneficial microbial species in the gut, thus maximizing health benefits (Gibson et al. 2004). In the modern era, such beneficial microorganisms are referred to as probiotics, while the carbohydrates that support their growth are referred to as prebiotics. The list of microbes used as probiotics extends from bacteria to fungi. Probiotics generally belong to the species of *Lactobacillus*, *Bifidobacterium*, *Bacillus*, *Propionibacterium*, *Enterococcus*, and *Saccharomyces* (Kerry et al. 2018). However, lactobacilli and bifidobacteria form a significant part of most probiotic food formulations available commercially, due to their well-known beneficial effects on the health of the host (Goldin and Gorbach 1992). At present, several carbohydrate molecules are marketed worldwide as prebiotics, with fructooligosaccharides (FOS), galactooligosaccharides (GOS), and inulin widely accepted. Inclusion of prebiotics and probiotics into food product formulations offers an attractive option for manufacturers interested in novel and healthy foods (Rolim 2015). With respect to suitable food matrices as carriers for probiotics and prebiotics, the majority of reports evaluate the potential of various dairy products while other studies examine fermented oat drinks, cereal bars, chocolate, and juices (Ranadheera et al. 2017). This chapter summarizes various prebiotic and probiotic food formulations that have been developed and their functional/health benefits.

12.2 Need for Probiotics and Prebiotics

Humans have become more susceptible to food-borne infections, allergies, and metabolic diseases due to dynamic changes in food habits and a more sedentary life style (Broussard and Devkota 2016). Diet may modulate various physiological functions and play detrimental or beneficial roles in some diseases (Koletzko et al. 1998). Studies have shown that food-borne infections are most likely to occur in individuals consuming contaminated foods, while alteration in the gut microbial population may influence metabolic disorders such as type 2 diabetes and cardiovascular diseases (CVDs) (David et al. 2014; Kasubuchi et al. 2015). Differences in gut bacteria can be seen between individuals in various countries, while similarities of faecal microbiomes among family members extend across countries and cultures (Yatsunenko et al. 2012). Unhealthy food creates a vicious cycle of disease and malnutrition affecting all age groups, particularly infants, young children, and the elderly. Over 200 diseases ranging from diarrhoea to cancers occur due to consumption of contaminated foods, which might comprise harmful microorganisms or chemical substances. As per the FAO global report of 2019, every year an estimated 600 million people become ill and 0.42 million die after eating contaminated food.

In spite of remarkable research and development in treatment and prevention procedures, infectious diseases still remain the foremost cause of death worldwide (Sarmah et al. 2018). Most infectious diseases are treated and controlled by using broad-spectrum antibiotics. However, these agents are becoming increasingly ineffective due to the rapid emergence of resistant microbial strains (Hancock and Sahl 2006). Moreover, according to the World Health Organization, most antibiotics will no longer be effective which makes things worse for the human community, since only a handful of antibiotics will be available to protect them from the dreadful pathogens. This suggests a high requirement for novel and improved alternative therapeutic and prophylactic strategies to manage several diseases. The introduction of probiotics and prebiotics to the diet favours a healthy microbial ecology in the gut, thus helping to maintain better health.

Probiotic and prebiotic foods could be used for improvement of a number of physiological functions in the human body, including gastrointestinal function, balancing of colonic microflora, control of nutrient bioavailability, carrier for antioxidants and vitamins, metabolism of macronutrients, and the related hormonal regulation. Dietary modification towards a balanced intake of nutrients along with ingestion of pre- and probiotics will help in preventing disease and to a promote healthier life (Cerdo et al. 2019).

12.3 Probiotic Food Formulations

The word 'probiotic' is translated from the Greek meaning 'for life'. Probiotics have a long historical record. The term probiotic was first used by Lilly and Stillwell (1965) to describe 'substances secreted by one microorganism which stimulated the growth of another'. The definition of probiotic was revised over time with advances in research and an increased understanding of the role of these microbes in human health. An early definition of probiotic was given by Parker (1974) as 'organisms and substances which contribute to intestinal microbial balance.' However, this definition was subsequently modified by Fuller (1989) to 'a live microbial feed supplement which beneficially affects the host animal by improving its intestinal microbial balance'. A probiotic would therefore incorporate living microorganisms into the diet. The most widely accepted definition was given by the FAO and WHO (2002), which concentrated exclusively on their health purpose as 'live microorganisms which when administered in adequate amounts confer a health benefit on the host'.

A probiotic organism must remain viable during processing and storage of food products to exert health benefits. They should also be able to survive in the intestinal ecosystem, and exert a physiological benefit on the host. Therefore, it is essential that the ingested probiotics bacteria reach the intestine in viable form, and become established there to exert these positive health effects. In order to perform its physiological functions in the gut, the probiotic strain must overcome a number of physical and chemical barriers in the gastrointestinal tract such as gastric acidity, pH, hydrolytic enzymes, and bile acid secretion. In addition, the probiotic strain should also not induce an immune response. It must be harmless to the host, and there must be no local or general pathogenic, allergic, or mutagenic/carcinogenic reactions provoked by the microorganism themselves, their fermentation products, or their cell components. The strain must also be genetically stable for the maintenance of its favourable properties. For the production of probiotics, it is important that the microorganisms multiply rapidly and densely on cost-effective media and that they remain viable during processing and storage. A number of these requirements viz. viability during processing and storage; sensitivity to low pH, gastric juice, bile, pancreas, intestinal juice, and intestinal or respiratory mucus; adherence to isolated cells or cell cultures; and interactions with other microorganisms; can be screened during *in vitro* experiments. After successful *in vitro* studies, *in vivo* experiments and clinical trials are conducted to establish probiotic efficacy. Some of the major requirements of probiotics that are vital for their use in development of foods for human consumption are:

- Preferably normal inhabitants of human intestinal tract (although some probiotics from other sources like dairy, non-dairy, and plant origin are available)
- Generally regarded as safe (GRAS) status
- Should be non-allergic and non-carcinogenic
- Proliferation and/or colonization in the colon
- Should not initiate any immune reaction in the body
- Non-pathogenic, toxic, allergic, mutagenic, or carcinogenic reaction by the probiotic strain itself.

- Genetically stable and should not contain antibiotic resistance transferrable plasmid
- Should be fast growing
- Must survive food processing and storage conditions.

In general, probiotic organisms intended for human consumptions should be safe and have established health benefits. For successful application in foods, the probiotic should also be technologically compatible with the food matrix and manufacturing process. It should survive the processing stages and remain stable during storage. The probiotic organism used should not deteriorate the quality attributes of food products and the product must be organoleptically acceptable. With respect to dosage, it has been suggested that approximately 10^9 cfu/day of probiotic microorganisms is necessary to elicit health effects. Based on daily consumption of 100 g or ml of probiotic food, a product should contain at least 10^7 cells/g or ml of probiotic food.

Probiotic fermented products are manufactured by adding the probiotic strains simultaneously with the starter cultures during fermentation in the tank. Alternatively, fermentation is carried out separately and probiotic cultures are added later to the final product. Incorporation of probiotic organisms into food products provides a potential to improve the quality of the product and the health status of consumers. There are many food formats that can successfully incorporate probiotics and deliver them at the required dose with optimal strain combinations. Probiotics have been incorporated in a number of foods, including the traditional vehicles of fermented milks and non-fermented foods beyond the dairy sector. Yoghurt, cheese, frozen dairy desserts, spray dried milk powder, fruit juices, confectionary, and bakery items have been employed as delivery vehicles for probiotics. Table 12.1 shows some probiotic products developed along with the probiotic survival rate at the end of storage period. The main limiting factor for probiotic stability in these food formats is water activity. Generally the water activity should not be more than 0.25 to achieve a year's shelf life at 25 °C, with the exception of fat-based foods such as chocolate and peanut butter, where good probiotic stability can be obtained despite higher water activities of up to 0.4 (Fenster et al. 2019). Other important factors that influence successful incorporation of probiotics are fermentation and storage temperature, packaging type, processing steps, and interaction with other ingredients (Cruz et al. 2013; Masoud et al. 2015). In addition, sensory profile of the probiotic products must also be taken into consideration as this can affect consumer acceptability. Amongst all the food categories, fermented milk products are regarded as the best matrix for carrying probiotics.

12.3.1 Dairy-Based Probiotic Foods

Probiotics can be found in a wide variety of commercial dairy products including fresh milk, yoghurt, cheese, ice cream, etc. Examples of some established probiotic organisms and major dairy product classes are listed in Table 12.2. Dairy products play an important role in delivering probiotic bacteria to humans, as these products provide a suitable environment for probiotic bacteria that support their growth and viability.

Probiotic dairy beverages were the first commercialized products that are still consumed in larger quantities than other probiotic products. *Lactobacillus rhamnosus* GG is the most widely used strain in the manufacture of dairy beverages. Other reported commercial probiotics in dairy beverages include *Lactobacillus acidophilus* La-5, *Lactobacillus casei* DN 114001, *L. rhamnosus* GG, *Lactobacillus plantarum* 299v, *Bifidobacterium bifidum* BB536, and *Bifidobacterium longum* CKL 1969. Acidophilus and bifidus milk are the best examples of probiotic fermented milk, prepared by inoculation of *L. acidophilus* and *B. bifidum* species, respectively. Since most lactobacilli are slow growers and

Table 12.1 Probiotic food products and their viability at the end of storage period.

Probiotic products	Probiotic strain/s	Storage period	Count at the end of storage (cfu/g or cfu/ml)
Dairy-based			
Fermented cow's milks	*L. acidophilus* *L. rhamnosus*	7 days	10^7
Fermented dairy drink from goat's milk	*L. acidophilus* *B. animalis* subsp. *lactis*	21 days	10^7
Cow's milk fruit yoghurt	*L. acidophilus* *B. animalis* subsp. *lactis*	35 days	10^6–10^7
Ewe's milk yoghurt	*L. acidophilus* *B. bifidum* *L. casei*	14 days	10^7
Ice cream	*L. johnsonii*	8 months	10^7
Goat's milk ice cream	*L. acidophilus, B. animalis subsp. lactis, Propionibacterium jensenii*	52 weeks	10^7–10^8
Cheddar cheese	*L. paracasei, Lactococcus lactis* subsp. *cremoris*	90 days	10^7
Semi-hard goat's cheese	*L. acidophilus* *B. lactis*	70 days	10^6
Gouda cheese	*L. acidophilus* NCFM, *L. paracasei* Lpc-37, *L. rhamnosus* HN001	200 days	10^8
Fresh minas cheese	*L. paracasei*	21 days	10^8
Milk powder	*L. acidophilus* *B. animalis* subsp. *lactis, Propionibacterium jensenii*	24 weeks	10^5–10^7
Soya-based			
Soya frozen dessert	*L. acidophilus* *L. paracasei* *B. lactis* *L. rhamnosus* *S. boulardi*	28 weeks	10^5–10^7
Soy milk	*B. breve*	20 days	10^9
Cereal-based			
Oat bars	*B. lactis*	7–14 days	~10^7
Milk-based maize/rice pudding	*B. animalis* *L. acidophilus* *L. rhamnosus*	21 days	10^8–10^9
Oat meal gruel mixed with fruit drinks	*L. plantarum*	30 days	10^{10}
Fruit and fruit juice			
Dried apple fruits	*L. casei*	–	10^6
Pineapple juice	*L. casei*	42 days	10^6
Cashew apple juice powder (spray dried)	*L. casei* NRRL B-442	35 days	~10^6

(*Continued*)

Table 12.1 (Continued)

Probiotic products	Probiotic strain/s	Storage period	Count at the end of storage (cfu/g or cfu/ml)
Orange juice powder (spray and freeze dried)	*L. plantarum* 299v	180 days	10^6
Vegetable-based			
Carrot blended with orange juice	*L. plantarum* CECT220	30 days	10^8–10^9
Tomato juice	*L. plantarum* *L. acidophilus* *L. casei* *L. delbrueckii*	30 days	10^4–10^8
Vegetable pickle products	*L. casei* LA284	70 days	10^4–10^8
Fermented kale juice	*L. plantarum* *L. acidophilus* *L. casei* *L. brevis*	4 weeks	10^8
Meat products			
Dry sausages (beef and pork)	*L. rhamnosus*	28 days	10^8
Pork Ioins	*L. casei* LOCK 0900	21 days	10^7–10^8
Miscellaneous			
Chocolate	*L acidophilus* LA3, *B. animalis* subsp. *lactis* BLC1	120 days	10^7–10^8
Probiotic dark chocolate	*L. plantarum* 564 *L. plantarum* 299v	360 days	10^5–10^6
Chocolate	*L. casei* 01 *L. acidophilus* LA5	60 days	$>10^6$
Potato chips	*L. acidophilus* *L. rhamnosus*	14 days	$>10^6$

Source: Adopted and modified from Ranadheera et al. (2017).

Bifidobacterium species favours anaerobic conditions, great precautions are required during the probiotic product manufacture to facilitate growth of these cultures. Higher inoculum levels (2–10%) and optimum growth conditions (37 °C) are suggested in the literature. Milk also needs to be severely heated from 80 to 125 °C for 5–60 minutes to kill all the unwanted microorganisms, including spores. Several factors have been reported to affect the viability of probiotic cultures in fermented milks. Acidity, pH, dissolved oxygen content, redox potential, hydrogen peroxide, presence of starter cultures, and various additives (including preservatives) affect the viability of probiotic bacteria.

Yoghurt is known for its nutritional value and health benefits and is one of the original sources of probiotics. Yoghurt is produced using a symbiotic culture combination of *Lactobacillus delbrueckii* subsp. *bulgaricus*, and *Streptococcus thermophilus*. The probiotic characteristics of the bacterial strains used for yoghurt production are still debatable. In addition, other lactobacilli and bifidobacteria are also sometimes added to the yohurt during or after fermentation. The viability of probiotics in yoghurt depends on the availability of nutrients, inhibitors and growth promoters, incubation

Table 12.2 Examples of commercial probiotic dairy products available in the market.

Product name	Probiotic organisms used
Muller Vitality Probiotic Yoghurt, UK.	*L. acidophilus* and *Bifidobacterium* spp.
Crunch n Yoghurt™, General Mills, USA	*L. acidophilus, L. bulgaricus*, and *Streptococcus thermophilus*
Classic flavour, France	*L. acidophilus*
Lactinex powder, Hyson, Westcott and Dunning	*L. acidophilus* and *L. bulgaricus*
LGG capsules, Japan	*L. rhamnosus* GG
LGG Milk, Japan	*L. rhamnosus* GG
LGG+, Iceland and Greenland	*L. rhamnosus* GG
Dukat BioAktiv Dairy Product, Slovenia	*L. rhamnosus* GG
Tnuva LGG1, Israel	*L. rhamnosus* GG
Beautiful Day LGG, Taiwan	*L. rhamnosus* GG
Culturelle capsules, USA	*L. rhamnosus* GG
Yakult, Japan	*L. casei* strain Shirota
Bacilac, Belgium	*L. acidophilus* plus *L. rhamnosus*
VSL-3, Italy	Eight LAB species
Vitacel® prolac, Rosenberg	*Lactobacillus reuteri*
Amul Prolife probiotic ice cream, India	*L. cremoris, L. lactis, L. bulgaricus*, and *L. acidophilus*
Symbalance yoghurt, Switzerland	*L. reuterii, L. acidophilus*, and *L. casei*
Danone (Bio Activia)	*B. esselnsis*

temperature, fermentation time, and storage temperature. Often growth promoters are added to probiotic products to maintain the cell viability. Hussain et al. (2015) developed probiotic *lassi* (a sweetened stirred yoghurt-like beverage) supplemented with *Aloe vera*. The authors reported that *Aloe vera* sustained the growth of probiotic *Lactobacillus paracasei* ssp. *paracasei* NCDC 627 in *lassi*. A cell count of >8 log CFU/ml was observed at the end of a 12-days storage period. Packaging material also plays an important role in probiotic survivability in food products. Application of high barrier packaging materials resulted in better survivability of probiotics in food products. Higher probiotic count was reported for *Aloe vera* supplemented probiotic *lassi* samples when packaged in glass (7.17 ± 0.15 cfu/ml) and ethylene-vinyl alcohol copolymer (EVOH) (6.97 ± 0.34 cfu/ml) materials compared to low density polyethylene (LDPE) (6.70 ± 0.12 log cfu/ml) at the end of 30 days storage period (Kumar et al. 2017).

Cheese is also considered a suitable vehicle for the delivery of probiotic cultures in fermented dairy foods. Cheeses have a number of advantages over yoghurt and other fermented milks due to the higher pH and buffering capacity, higher nutritional value, more solid consistency, and longer shelf life. Probiotics can be added to cheese either before fermentation (together with a starter culture) or after the fermentation is carried out by a starter culture. Several studies (Blaiotta et al. 2017; Cuffia et al. 2018) reported better survival rates of probiotics in cheese at the end of their shelf life. Probiotics in cheese were found to survive the passage through the simulated human gastrointestinal tract. However, comparing the serving size of yoghurt to that of cheese, cheese needs to have a higher density of probiotic cells and higher viability to provide the same health benefits. Fresh cheese, such as cottage cheese, has a high recommended daily intake, and a limited shelf life at refrigerated storage temperature. It may, thus, serve as a food with a high potential to be applied as a carrier of probiotics. Ripened cheeses, including Cheddar cheeses, however, have long ripening times, hence development of a probiotic cheese requires careful examination of the suitability of particular strain(s) to maintain viability throughout its shelf life (Stanton et al. 1998).

Other dairy products including quarg, chocolate mousse, frozen fermented dairy desserts, sour-cream, and ice cream can act as suitable vehicles for probiotic supplementation. Addition of probiotic culture combination DVS-Probio-Tec™ ABT-1 (*L. acidophilus*-5, *Bifidobacterium*-12, *S. thermophilus*) to Quarg cheese improved its nutritional characteristics (Mirjana et al. 2007). An appreciable quality probiotic chocolate mousse was also developed by supplementing with *L. paracasei* subsp. *paracasei* LBC82, solely or in combination with inulin (Aragon-Alegro et al. 2007). Sour-cream was investigated as a probiotic vehicle and proved a feasible probiotic carrier by Wilson et al. (2004). Ice cream also has good potential as a probiotic carrier. Neutral pH and high total solids in ice cream aid probiotic stability. In contrast to fermented dairy products, where probiotic viability is measured in days or weeks, ice cream is able to accommodate probiotic strains for more than a year (Alamprese et al. 2005). However, several factors that need to be considered include appropriate culture selection (low-temperature resistant strains), stage of probiotic addition, transport conditions, and storage temperatures. Overrun, which aerates the ice cream mix prior to packaging, may cause a drop in viable cell counts due to increased oxygen incorporation. This can be compensated for by adding excess probiotics or by culturing the probiotic population during a potential fermentation step prior to the actual ice cream manufacture. Additionally, the toxic activity of oxygen might be accommodated by utilizing aero-tolerant species such as lactobacilli as compared to strictly anaerobic species.

12.3.2 Non-dairy Based Probiotic Products

According to the National Institute of Diabetes and Digestive and Kidney Diseases (NIDDK) of the US National Institutes of Health, about 75% of the world population is lactose intolerant. Non-dairy probiotic products received considerable interest amongst vegetarians and lactose intolerance consumers. Some non-dairy-probiotic products developed are listed in Table 12.3.

Fermentation of vegetables as a mode of preservation has been known since ancient times. Fermented vegetables can offer a suitable medium to deliver probiotics. *L. rhamnosus*, *L. casei*, and *L. plantarum* are better adapted to the vegetable products during fermentation. However, a low incubation temperature of typical vegetable fermentations is a big challenge for the introduction of the traditional probiotic *L. acidophilus* and *Bifidobacterium* bacteria. Nevertheless, when the temperature is adjusted to 37 °C, probiotic bacteria grow rapidly in plant-based substrates.

Fruit juice-based probiotic products have gained increased popularity in recent times. The fruit juices contain beneficial nutrients that can be an ideal medium for probiotic cultures. Fruit juices have pleasing taste profiles liked by all age groups and they are perceived as being healthy and refreshing. The fruits are rich in several nutrients such as minerals, vitamins, dietary fibres, and antioxidants. These characteristics allow the selection of appropriate strains of probiotics to manufacture refreshing healthy fruit juice. Addition of probiotics may impart different taste profiles to probiotic juices when compared to their conventional counterparts. Different attempts have been made to reduce the sensations of unpleasant aromas and flavours in probiotic fruit

Table 12.3 Examples of non-dairy probiotic products developed.

Category	Product
Fruit- and vegetable-based	Fermented banana pulp
	Fermented banana
	Beets-based drink
	Tomato-based drink
	Dried fruits
	Green coconut water
	Peanut milk
	Cranberry, pineapple, ginger, blackcurrant, passion fruit, cabbage, carrot, Noni and orange juices
	Onion
	Probiotic banana puree
	Non-fermented fruit juice
Soy-based	Non-fermented frozen desserts
	Fermented soy milk
	Soy-based stirred yoghurt-like drinks
Cereal-based	Cereal-based puddings
	Rice-based yoghurt
	Oat-based drink
	Yosa (oat-bran pudding)
	Mahewu (fermented maize beverage)
	Maize-based beverage
	Wheat, rye, millet, maize and other cereals fermented probiotic beverages Malt-based drink
	Boza (fermented cereals)
	Millet or sorghum flour fermented probiotic beverage
Others	Starch-saccharified probiotic drink
	Probiotic cassava-flour product
	Meat products
	Dosa (rice and Bengal gram-based fermented product)

juice, which includes incorporation of 10% (v/v) of tropical fruit juices, mainly pineapple, mango, and passion fruit. Other probiotic cultures used in the manufacture of fruit-based probiotics are *B. longum* and *L. plantarum, L. acidophilus,* and *L. casei.*

Cereal-based products offer many possibilities. Cereals are good substrates for the growth of probiotic strains as they contain essential nutrients. Also, the presence of non-digestible components in the cereal matrix may serve as prebiotic source. Due to the nutritional complexity of cereals, a systematic approach is required to identify the factors that enhance the growth of probiotics. A multitude of fermented cereal products have been developed, but only recently, the presence of probiotic microorganisms in traditional fermented cereal foods have been reported (Ganguly et al. 2019; Mridula and Sharma 2015; Salmerón et al. 2015). Oats are widely reported with respect to probiotic-based cereal products. Several studies (Nair and Shah 2017; Sangwan and Singh 2018; Sharma et al. 2017) have demonstrated the potential of oats as substrates for the development of probiotic products containing *L. plantarum*, *Lactobacillus reuteri*, *L. acidophilu*, *L. rhamnosus*, *Lactobacillus reureri*, *Bacillus*, and *Bifidobacteria* strains.

It is believed that meat matrices might support the survival of probiotic bacteria through the gastrointestinal tract. However, manufacturing probiotic meat products is difficult compared to the manufacturing other probiotic products because of the nature of the raw material. Probiotic cultures should be able to adapt to the conditions in fermented meat products and become dominant in the final products which contain high natural background microbiota. Furthermore, high content of curing salt, low pH, and water activity due to acidification and drying might affect cell viability. Probiotics can be added to meat products such as sausages, hams, and pork loins after the curing steps, which then allow for fermentation (2–4 days at 22–24 °C) and ripening (21 days at 16–24 °C) followed by storage between 4–6 °C until maturation (6 months) (Kolozyn-Krajewska and Dolatowski 2012). The selection of appropriate probiotic strains which can remain viable in the meat environment is important. Lactic acid bacteria species isolated from similar sources may serve as useful probiotics for meat products. Lactobacilli, mainly *L. plantarum, Lactobacillus pentosus, Lactobacillus curvatus, L. acidophilus, L. rhamnosus, L. casei, L. paracasei*, and *Bifidobacterium* spp. isolated from meat and sausages are found suitable for meat fermentation and as probiotics. In order to support and maintain growth and viability in meat products, probiotic strains are encapsulated. Muthukumarasamy and Holley (2006) used alginate microencapsulated *L. reuteri* and *B. longum* in the preparation of a probiotic meat-based sausage.

12.4 Prebiotic Food Formulations

Prebiotics are non-digestible food ingredients that beneficially affect the host by selectively stimulating growth, activity, or both, of one or a limited number of bacterial species resident in the colon (Gibson and Roberfroid 1995). The definition of prebiotic has been revised multiple times during 2003–2017. Bindels et al. (2015) suggested a more comprehensive definition for prebiotics as non-digestible compounds that, through their metabolization by microorganisms in the gut, modulate composition and/or activity of the gut microbiota, thus conferring a beneficial physiological effect on the host. Recently, Gibson et al. (2017) defined prebiotics as substrates that are selectively utilized by host microorganisms conferring health benefits. To exhibit such effects, a prebiotic must neither be hydrolyzed nor absorbed in the upper part of the gastrointestinal tract, and must selectively be fermented by one or a limited number of potentially beneficial bacteria residing in the colon. Prebiotics may be classified using five basic criteria such as: (i) resistance to digestion in the upper sections of the alimentary tract; (ii) fermentation by intestinal microbiota; (iii) beneficial effect on host health; (iv) selective stimulation of growth of probiotics; and (v) stability in various food/feed processing conditions. Primarily, prebiotics are added to food products to support the growth of probiotics to maximize health effects, although prebiotics have been known to possess inherent capabilities to influence product properties. The numbers of probiotics in the human gut tend to decrease with age. Supplementation of food with prebiotics has been suggested to maintain a high level of probiotics, thus maintaining good health.

Prebiotic compounds can be classified on the basis of their chemical nature, chain length or degree of polymerization, mode of usage, etc. Depending on their chemical nature, prebiotic compounds are categorized into three types such as saccharide derivatives (disaccharides, oligosaccharides, and polysaccharides), proteins, and lipids (Ishwarya and Prabhasankar 2014). Most of the prebiotics used as food adjuncts are saccharide derivatives and mainly obtained from plants. A range of dietary compounds have been

Table 12.4 Prebiotic ingredients used in food products.

Prebiotic ingredient	Concentration used	Product	Role
Lupine seed (Raffinose family oligosaccharides)	2%	Fermented milk	Reduced fermentation time and increased probiotic count *B. lactis* Bb-12 and *L. acidophilus* La5
Oat bran, green banana flour (β-glucan)	9%	Dairy fruit beverage	Increased survival of *L. casei* LC-1
Quinoa flour (dietary fibre and unsaturated fatty acids)	3% (Dietary fibre 7%)	Fermented milk	Increased nutritive value
Lentil flour (Raffinose family oligosaccharides)	1–2% (1–1.3% prebiotic carbohydrates)	Yoghurt	Improved physical and rheological properties
Lemon and orange fibres	3%	Fermented milk	Enhanced bacterial growth, survival of probiotic and sensorial properties
β-glucan and banana	1.1% dietary fibre and 0.5% β-glucan	Oat fermented beverage	Improved palatability and viscosity
Wheat	0.25%	Symbiotic wheat milk beverage	Improved physicochemical, microbiological and sensory characteristics
Inulin	3%	Yoghurt	Increase viable count
Inulin	4%	Soy milk yoghurt	Growth stimulation

suggested as prebiotics (Table 12.4), that includes mainly inulin, fructo-oligosaccharides (FOS), galacto-oligosaccharides (GOS), β-glucan, and lactulose, although other compounds from various plant sources have been used in food products.

12.5 Functional Properties

Probiotics and prebiotics provide a multitude of functional properties when incorporated into food systems. The incorporation of prebiotics helps in texture development, improved sensory acceptance, and enhanced shelf life of food products due to their water holding capacity, gel-forming ability, fat mimetic roles, anti-sticking properties, texturizing and thickening effects (Rakhesh et al. 2015). It was observed that probiotics also have been used to improve textural attributes of several food products. Probiotic cheese with the addition of *L. acidophilus* was firmer at the end of storage due to higher values of pH and hardness (Buriti et al. 2005). Percentage of syneresis and proteolytic index showed no difference in cheese made with probiotic and traditional cultures. *Dahi* made by exopolysaccharides producing probiotic *Lactobacillus fermentum* NCDC400 (formerly known as V10) exhibited higher rheological and sensory properties (Behare et al. 2013).

Textural analysis indicated that adhesion and stickiness values (parameters representing force necessary to remove the material that adheres to the mouth during eating) were higher for *L. fermentum* NCDC400 *dahi* than for the control *dahi*, indicating that the probiotic strain contributed to the mouth-feel of the product. Van Ba et al. (2018) observed that addition of probiotic *L. plantarum* to fermented pork sausage resulted in a significantly lower spoilage bacteria count and lipid oxidation levels. Also, the probiotic addition improved the odour, taste, and acceptability scores of the sausage.

Compared to probiotics, prebiotics play a major role in the techno-functional properties of food. Inherent characteristics of prebiotics viz. source, molecular weight, chemical structure, and tendency to react with other food components play a major role in functionality of prebiotics. The majority of prebiotics are dietary fibres that pass through the digestive tract largely undigested. In the colon, prebiotics are selectively fermented by the beneficial flora, which stimulates their growth and reinforces its action against putrefactive microorganisms. Prebiotic interaction with casein micelle or other carbohydrate can change the gel structure of food products strongly. Prebiotics, such as inulin, were shown to decrease whey syneresis and improve the firmness and cohesiveness of low-fat fermented milk. However, this depends on the concentration and level of prebiotic added to these products. The use of a higher concentration of prebiotics decreases the firmness of the protein gel due to an increase in viscosity of the continuous phase and trapping the casein aggregates within, while sub-optimal levels of prebiotics can lower the apparent viscosity due to more compaction of casein aggregates. Some studies indicated that prebiotic addition to fermented milk products increases the sweetness and creamy mouth feel, enhances aroma, and improves appearance. Isik et al. (2011) reported that addition of inulin in low-fat frozen yoghurt increased the textural attributes, melting rate, and decreased fluffiness. Glibowski and Kowalska (2012) reported that addition of inulin to kefir increased the firmness and cohesiveness.

Ice cream supplemented with 2.5% inulin supported the growth of *L. casei* and improved nutritional and sensory properties (Criscio et al. 2010). Morais et al. (2014) reported that addition of the prebiotic inulin at 7.5% (w/w) to low-fat chocolate dairy dessert resulted in a similar sensory acceptability to that of its high-fat counterpart. Aidoo et al. (2014) developed a sugar-free chocolate formulation consisting of 75.36% polydextrose and 25.64% inulin with acceptable rheological and physical quality characteristics. In some cases, prebiotics are recommended as an encapsulating material to deliver probiotics and other functional ingredients. Encapsulation of *Lactobacillus* and *Bifidobacterium* species in yoghurt by FOS improves bacterial count without any undesirable effects on the chemical properties of the food. Cruz et al. (2013) evaluated the effect of an increasing concentration of oligofructose addition on physicochemical, rheological, and microbiological characteristics of yoghurt. Addition of the prebiotic did not affect pH, proteolysis, or the viability of *S. thermophilus* or *Lactobacillus delbreuckii* subsp. *bulgaricus* during the 28 days of refrigerated storage. The sensory properties of a peach-flavoured drinkable yoghurt made with FOS were comparable to the control product, indicating that a prebiotic can be added without impacting sensorial attributes (Gonzalez et al. 2011). On the other hand, inulin concentration above 5% in bread affected sensory characteristics and resulted in smaller loaves, harder crumb, and darker colour (Morris and Morris 2012). Sheep's milk ice cream containing 10% inulin or FOS was creamier and brighter in appearance than the control full-fat ice cream (Balthazar et al. 2017). Authors suggested these prebiotics as alternative fat replacers for manufacture of low-fat sheep's milk ice cream. Table 12.5 and Figure 12.1 present functional properties of probiotic and prebiotic foods.

Table 12.5 Functional properties of probiotics and prebiotics in food.

Functional properties	Food products
A. Probiotics	
Physicochemical and sensory properties	Wheat symbiotic beverages
Improved rheological and sensory properties	Flavoured probiotic beverages
Rich, smooth, and traditional taste	Greek yoghurt
Improved aroma, appearance, mouthfeel, texture, and taste	Probiotic goat milk drink
Proteolysis, ripening time, texture, and sensory characteristics	Probiotic cheese
Creamy sensation, high textural quality, and enhanced flavour	Probiotic ice cream
B. Prebiotics	
Sugar replacement, mouthfeel and texture, fibre	Yoghurt and dessert
Sugar replacement, mouth feel, foam stabilization, and prebiotic	Beverages and drinks
Fat or sugar replacement, texture, fibre, and prebiotic	Breads and fillings
Meat products	Fat replacement, texture, stability, and fibre
Fat or sugar replacement, fibre, and prebiotic	Dietetic products
Sugar replacement, moisture retention, rheological characteristics, and fibre	Chocolate
Sugar replacement, fibre, and prebiotic	Sugar confectionary
Sugar replacement and prebiotic	Soups and sauces
Texture, body, and mouth feel, fibre, stability, and prebiotics	Baby foods
Improved rheological and sensory properties	Low fat custard
Colour, taste, creaminess, appearance, sensory characteristics	Soy cream dessert
Fat or sugar replacement, mouth feel, and appearance	Sheep's milk ice cream

12.6 Health Effects

A number of health benefits have been claimed for probiotics and prebiotics. Inclusion of these agents in the diet is considered to be important in maintaining good health. Consumption of probiotic cultures and prebiotics regularly in the diet seems an ideal method to re-establish the intestinal flora balance after antibiotic treatment. There is a growing agreement relating to the beneficial aspects of specific dairy products such as fermented milk and yoghurt and of bacterial cultures that ferment the dairy products. In addition, the use of specific prebiotic substances that are compatible with the product and enhance probiotic growth further maximizes health properties. Some of the physiological benefits attributed to probiotics and prebiotics are shown in Figure 12.2 and discussed in the following sections.

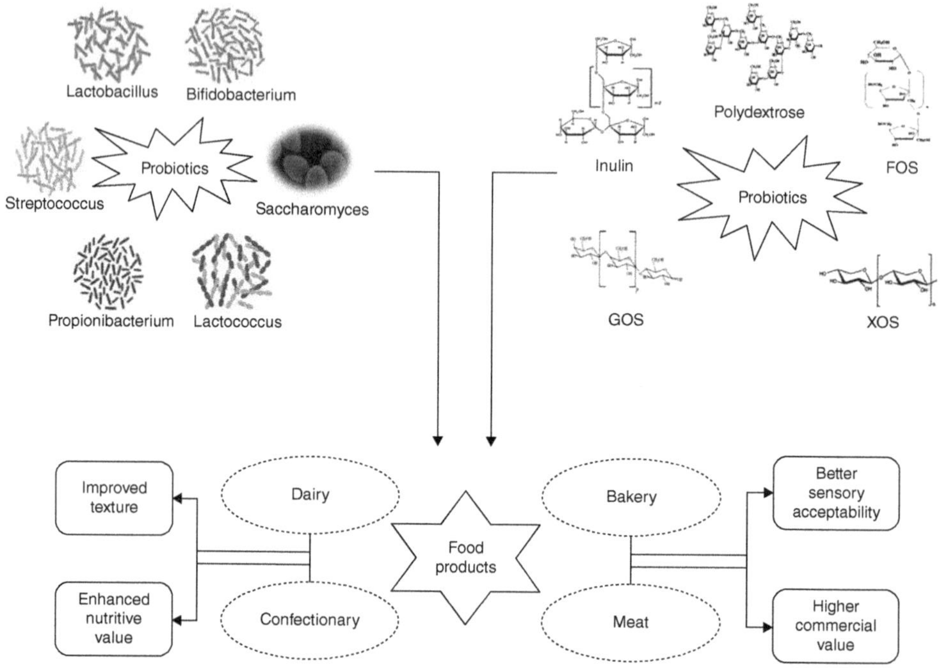

Figure 12.1 Functional roles of probiotics and prebiotics in food products.

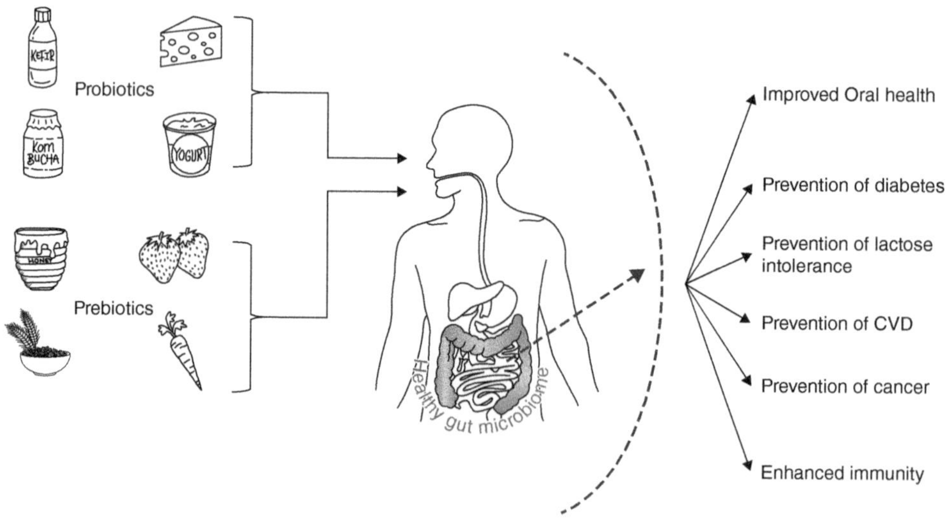

Figure 12.2 Probiotics and prebiotics in human health.

12.6.1 Reduced Risk of Lactose Intolerance

Most humans, commonly non-Caucasians, become lactose intolerant after weaning. Lactose intolerant people cannot metabolize lactose due to the lack of an essential enzyme, β-galactosidase. If lactose passes through the small intestine, it is converted to acid and gas in the large intestine by the colonic microflora. In addition to this, the presence of hydrogen in the breath is a signal for lactose mal-digestion. The major factor that improves the digestibility of lactose by hydrolysis is the bacterial enzyme β-galactosidase. Use of certain starter cultures along with probiotics in fermented milk products alleviates the problem of bad breath and associated symptoms. The beneficial effects of probiotics on lactose intolerance are due to the lower lactose concentration and high lactase activity of bacterial preparations in the fermented dairy products. Since lactose is converted to lactic acid by the β-galactosidase enzyme in fermented milk products, such as yoghurt, they are more suitable than milk for individuals suffering with lactose intolerance. However, yoghurt microflora, consisting of *L. delbrueckii* subsp. *bulgaricus* and *S. thermophilus*, are not generally resistant to gastric acidity. Hence, milk products that contain probiotic bacteria are more effective for lactose intolerant humans. Therefore, the β-galactosidase activity of probiotic strains and other lactic acid bacteria used in dairy products is really important. β-galactosidase activity within probiotics varies remarkably. Both the enzyme activity of probiotic strain and the activity remaining in the final product for efficacy in lactose intolerant subjects needed to be considered. The common probiotics used in yoghurt to overcome lactose intolerance and to provide other health benefits are *L. acidophilus*, *L. rhamnosus* GG, *L. casei* strain Shirota, *B. bifidum*, *B. longum*, etc. The combination of *L. casei* strain *Shirota* and *Bifidobacterium breve* Yakult survived gastrointestinal transit and improved the symptoms of lactose intolerance significantly in the patients (Almeida et al. 2012). This effect persisted even after the intervention ceased. Consumption of Kefir comprising a mixture of starter and probiotic cultures improved lactose digestion and tolerance in adults with lactose maldigestion (Hertzler and Clancy 2003). A mix of probiotics in yoghurt together with *B. longum* capsules increased the β-galactosidase activity in faeces and alleviated the complaints of lactose intolerance (He et al. 2008). In a randomized placebo-controlled clinical trial, Roskar et al. (2017) evaluated the efficacy of a probiotic product comprising *Bifidobacterium animalis* subsp. *animalis* IM386 and *L. plantarum* MP2026 in lactose intolerant individuals. Results suggested improvement in lactose intolerance symptoms in the probiotic group but observed great variability amongst individuals due to a smaller sample size.

GOS feeding to lactose intolerant participants for 36 days showed improvements in at least one of the disease symptoms (pain, bloating, diarrhoea, cramping, or flatulence) in 71% of the subjects (Azcarate-Peril et al. 2017). Furthermore, at the end of the study, the bifidobacteria count significantly increased by 90% in non-lactose tolerant participants consuming GOS. Growth of lactose fermenting *Bifidobacterium*, *Faecali bacterium*, and *Lactobacillus* increased significantly. Oat fermented beverage supplemented with high glucan levels and banana suggested as an alternative for people suffering from intolerance to milk proteins and lactose (Goncerzewicz et al. 2016).

12.6.2 Antimicrobial Properties

The intestinal microflora is a complex ecosystem, and newly-introduced organisms face an extremely competitive environment. Organisms that can produce a product that will inhibit the growth or kill existing unwanted organisms in the intestinal environment have a characteristic advantage. The possibility that probiotics will establish in the gastrointestinal tract is enhanced by their ability to eliminate competitors. Prebiotics influence the gut microbiota that affect growth of pathogenic organisms by facilitating production of acidic metabolic end products, consequently lowering the colonic pH, competitive effects due to limitations in number of colonization sites, and enhancement of the immune system. The antagonistic conditions are produced by probiotic organisms either by changing the conditions at the site of action or production of bacteriocins or antibiotic-like substances. Bifidobacteria produce acetic and lactic acids in a molar ratio of 3:2. *L. acidophilus* and *L. casei* produce lactic acid as the main end product of fermentation. Certain lactobacilli also produce hydrogen peroxide and bacteriocins as antimicrobial substances. These inhibitory substances create antagonistic environments for food-borne pathogens and spoilage organisms. Bacteriocins are the ribosomally synthesized proteinaceous molecules produced by certain LAB and probiotic strains that show bacteriostatic or bactericidal action against closely-related or unrelated species of bacteria yeast and moulds. Pathogenic microorganisms are able to penetrate and/or colonize body surfaces, due to a massive attack of the pathogens or to a (temporarily) reduced colonization resistance. Probiotic organisms are used to improve the colonization resistance on body surfaces, such as the gastrointestinal, urogenital, and respiratory tract. Consumers taking yoghurt had lower *Enterobacteriaceae* count and increased galactosidase activity in the gastrointestinal tract than those that did not take yoghurt (Alvaro et al. 2007). Tejero-Sarinena et al. (2013) studied the effect of probiotics on the survival of *Salmonella enterica*, *Serover typhimurium*, and *Clostridium difficile* in an *in vitro* model and postulated that probiotics inhibit pathogens by the production of short-chain fatty acids (SCFAs) such as acetic, propionic, and butyric acid. Acidophilin, a dietary fermented milk product made by fermentation of pasteurized milk with *L. acidophilus*, *Lactococcus lactis* subsp. lactis, and kefir starter is used in the treatment of colitis, enterocolitis, and other intestinal diseases as a supportive medicinal product (Ozer and Kirmaci 2010). A probiotic whey beverage fermented with a combination of *L. acidophilus* NCDC 15 and *L. casei* NCDC 12 exhibited antagonistic effects against *Escherichia coli*, *Klebsiella pneumonia*, *Salmonella typhi*, and *Staphylococcus aureus* (Tripathi and Jha 2004). Drinking yoghurt comprising a mixture of *L. casei*, *Lactobacillus bulgaricus*, and *S. thermophilus* prevented antibiotic-associated diarrhoea in adults (Hickson et al. 2007).

Sachdeva and Nagpal (2009) performed meta-analysis of randomized trials on *Helicobacter pylori* eradication by prebiotics in fermented milk products. Preliminary comparisons showed increased efficacy of fermented milk-based products over capsule/sachet-based probiotics for eradicating intestinal *H. pylori*. A prebiotic formula comprising GOS and FOS fed to 160 healthy bottle-fed infants within 0–14 days of birth was well tolerated (Costalos et al. 2008). The study showed higher percentages of *Bifidobacterium* and lower percentages of *E. coli*, while showing suppression of *Clostridium* in the stool. Chen et al. (2016) investigated the effect of prebiotic formula comprising inulin, FOS, and GOS on toddler gut health by measuring faecal microbiota. Consumption of prebiotic formula three times per day, giving a total intake of 1.8 g prebiotic ingredients significantly increased the number of *Bifidobacterium* spp. and reduced the population of *Clostridium perfringens* and total anaerobic bacteria. Oral feeding of *lassi* (Hussain et al. 2017) and ice cream (Yadav et al. 2017) supplemented with probiotic strain *L. paracasei* ssp. *paracasei* NCDC 627 prevented *Shigella dysenteriae*-induced infection in mice.

12.6.3 Anticarcinogenic Properties

In the last two decades, the number of people suffering from colon cancer has been gradually increasing, particularly in industrialized countries (Colace et al. 2017). Earlier studies indicated that diet containing probiotic organisms and antibiotics can lower the generation of carcinogens in the colon and reduce chemically-induced tumours. These effects appear to be mediated through the intestinal microflora. A possible mechanism for these anticancer effects relies on inhibiting intestinal bacterial enzymes that convert pro-carcinogens to more proximal carcinogens. This is due to the ability of beneficial organisms to inhibit the growth of undesirable flora that have high activities of enzymes such as β-glucuronidase, nitroreductase, azoreductase, and β-glycosidase or the capability for nitrosation.

Liver cancer is the sixth-most commonly diagnosed cancer in the world, and third-most common cause of death from cancer, according to Cancer Research UK. The biggest risk factor for liver cancer is said to be chronic hepatitis B virus infection and consumption of foods contaminated with mycotoxins. The mycotoxin aflatoxin B1 (AFB1) is classified as a group I carcinogen in humans (Saini and Kaur 2012). Some studies have suggested that oncogenes are critical molecular targets for AFB1, and AFB1 causes characteristic genetic changes in the p53 tumour suppressor gene and ras proto-oncogenes. Probiotics may exert anticancer effects through modulation of gut microbiota composition and antimicrobial compound production, improvement of gut barrier function, and modulation of local and systemic immunity. Some probiotic bacterial strains have been successfully shown to bind mycotoxins; *L. rhamnosus* binds AFB1 *in vivo*, thus reducing bio-absorption of the toxin from the gut, while *L. acidophilus* and *B. longum* neutralize AFB1 and aflatoxin M1 through a similar binding mechanism. Li et al. (2016) reported that feeding the probiotic mixture comprising *L. rhamnosus* GG, *E. coli* Nissle 1917, and heat-inactivated VSL#3 to mice shifted the gut microbiota composition and reduced the size of liver tumours. Dietary supplementation using probiotics *L. rhamnosus* LC705 and *Propionibacterium freudenrichii* subsp. *Shermani* effectively reduced the excretion of aflatoxin DNA adducts (AFB1-N7-guanine) in urine (de LeBlanc et al. 2007). Probiotic fermented milk and chlorophyllin reduced tumour incidences, suggesting a protective effect of the probiotic against AFB1-induced hepatocarcinogenesis (Kumar et al. 2011).

Colorectal cancer is the one of the leading causes of cancer-related deaths worldwide. The global burden of colorectal cancer is expected to increase by 60% to more than 2.2 million new cases and 1.1 million deaths by 2030 (Arnold et al. 2017). Dietary interventions and natural bioactive supplements have now been extensively studied to reduce the risks of colon cancer as preventive measures. Probiotics have been found to decrease faecal concentrations of pro-cancer enzymes, and secondary bile salts and to reduce absorption of harmful mutagens that may contribute to colon carcinogenesis. Normal intestinal flora can also influence carcinogenesis by producing enzymes (glycosidase, B-glucuronidase, azoreductase, and nitroreductase) that transform pre-carcinogens into active carcinogens. Use of the probiotics *L. acidophilus* and *L. casei* in humans helped to decrease levels of these enzymes. The mechanisms proposed for probiotic organisms to prevent colon cancer includes enhancing the host's immune response, altering the metabolic activity of the intestinal microflora, binding and degrading carcinogens, producing antimutagenic compounds, and altering the physiochemical conditions in the colon. *In vitro* studies have demonstrated that probiotic strains *L. fermentum* NCIMB-5221 and 8829 are highly potent in suppressing colorectal cancer cells and promoting normal epithelial colon cell growth through the production of ferulic acid (SCFAs) (Kahouli et al. 2015). *L. acidophilus* supplemented with fermented milk products is known to prolong the induction of colon tumours. Daily oral administration of microencapsulated *L. acidophilus* bacterial cells in the yoghurt resulted in significant suppression of colon tumour incidence, tumour multiplicity, and reduced tumour size in mice (Urbanska et al. 2009). Feeding milk and colostrum fermented with *L. acidophilus* resulted in a 16–41% reduction in tumour proliferation (Andrews and Tan 2012). Munjal et al. (2009) reported that a fermentation product of the inulin type fructan showed growth inhibition and induction of apoptosis in human colorectal carcinoma. Feeding prebiotic and soybean in combination significantly reduced incidences (40–70%) of azoxymethane (AOM) induced colon tumours (Gourineni et al. 2011).

12.6.4 Immunomodulatory Activities

Lactobacilli used in dairy products can enhance the immune responses of the host. Modulation of the immune system is one of the mechanisms through which probiotics are thought to mediate health effects. Enhanced phagocytic activity of granulocytes, cytokine excretion in lymphocytes, and increased immunoglobulin secreting cells in the blood, are typical responses to probiotics, all of which indicate changes in the immune system. Because immune function declines with age, enhancing immunity in the elderly with probiotics would be of great importance. Probiotic cultures stimulate both non-specific (innate) and specific (adaptive) immunity. Possible stimulation of an immune response by probiotic bacteria may explain potential therapeutic and prophylactic applications of such cultures in the treatment of infections and carcinogenesis. Inhibition of carcinogenesis correlates with changes in immune activity in response to probiotic consumption. Organisms that have been identified as having this property include *B. longum*, *L. acidophilus*, *L. casei* strain Shirota, *L. rhamnosus*, *Lactobacillus helveticus*, *L. fermentum*, etc. Immunological responses of the host can be measured by lymphocyte proliferation, secretion of interleukins 1, 2, and 6, tumour necrosis factor, prostaglandin E production and serum total protein, albumin, globulin, and gamma interferon levels. Vinderola et al. (2005) determined the immunomodulation capacity of Kefir and pasteurized Kefir on the intestinal mucosal immune response in mice. Both Kefir types modulated the mucosal immune system in a dose-dependent manner. A 10-times diluted sample of control Kefir (not pasteurized) induced a similar magnitude of immunomodulation when compared to pasteurized Kefir, thus indicating the importance of cell viability. An improvement in the nutritional and immunological status of enterally-fed elderly subjects was observed by the administration of fermented milk containing *Lactobacillus johnsonii* La1 (Fukushima et al. 2007). A probiotic cheese containing *L. acidophilus* NCFM increased the cytotoxicity of NK cells in elderly volunteers (Ibrahim et al. 2010). In a study conducted by Lollo et al. (2013), on the effect of probiotic yoghurt and a probiotic whey beverage comprising *S. thermophilus* TA040, *L. bulgaricus* LB340, *L. acidophilus* LA14, and *B. longum* BL05 on immunomodulation after exhaustive treadmill exercise in Wistar rats, results demonstrated that the probiotic yoghurt outperformed the probiotic whey beverage in blood cell indicators (neutrophils and lymphocytes), cytokines (TNF-α and IL-1β), and various standard parameters.

12.6.5 Prevention of Cardiovascular Disease

CVD is a condition in which the main coronary arteries supplying the blood are no longer able to supply sufficient blood and oxygen to the heart muscle (myocardium). The main cause of the reduced blood flow is accumulation of plaques, mainly in the intimae of arteries, a disease known as atherosclerosis. This is a slow but progressive disease, which usually begins in childhood, but does not usually manifest itself until later life. The risk factors known to predispose an individual to CVD have been categorized into those that are not modifiable, such as age, sex, race, and family history, and those that are modifiable, such as hyperlipidaemia, hypertension, obesity, cigarette smoking, and lack of exercise. Epidemiological studies examining risk factor in different populations have observed a positive correlation between an individual's cholesterol levels and development of CVD. High blood concentrations of low density lipoproteins (LDL) and lipoproteins are associated with increased risk of CVD; also small high density lipoproteins (HDL) and large very low density lipoproteins (VLDL) particles may also play a role in the development of CVD. HDL lipoprotein is involved in reverse cholesterol transport, carrying cholesterol from cells to the liver for removal from the body and a low level of HDL cholesterol reflects impairment in this process.

It has been estimated that every 1% increase in LDL cholesterol level leads to a 2–3% increase in CVD risk. HDL cholesterol levels may influence the relationship between total cholesterol levels and CVD risk. Current dietary strategies for the prevention of CVD advocate adherence to low-fat/low saturated fat diets. Furthermore, supplementation of diet with fermented dairy products or probiotics containing dairy products has the potential to reduce serum cholesterol levels in humans and animals. The mechanism of action of probiotics on cholesterol is unclear; the possible mechanisms are cholesterol assimilation by bacteria, deconjugation of bile salts, cholesterol binding to bacterial cell walls, and reduction in cholesterol biosynthesis. Nevertheless, several reports showed the positive role of probiotic-containing foods on managing cholesterol levels in animal and human studies. Rats fed with yoghurt, probiotic dahi, and probiotic yoghurt containing *L. acidophilus* and *B. bifidum* exhibited significant reduction of 2.63, 4.1, and 4.68 mg/100 ml of serum cholesterol level at the end of 30 days (Vijayendra and Gupta 2012). Xiao et al. (2003) reported that consumption of low-fat yoghurt containing *B. longum* BL1, in a trial involving 32 hypercholesterolemic patients, showed a significant reduction in triglycerides, total serum and LDL cholesterol, and 14.5% increase in HDL cholesterol. Consumption of probiotic yoghurt containing *L. acidophilus* La5 and *B. breve* subsp. *lactis* Bb12 up to 8 weeks by 72 patients with non-alcoholic fatty liver disease, reduced serum levels of alanine transaminase/alanine aminotransferase (ALT), aspartate aminotransferase (ASP), total cholesterol (TC), and low-density-lipoprotein cholesterol (LDL-C) (Nabavi et al. 2014). Incorporation of the probiotic *L. paracasei* subsp. *paracasei* VC213 and *Enterococcus lactis* BT161 in cheese manufacture decreased cholesterol content during cheese ripening by up to 60 days (Albano et al. 2018). The authors suggested these strains as good candidates for novel probiotic-containing formulations and could be used to prepare functional dairy products.

Regular prebiotic intake regularly has been associated with reduced risk of vascular problems. A soy food diet enriched with prebiotic resulted in greater reductions in LDL cholesterol and increase in HDL cholesterol in hyperlipidemic adults (Wong et al. 2010). Symbiotic treatment of *Pediococcus acidilactici* LAB 5 with sorbitol for one month lowered the plasma cholesterol levels of Swiss albino mice (Mandal et al. 2009).

12.6.6 Urogenital Tract Infection (UTI)

Excluding sexually-transmitted diseases, almost all infections of the vagina and bladder are caused by microorganisms that originate in the bowel. A strong correlation has been reported between the presence of commensals, particularly lactobacilli in the vagina, with health, and an absence of these microorganisms in patients with urogenital infections. Disruption of the normal vaginal flora can be caused by broad-spectrum antibiotics, spermicides, hormones, dietary substances, etc. Probiotic microorganisms delivered as foods and topical preparations can have a role in preventing urogenital tract disorders. Some studies have shown that the daily oral intake of *L. fermentum* and *L. rhamnosus* can modify vaginal flora (Reid et al. 2001). Barrons and Tassone (2008) suggested that probiotics and fermented milk products consumption supresses urogenital tract infections. Consumption of fermented milk products containing probiotic bacteria greater than three times a week appeared to be associated with lower risks of UTI in fertile women (Kontiokari et al. 2003).

12.6.7 Enhancement of SCFA Production

Since the enhancement of intestinal microflora with probiotics primarily involves the stimulation of intestinal fermentation, the stimulation of SCFA production is one of the essential factors for the beneficial effects exerted by probiotics. Although increases in lactobacilli numbers stimulate lactate production, however, lactate does not accumulate in the large intestine, except in those patients with short bowel syndrome and dyspeptic diarrhoea. Rather, lactate is normally metabolized to acetate, propionate, or butyrate by lactate-utilizing bacteria. Thus, the administration of probiotics with other lactate-utilizing bacteria, butyrate-producing bacteria in particular, could be a more effective way to achieve maximum health benefits. The type and composition of the diet has an influence on the gut microbiota and, therefore, on SCFA production. The production of SCFA, such as acetate, propionate, and butyrate, is one of the functions of the gut microbiota and several bacteria may also produce them during fermentation of food matrices. Untreated barley has been shown to increase probiotic species as well as butyrate concentrations (Moen et al. 2016). Intake of probiotic fortified fermented milk products directly improved production of SCFA (Veiga et al. 2014). Asarat et al. (2015) investigated the effect of skimmed milk supplemented prebiotics (inulin, hi-maize, or β-glucan) and probiotics on SCFAs production. A combination of inulin and *L. rhamnosus* GG ATCC 53013 released the highest concentration of SCFAs, compared to LGG and hi-maize or β-glucan. The purified SCFAs, particularly butyrate, regulated cytokine milieu in LPS-stimulated human peripheral blood mononuclear cells to anti-inflammatory cytokines.

12.6.8 Antidiabetic Activities

Several authors have postulated the connection between the composition of the intestinal microbiota and metabolic diseases such as obesity and diabetes. Recent research has also proved that changes in the Firmicutes species can be associated with type-2 diabetes. Ratio of Bacteroidetes/Firmicutes has been positively correlated with plasma glucose concentrations. A similar pattern was also observed in the development of autoimmune diseases such as type-1 diabetes. Management of diabetes by modulating gut hormones via probiotic and prebiotic food product could be a promising strategy. Skimmed milk and a control *dahi* conferred protection against streptozotocin (STZ) induced diabetes in male Wistar rats. However, these products were less effective than dahi with *L. acidophilus* NCDC14 and *L. casei* NCDC 19 in delaying insulin resistance (Yadav et al. 2008). In a randomized, double-blind, placebo-controlled clinical trial on patients with type 2 diabetes, supplementation of *L. acidophilus* La-5, *B. animalis* subsp. *lactis* BB-12 into probiotic soy and probiotic soy milk up to six weeks, showed significant differences between groups concerning mean changes of HbA1c, total cholesterol, and

LDL cholesterol (Hariri et al. 2015). In another study with similar probiotics species in yoghurt, fasting blood glucose and antioxidant status in patients with Type-2 diabetes was reduced (Ejtahed et al. 2012).

Arabinoxylan fibres supplemented at 15 g/day can significantly improve glycaemic control in people with type 2 diabetes (Lu et al. 2004). Consumption of arabinoxylan in subjects with impaired glucose tolerance improved fasting serum glucose and triglycerides; however, this beneficial effect was not seen by changes in fasting adipokine concentrations (Garcia et al. 2007). Oligofructose-enriched inulin in young children between 8 and 17 years of age decreased endotoximia, reduced insulin resistance, improved glycaemic control, and changed gut microbiota permeability and inflammation (Ho et al. 2016). It has been suggested as a potential and novel agent for treating type 1 diabetes.

12.7 Conclusion

Consumers are increasingly managing their health by self-medication, which is generating expanding market opportunities for the functional food industry. Clinical studies have demonstrated the beneficial health effects of probiotics and prebiotics. As members of the autochthonous microflora of the gastrointestinal tract, probiotics offer considerable potential because of their history of safe use and the general body of evidence that supports their positive role. Prebiotics are used in conjunction with probiotics to explore targeted benefits. Dairy probiotic products have been widely researched due to their health appeal and incessantly increasing demand by consumers. Non-dairy probiotic products are also finding promising health applications and some are gaining market acceptability. Although ongoing research is further demonstrating the promising potential health effects of probiotics and prebiotics, more standardized and verifiable clinical studies are needed to demonstrate the safety, efficacy, and limitations of a putative probiotic, to determine effects on the diseased individuals and effects of long-term consumption, and to resolve whether they is superior to existing therapies.

Acknowledgements

First and second authors are thankful to IDP-NAHEP, ICAR-NDRI, Karnal, Haryana, India for providing a faculty development grant for undergoing research training at Teagasc Food Research Centre, Moorepark, Fermoy, Cork, Ireland.

References

Aidoo, R.P., Afoakwa, E.O., and Dewettinck, K. (2014). Optimization of inulin and polydextrose mixtures as sucrose replacers during sugar-free chocolate manufacture – rheological, microstructure and physical quality characteristics. *Journal of Food Engineering* 126: 35–42.

Alamprese, C., Foschino, R., Rossi, M. et al. (2005). Effect of *Lactobacillus rhamnosus* GG addition in ice cream. *International Journal of Dairy Technology* 58: 200–206.

Albano, C., Morandi, S., Silvetti, T. et al. (2018). Lactic acid bacteria with cholesterol-lowering properties for dairy applications: in vitro and in situ activity. *Journal of Dairy Science* 101: 10807–10818.

Almeida, C.C., Lorena, S.L., Pavan, C.R. et al. (2012). Beneficial effects of long-term consumption of probiotic combination of *Lactobacillus casei* Shirota and *Bifidobacterium breve* Yakult may persist after suspension of therapy in lactose-intolerant patients. *Nutrition in Clinical Practice* 27: 247–251.

Alvaro, E., Andriex, C., Rochet, V. et al. (2007). Composition and metabolism of the intestinal microbiota in consumers and non-consumers of yoghurt. *British Journal of Nutrition* 97: 126–133.

Andrews, J.M. and Tan, M. (2012). Probiotics in luminal gastroenterology: the current state of play. *Internal Medicine Journal* 42 (12): 1287–1291.

Aragon-Alegro, L.C., Alegro, J.H.A., Cardarelli, H.R. et al. (2007). Potentially probiotic and symbiotic chocolate mousse. *LWT – Food Science and Technology* 40: 669–675.

Arnold, M., Sierra, M.S., and Laversanne, M. (2017). Global patterns and trends in colorectal cancer incidence and mortality. *Gut* 66: 683–691.

Asarat, M., Apostolopoulos, V., Vasiljevic, T., and Donkar, O. (2015). Short chain fatty acids produced by symbiotic mixtures in skim milk differentially regulate proliferation and cytokine production in peripheral blood mononuclear cells. *International Journal of Food Sciences and Nutrition* 66: 755–765.

Azcarate-Peril, M.A., Ritter, A.J., Savaiano, D. et al. (2017). Impact of short-chain galactooligosaccharides on the gut microbiome of lactose-intolerant individuals. *Proceedings of the National Academy of Sciences of the United States of America* 114: E367–E375.

Backhed, F. (2005). Host-bacterial mutualism in the human intestine. *Science* 307: 1915–1920.

Balthazar, C.F., Silva, H.L.A., Cavalcanti, R.N. et al. (2017). Prebiotics addition in sheep milk ice cream: a rheological, microstructural and sensory study. *Journal of Functional Foods* 35: 564–573.

Barrons, R. and Tassone, D. (2008). Use of *Lactobacillus* probiotics for bacterial genitourinary infections in women: a review. *Clinical Therapeutics* 30: 453–468.

Behare, P.V., Singh, R., Nagpal, R., and Rao, K.H. (2013). Exopolysaccharides producing *Lactobacillus fermentum* strain for enhancing rheological and sensory attributes of low-fat Dahi. *Journal of Food Science and Technology* 50: 1228–1232.

Bindels, L.B., Delzenne, N.M., Cani, P.D., and Walter, J. (2015). Towards a more comprehensive concept for prebiotics. *Nature Reviews. Gastroenterology & Hepatology* 12: 303–310.

Blaiotta, G., Murru, N., Di Cerbo, A. et al. (2017). Commercially standardized process for probiotic 'Italico' cheese production. *LWT – Food Science and Technology* 79: 601–608.

Broussard, J.L. and Devkota, S. (2016). The changing microbial landscape of Western society: diet, dwelling and discordance. *Molecular Metabolism* 5: 737–742.

Buriti, F., Da Rocha, J., and Saad, S. (2005). Incorporation of *Lactobacillus acidophilus* in Minas fresh cheese and its implication for textural and sensorial properties during storage. *International Dairy Journal* 15: 1279–1288.

Cerdo, T., Garcia-Santos, J.A., Bermudez, M.G., and Campoy, C. (2019). The role of probiotic and prebiotics in the prevention and treatment of obesity. *Nutrients* 11: 635. https://doi.org/10.3390/nu11030635.

Chen, Y., Liao, F.H., Lin, S.H., and Chien, Y.W. (2016). A prebiotic formula improves the gastrointestinal bacterial flora in toddlers. *Gastroenterology Research and Practice* https://doi.org/10.1155/2016/3504282.

Colace, L., Boccia, S., De Maria, R., and Zeuner, A. (2017). Colorectal cancer: towards new challenges and concepts of preventive healthcare. *Ecancermedical Science* 11: ed74.

Conlon, M.A. and Bird, A.R. (2015). The impact of diet and lifestyle on gut microbiota and human health. *Nutrients* 7: 17–44.

Costalos, C., Kapiki, A., Apostolou, M., and Papathoma, E. (2008). The effect of a prebiotic supplemented formula on growth and stool microbiology of term infants. *Early Human Development* 84: 45–49.

Criscio, T.D., Fratianni, A., Mignogna, R. et al. (2010). Production of functional probiotic, prebiotic and symbiotic ice cream. *Journal of Dairy Science* 93: 4555–4564.

Cruz, A.G., Cavalcanti, R.N., Guerreiro, M.R. et al. (2013). Developing a prebiotic yoghurt: rheological, physic-chemical and microbiological aspects and adequacy of survival analysis methodology. *Journal of Food Engineering* 114: 323–330.

Cuffia, F., Bergamini, C., and Candioti, M. (2018). Probiotic soft sheep's cheese: evaluation of probiotic survival and its influence on proteolysis and organoleptic characteristics. *International Food Research Journal* 25 (1): 399–407.

David, L.A., Maurice, C.F., Carmody, R.N. et al. (2014). Diet rapidly and reproducibly alters human gut microbiome. *Nature* 505: 559–563.

Ejtahed, H.S., Mohtadi-Nia, J., Homayounti-Rad, A. et al. (2012). Probiotic yoghurt improves antioxidant status in type 2 diabetic patients. *Nutrition* 28: 539–543.

FAO and WHO (2002). *Guidelines for the Evaluation of Probiotics in Food: Report of a Joint FAO/WHO Working Group on Drafting Guidelines for the Evaluation of Probiotics in Food*. London, Ontario, Canada: Food and Agriculture Organization of the United Nations, World Health Organization.

Fenster, K., Freeburg, B., Hollard, C. et al. (2019). The production and delivery of probiotics: a review of practical approach. *Microorganisms* 7 (3): 83. https://doi.org/10.3390/microorganisms7030083.

Fukushima, Y., Miyaguchi, S., Yamano, T. et al. (2007). Improvement of nutritional and incidence of infection in hospitalized, enterally fed elderly by feeding of fermented milk containing probiotic *Lactobacillusjohnsonii* La1 (NCC533). *British Journal of Nutrition* 98: 969–977.

Fuller, R. (1989). Probiotics in man and animals. *Journal of Applied Bacteriology* 66: 365–378.

Ganguly, S., Sabikhi, L., and Singh, A.K. (2019). Effect of whey-pearl millet-barley based probiotic beverage on Shigella-induced pathogenicity in murine model. *Journal of Functional Foods* 54: 498–505.

Garcia, A.L., Otto, B., and Reich, S.C. (2007). Arabinoxylan consumption decreases postprandial serum glucose, serum insulin and plasma, total ghrelin response in subjects with impaired glucose tolerance. *European Journal of Clinical Nutrition* 61: 334–341.

Gibson, G.R. and Roberfroid, M.B. (1995). Dietary modulation of the human colonic microbiota: introducing the concept of prebiotics. *Journal of Nutrition* 125: 1401–1412.

Gibson, G.R., Probert, H.M., Van Loo, J. et al. (2004). Dietary modulation of the human colonic microbiota: introducing the concept of prebiotics. *Nutrition Research Reviews* 17: 259–275.

Gibson, G.R., Hutkins, R., and Sanders, M.E. (2017). Expert consensus document: the International Scientific Association for Probiotics and Prebiotics (ISAPP) consensus statement on the definition and scope of prebiotics. *Nature Reviews. Gastroenterology & Hepatology* 14: 491–502.

Glibowski, P. and Kowalska, A. (2012). Rheological, texture and sensory properties of kefir with high performance and native inulin. *Journal of Food Engineering* 111: 299–304.

Goldin, B.R. and Gorbach, S.L. (1992). Probiotics for humans. In: *Probiotics: the Scientific Basis* (ed. R. Fuller), 355–376. London: Chapman & Hall.

Goncerzewicz, A., Misiewicz, A., Owczarek, I. et al. (2016). The effect of newly developed oat-banana fermented beverage with a beta-glucan additive on ldhL gene expression in *Streptococcus thermophilus* TKM3KKP 2030p. *Current Microbiology* 73: 773–780.

Gonzalez, N.J., Adhikari, K., and Sancho-Madriz, M.F. (2011). Sensory characteristics of peach-flavoured yoghurt drinks containing prebiotics and symbiotics. *LWT – Food Science and Technology* 44: 158–163.

Gourineni, V.P., Verghese, M., Boateng, J. et al. (2011). Combinational effects of prebiotics and soybean against azoxymethane induced colon cancer *in vivo*. *Journal of Nutrition and Metabolism*: 868197. https://doi.org/10.1155/2011/868197.

Hancock, R.E. and Sahl, H.G. (2006). Antimicrobial and host defense peptides as new anti-infective therapeutic strategies. *Nature Biotechnology* 24: 1551–1557.

Hariri, M., Salehi, R., Feizi, A. et al. (2015). A randomized, double-blind, placebo-controlled, clinical trial on probiotic soy milk and soy milk: effects on epigenetics and oxidative stress in patients with type II diabetes. *Genes & Nutrition* 10: 52. https://doi.org/10.1007/s12263-015-0503-1.

He, T., Priebe, M.G., Zhong, Y. et al. (2008). Effects of yoghurt and bifidobacteria supplementation on the colonic microbiota in lactose-intolerant subjects. *Journal of Applied Microbiology* 104: 595–604.

Hertzler, S.R. and Clancy, S.M. (2003). Kefir improves lactose digestion and tolerance in adults with lactose maldigestion. *Journal of the American Dietetic Association* 103: 582–587.

Hickson, M., Souza, A.L.D., Muthu, N. et al. (2007). Use of probiotic lactobacillus preparation to prevent diarrhoea associated with antibiotics: randomized double blind placebo controlled trial. *BMJ* 335: 80. https://doi.org/10.1136/bmj.39231.599815.55.

Ho, J., Reimer, R.A., Doulla, M., and Huang, C. (2016). Effect of prebiotic intake on gut microbiota, intestinal permeability and Glycemic control in children with type 1 diabetes: study protocol for a randomized controlled trial. *Trials* 17: 347. https://doi.org/10.1186/s13063-016-1486-y.

Hussain, S.A., Patil, G.R., Yadav, V., and Singh, R.R.B. (2015). Effect of storage on sensory quality, pH, wheying-off and probiotic count of lassi supplemented with *Aloe barbadensis* Miller juice. *Indian Journal of Dairy Science* 68 (2): 105–110.

Hussain, S.A., Patil, G.R., Reddi, S. et al. (2017). Aloe vera (*Aloe barbadensis* Miller) supplemented probiotic lassi prevents Shigella infiltration from epithelial barrier into systemic blood flow in mice model. *Microbial Pathogenesis* 102: 143–147.

Ibrahim, F., Ruvio, S., Granlund, L. et al. (2010). Probiotic and immunosenescence: cheese as a carrier. *FEMS Immunology and Medical Microbiology* 59: 53–59.

Ishwarya, S.P. and Prabhasankar, P. (2014). Prebiotics: application in bakery and pasta products. *Critical Reviews in Food Science and Nutrition* 54 (4): 511–522.

Isik, U.D., Boyacioglu, D., Capanoglu, E., and Erdil, D.N. (2011). Frozen yoghurt with added inulin. *Journal of Dairy Science* 94: 1647–1656.

Kahouli, I., Malhotra, M., Alaoui-jamali, M.A., and Prakash, S. (2015). In vitro characterization of the anticancer activity of the probiotic bacterium *Lactobacillus fermentum* NCIMB 5221 and potential against colorectal cancer cells. *Journal of Cancer Sciences and Therapy* 2015: 224–235.

Kasubuchi, M., Hasegawa, S., Hiramatsu, T. et al. (2015). Dietary gut microbial metabolites, short chain fatty acids and host metabolic regulation. *Nutrients* 7: 2839–2849.

Kerry, R.G., Patra, J.K., Gouda, S. et al. (2018). Benefaction of probiotics for human health: a review. *Journal of Food and Drug Analysis* 26: 927–939.

Koletzko, B., Aggett, P.J., and Bindels, J.G. (1998). Growth development and differentiation: a functional food science approach. *Brazilian Journal of Nutrition* 80: 35–45.

Kolozyn-Krajewska, D. and Dolatowski, Z. (2012). Probiotic meat products and human nutrition. *Process Biochemistry* 47: 1761–1772.

Kontiokari, T., Laitinen, J., Jarvi, L. et al. (2003). Dietary factors protecting women from urinary tract infection. *The American Journal of Clinical Nutrition* 77: 600–604.

Kumar, M., Verma, V., Nagpal, R. et al. (2011). Effect of probiotic fermented milk and chlorophyllin on gene expression and genotoxicity during AFB1-induced hepatocellular carcinoma. *Gene* 490: 54–59.

Kumar, A., Hussain, S.A., Raju, P.N. et al. (2017). Packaging material type affects the quality characteristics of Aloe-probiotic lassi during storage. *Food Bioscience* 19: 34–41.

de LeBlanc, A.M., Matar, C., and Perdigon, G. (2007). The application of probiotics in cancer. *British Journal of Nutrtion* 98: S105–S110.

Li, J., Sung, C.Y., and EI-Nezami, H. (2016). Probiotics modulated gut microbiota suppresses hepatocellular carcinoma growth in mice. *Proceedings of the National Academy of Sciences of the United States of America* 113: E1306–E1315.

Lilly, D.M. and Stillwell, R.H. (1965). Probiotics: growth promoting factors produced by microorganisms. *Science* 147: 747–748.

Lollo, P.C., de Moura, C.S., Morato, P.N. et al. (2013). Probiotic yoghurt offers higher immune-protection than probiotic whey beverage. *Food Research International* 54: 118–124.

Lu, Z.X., Walker, K.Z., Muir, J.G., and Dea, O. (2004). Arabinoxylanfiber improves metabolic control in people with type II diabetes. *European Journal of Clinical Nutrition* 58: 621–628.

Mandal, V., Sen, S.K., and Mandal, N.C. (2009). Effect of prebiotics on bacteriocin production and cholesterol lowering activity of *Pediococcusacidilactici* LAB 5. *World Journal of Microbial Biotechnology* 25: 1837–1841.

Markowiak, P. and Slizewska, K. (2017). Effects of probiotics, prebiotics and symbiotics on human health. *Nutrients* 9: 1021. https://doi.org/10.3390/nu9091021.

Masoud, R., Fadaei, V., Khosravi-Darani, K., and Nikbakht, H.R. (2015). Improving the viability of probiotic bacteria in yoghurt by homogenization. *Journal of Food Processing & Preservation* 39: 2984–2990.

Mirjana, S., Ilicic, M.I., Milan, S. et al. (2007). Nutritive characteristics of probiotic quarg as influenced by type of starter. *Acta Periodica Technologica* 38: 11–19. https://doi.org/10.2298/APT0738011D.

Moen, B., Berget, I., Rud, I. et al. (2016). Extrusion of barley and oat influence the fecal microbiota and SCFA profile of growing pigs. *Food & Function* 7: 1024–1032.

Morais, E.C., Morais, A.R., Cruz, A.G., and Bolini, H.M. (2014). Development of chocolate dairy dessert with addition of prebiotics and replacement of sucrose with different high intensity sweeteners. *Journal of Dairy Science* 97: 2600–2609.

Morris, C. and Morris, G.A. (2012). The effect of inulin and fructooligosacchride supplementation on the textural, rheological and sensory properties of bread and their role in weight management: a review. *Food Chemistry* 133: 237–244.

Mridula, D. and Sharma, M. (2015). Development of non-dairy probiotic drink utilizing sprouted cereals, legume and soymilk. *LWT – Food Science and Technology* 62 (1): 482–487.

Munjal, U., Glei, M., Pool-Zobel, B.L., and Scharlau, D. (2009). Fermentation products of inulin-type fructans reduce proliferation and induce apoptosis in human colon tumor cells of different stages of carcinogenesis. *British Journal of Nutrition* 27: 1–9.

Muthukumarasamy, P. and Holley, R.A. (2006). Microbiological and sensory quality of dry fermented sausages containing alginate microencapsulated *Lactobacillus reuteri*. *International Journal of Food Microbiology* 111: 164–169.

Nabavi, S., Rafraf, M., Somi, M.H. et al. (2014). Effects of probiotic yoghurt consumption on metabolic factors in individuals with non-alcoholic fatty liver disease. *Journal of Dairy Science* 97: 7386–7393.

Nair, B.M. and Shah, K. (2017). 'Hildur Lassi': a probiotic fermented food with oats and milk. *International Journal of Fermented Foods* 6 (2): 95–95.

Ozer, B. and Kirmaci, H.A. (2010). Functional milks and dairy beverages. *International Journal of Dairy Technology* 63: 1–15.

Parker, R.B. (1974). Probiotics, the other half of the antibiotic story. *Animal Nutrition and Health* 29: 4–8.

Rakhesh, N., Fellows, C.M., and Sessions, M. (2015). Evaluation of the technological and sensory properties of durum wheat spaghetti enriched with different dietary fibers. *Journal of the Science of Food and Agriculture* 95: 2–11.

Ranadheera, C.S., Vidanarachi, J.K., Rocha, R.S. et al. (2017). Probiotic delivery through fermentation: dairy vs. non-dairy beverages. *Fermentation* 67: 1–17. https://doi.org/10.3390/fermentation3040067.

Reid, G., Bruce, A.W., Fraser, N. et al. (2001). Oral probiotics can resolve urogenital infections. *FEMS Immunology and Medical Microbiology* 30: 49–52.

Rolim, P.M. (2015). Development of prebiotic food products and health benefits. *Food Science and Technology* 35: 3–10.

Roskar, I., Svigelj, K., Stempelj, M. et al. (2017). Effects of probiotic product containing *Bifidobacteriumanimalis* subsp. *animalis* IM386 and *Lactobacillus plantarum* MP2026 in lactose intolerant individuals: randomized, Placebo-controlled clinical trial. *Journal of Functional Foods* 35: 1–8.

Sachdeva, A. and Nagpal, J. (2009). Effect of fermented milk-based probiotic preparations on *Helicobacter pylori* eradication: a systematic review and meta-analysis of randomized-controlled trials. *European Journal of Gastroenterology & Hepatology* 21: 45–53.

Saini, S.S. and Kaur, A. (2012). Aflatoxin B1: toxicity, characteristics and analysis: mini review. Glob. *Advanced Research Journal of Chemistry Materials Science* 1: 63–70.

Salmerón, I., Thomas, K., and Pandiella, S.S. (2015). Effect of potentially probiotic lactic acid bacteria on the physicochemical composition and acceptance of fermented cereal beverages. *Journal of Functional Foods* 15: 106–115.

Sangwan, S. and Singh, R. (2018). Synergistic effect of oats and LGG fermented milk on lowering hypercholesterolemia in rats. *Journal of Cereal Science* 82: 164–169.

Sarmah, P., Dan, M.M., Adapa, D., and Sarangi, T.K. (2018). A review on common pathogenic microorganisms and their impact on human health. *Electronic Journal of Biology* 14 (1): 50–58.

Sharma, P., Trivedi, N., and Gat, Y. (2017). Development of functional fermented whey–oat-based product using probiotic bacteria. *3 Biotech* 7: 272. https://doi.org/10.1007/s13205-017-0906-3.

Stanton, C., Gardiner, G., Lynch, P.B. et al. (1998). Probiotic cheese. *International Dairy Journal* 8 (5–6): 491–496.

Tejero-Sarinena, S., Barlow, J., Costabile, A. et al. (2013). Antipathogenic activity of probiotic against *Salmonella Typhimurium* and *Clostridium difficile* in anaerobic batch culture systems: is it due to synergies in probiotic mixtures or the specificity of single strains? *Anaerobe* 24: 60–65.

Tripathi, V. and Jha, Y.K. (2004). Development of whey beverage with antagonistic characteristics and probiotics. *International Journal of Food Properties* 7: 261–272.

Urbanska, A.M., Bhathena, J., Martoni, C., and Prakash, S. (2009). Estimation of the potential antitumor activity of microencapsulated *Lactobacillus acidophilus* yoghurt formulation in the attenuation of tumorigenesis in APC (Min/+) mice. *Digestive Diseases and Sciences* 54: 264–273.

Van Ba, H., Seo, H.W., Seong, P.N. et al. (2018). *Lactobacillus plantarum* (KACC 92189) as a potential probiotic starter culture for quality improvement of fermented sausages. *Korean Journal for Food Science of Animal Resources* 38 (1): 189–202.

Veiga, P., Pons, N., Agrawal, A. et al. (2014). Changes of the human gut microbiome induced by fermented milk product. *Scientific Reports* 4: 6328. https://doi.org/10.1038/srep06328.

Vijayendra, S.V.N. and Gupta, R.C. (2012). Assessment of probiotic and sensory properties of dahi and yoghurt prepared using bulk freeze-dried cultures in buffalo milk. *Annals of Microbiology* 62: 939–947.

Vinderola, C.G., Duarte, J., Thangavel, D. et al. (2005). Immunomodulating capacity of kefir. *Journal of Dairy Research* 72: 195–202.

Wilson, E., Seo, C., Shahbazi, A., and Ibrahim, S. (2004). Survival and growth of probiotic cultures in sour cream products. IFT Annual Meeting. 17A, Las Vegas, NV.

Wong, J.M.W., Kendall, C.W.C., de Souza, R. et al. (2010). The effect on the blood lipid profile of soy foods combined with a prebiotic: a randomized controlled trial. *Metabolism* 59: 1331–1340.

Xiao, J.Z., Kondo, S., Takahashi, N. et al. (2003). Effects of milk products fermented by *Bifidobacterium longum* on blood lipids in rats and healthy adult male volunteers. *Journal of Dairy Science* 86: 2452–2461.

Yadav, H., Jain, S., and Sinha, P.R. (2008). Oral administration of dahi containing probiotic *Lactobacillus acidophilus* and *Lactobacillus casei* delayed the progression of streptozotocin-induced diabetes in rats. *Journal of Dairy Research* 75: 474647–474655.

Yadav, V., Singh, R.R.B., Hussain, S.A., and Kapila, S. (2017). Immunomodulatory effects of oral administration of *Aloe vera* supplemented probiotic ice cream in mice. *Indian Journal of Dairy Science* 70 (1): 89–95.

Yatsunenko, T., Rey, F.E., Manary, M.J. et al. (2012). Human gut microbiome viewed across age and geography. *Nature* 486: 222–227.

13 Mathematical Tools for Food Formulation

Camila A. Perussello and Jesus M. Frias

Technological University Dublin, Environmental Sustainability and Health Institute, Dublin, Ireland

Summary

Mathematical modelling in food formulation design is a growing field that has been used by food scientists to find the effect of ingredients and their amount/rate in a formulation on the final product's technological and sensory properties and accordingly, consumer acceptance. This chapter outlines various mathematical models used in food formulation experimental design (factorial design and mixture designs) as well as statistical tools used to correlate the data from sensory panels and consumer studies (preference Mapping [PM] and multinomial mixed effect models). Furthermore, this chapter also discusses the nutrikinetic and nutridynamic characteristics of food formulations.

13.1 Introduction

Food formulations are typically developed from extensive, science-based laboratory tests and sensory analysis. Scaled-up trials in an industrial setup and consumer testing are mandatory before market launching. Most of the food industry's understanding of the drivers for consumer liking stems from studies related to the attitudes, habits, and preferences of target groups of people (Sidel and Stone 1993). Nevertheless, consumers are not always able to determine with precision and accuracy the intensity of the sensory descriptors of a food product (Ares and Varela 2017; Giménez et al. 2012). Consumer tests are directed to acceptance and preference scores of food samples, mainly through the use of hedonic scales. However, in these studies it is not always clear what specific attributes are responsible for driving liking/preference and to what extent these attributes determine the acceptability of a given formulation. That is why consumer tests are usually interpreted in parallel with the results from panel sensory sessions, where experts are trained to discriminate samples based on specific descriptors of a sample and their intensities

Food Formulation: Novel Ingredients and Processing Techniques, First Edition.
Edited by Shivani Pathania and Brijesh K. Tiwari.

(Lawless and Heymann 2010). Trained panel results are certainly also benefited by the correlation with consumer studies as they are unlikely to be representative of a target market, especially if small panels are used ($n<10$). One of the central challenges of food sensory analysis is how to match the results from trained and consumer tests, that is, how to correlate the food's attributes to linking and purchase intent (Ares and Varela 2017). Failing to couple these results accurately may incur a poor measure of the performance of a food formulation in the real food market. Empirical models (e.g. logit and multinomial logit) are commonly used to correlate consumer preference to food descriptors and/or sociodemographic factors (e.g. age, income, gender) (Roselli et al. 2018; Yin et al. 2010). Other mathematical tools (e.g. Principal Component Analysis – PCA) are employed to translate sensory panel results and instrumental data (i.e., Quantitative Descriptive Analysis – QDA) into an adequate formulation design (Rødbotten et al. 2015; Ares et al. 2015; Moussaoui and Varela (2010).

As opposed to inductive mathematical modelling, where models are built from experimental data, deductive models convert theories into mathematical equations that represent the problem under analysis. This is often the case in food encapsulation, where the release kinetics is a function of the formulation, the encapsulation conditions, and the type and geometry of the capsule. Depending on these factors, a variety of release mechanisms can be found, such as diffusion, particle fracture, and solvent-, fracturation-, or pressure-activated release (Ronald and Shahidi 2007). The release of encapsulated active compounds is frequently modelled by mechanistic models, for instance Fick's Law of diffusion (Crank 1975) and derivatives, such as the linear superposition model – LSM (Berens and Hopfenberg 1978), although empirical models (e.g. first-order kinetics, second-order kinetics, etc.) are also employed.

Both inductive and deductive modelling approaches are applied in the field of nutrition research, where compound delivery is as important as the technological and sensory properties of a food formulation. An example is the absorption kinetics of a nutrient of interest in the gastrointestinal system, which is typically modelled by zero-order or first-order absorption kinetics. Other models, such as Weibull-type absorption and Michaelis–Menten absorption, are applied to describe more complex profiles (Abuhelwa et al. 2017). The modelling approach choice is highly dependent on the typical mechanism(s) of absorption of a nutrient by the human organism, which in turn varies between bioactive compounds (Bao et al. 2019). Specific branches of nutrition research, such as Nutrikinetics (study of the absorption, distribution, metabolism, and excretion (ADME) of food compounds within the human body) and Nutridynamics (study of the effects of a living organism on human health) make use of sophisticated quantitative structure–activity relationship (QSAR) modelling (van Duynhoven et al. 2012; Lefebvre et al. 2015). QSAR models can predict the bioavailability of components by comparing their chemical structure with that of other substances with known activities (Lefebvre et al. 2015). Although QSAR models are much more common in pharmacokinetics and food risk assessment studies (Blaauboer et al. 2016), they can be used to complement earlier stages of food formulation design. This holds true when an optimal formulation or group of best formulations has been found in terms of technical and sensory characteristics, but their health effects are also relevant and, therefore, should be optimized.

This chapter outlines the main modelling techniques used in food formulation design, covering aspects from basic food formulation design to Nutrikinetics and Nutridynamics.

13.2 Food Formulation Experimental Design

There are three main classes of mathematical models: theoretical or deductive, empirical or inductive, and semi-empirical. Whereas the essence of theoretical modelling is the conversion of theories and concepts into mathematical formulations, the basis of inductive modelling is the extraction of information from the measured data. In this case, experimental data is the only information available and there is no need for consistent knowledge of the process mechanisms. In general, inductive modelling requires the application of statistical techniques to a data set, which means finding the best representation of the data behaviour by examining various relationships between these and potential models (Verdenius 2001). Importantly, adjusting a model does not imply adjusting the data to the phenomenon.

Mathematical modelling has several objectives, such as: (i) Forecast and prediction; (ii) Scenario analysis (what-if); (iii) Sensitivity analysis; (iv) Optimization/control; and (v) Metamodelling. Most commonly, however, the ultimate objective of mathematical modelling is to predict possible outcomes (dependent variables) from a set of inputs (independent variables) (Tijskens et al. 2001). In the context of food formulation design, this is equivalent to finding the effect of ingredients and their amount/rate in a formulation on the final product's technological and sensory properties and accordingly, consumer acceptance. The impact of process conditions can also be regarded as independent variables. In practise, many models are combinations of empirical and phenomenological approaches (Tijskens et al. 2001).

Most models in food formulation are either empirical or semi-empirical, such as regression models in mixture problems (Nikzade et al. 2012), linear and logistic correlations between sensory acceptability and trained sensory panel scores (Hough et al. 2002), and artificial intelligence network-built models in ingredient functionality determination (Goel et al. 2018). Table 13.1 shows a summary of the main mathematical tools employed in food formulation studies.

To obtain a good-quality inductive model, the model's hypotheses should reflect reasonably the behaviour of the data set, the data collection should cover the region of interest uniformly, contemplate a reasonable number of replicates, and avoid systematic error (Valentas et al. 1997). Being data-driven, the representation and prediction power of inductive models is directly dependent on the quality of the experimental data. Therefore, well-designed experiments are the base of good empirical models.

Experimental design or design of experiments (DOE) refers to the design of experimental tests to investigate the effect of input variables (e.g. process conditions, product composition) on output variables or responses (e.g. liking scores, quality characteristics, processing time). When it comes to the design and optimization of a mixture of ingredients to meet certain technological and quality characteristics, statistical tools such as full factorial, fractional factorial, and mixture designs (e.g. simplex-lattice, simplex-centroid, D-optimal, and crossed mixture) are traditionally used (Mason et al. 2003). Recently, the concept of food combinatoric principles (FCP) have been introduced by Musina et al. (2017), a technique to optimize the composition of food mixtures.

13.2.1 Factorial Designs

A full factorial design is a set of experiments that combines two or more factors, with a number of levels each, to yield all possible combinations of these levels across these

Table 13.1 Main approaches to food formulation empirical modelling.

Technique	Objective	Example application	References
	Design of experiments		
Factorial Design – full	Encapsulation of food bioactive	Ascorbic acid microcapsules	Da Cruz et al. (2019)
Factorial Design – incomplete (3^3)	Food formulation design; Ingredient substitution	Chocolate cake added with inulin and yacon	Moscatto et al. (2006)
Factorial Design – Full design at mixed levels	Food formulation design; Storage time effect	Yoghurt added with *cajeta* jam	Ramírez-Sucre and Vélez-Ruiz (2013)
Factorial Design – Taguchi method (3^4)	Food formulation design; Ingredient substitution	Thyme honey based-mixtures	Dimou et al. (2009)
D-optimal Mixture Design	Food formulation design; Ingredient substitution	Prebiotic sausage	Sarteshnizi et al. (2015);
Simplex-Centroid Mixture Design	Food formulation design; Ingredient substitution	Low cholesterol and low-fat mayonnaise	Nikzade et al. (2012)
	Mathematical methodologies		
Response Surface Methodology – RSM	Food formulation design; Process conditions optimization	Millet-legume extrudates	Chakraborty et al. (2011)
Food combinatoric principles – FCP	Food formulation design; Nutritional value optimization	Curd-grain blend	Musina et al. (2017)
Multinomial regression	Mixture design; Consumer preference	Cocktails containing mango juice, cassis and lime	Goos and Hamidouche (2019)
Preference Mapping/ Principal Component Analysis (PCA)	Consumer preferences mapping; Taste panel and consumer panel correlation	Goat and sheep fresh sausages	Paulos et al. (2015)
QSAR modelling	Ingredient functionality	Sweetness of a range of molecules	Goel et al. (2018).
Non-linear mixed effect model	Nutrikinetics and Nutridynamics	Polyphenol bioavailability from grapes and black tea extracts	van Velzen et al. (2014)

factors. Such an experiment allows the study of isolated effects of each factor on the responses, as well as the effects of interactions between factors. Full factorial designs, which are represented by Equation 13.1, may involve a large number of experiments if the number of factors is high (Yu et al. 2018):

$$n = m^k \tag{13.1}$$

where n is the total number of samples, k is the number of factors, and m is the number of levels of each factor.

Most factorial designs operate at only two levels, especially those employed to identify sensitivity and interactions (synergistic or antagonistic effects of factors). In turn, three-level designs are generally used for optimization and control (maxima or minima). The notation for full factorial designs is 2^n, where n is the number of levels of the factors (or variables) under study. A design comprising two levels on each factor ($n = 2$) is called a 2^2 or 2×2 factorial design, which is composed of four experiments. Designs with more than two levels are also possible, but are not as common as 2^n experiments. Whenever the number of combinations in a full factorial design is not feasible (e.g. $2^7 = 128$), some of the possible combinations are omitted, giving place to fractional factorial experiments. For practical reasons, in biological experimentation a full factorial design is employed for a reduced number of process parameters up to four (Antony 2014a). In fractional experiments, only the most important process parameters that influence critical quality characteristics are included in the experiment design (Antony 2014b). Further details on types of factorial designs can be found in Granato and Calado (2014).

Ramírez-Sucre and Vélez-Ruiz (2013) assessed the effects of formulation and storage time on the physicochemical, morphological, and rheological properties of a stirred yoghurt with the addition with *cajeta*, a Mexican caramel jam, as a flavourer. The full factorial design at mixed levels yielded 18 samples and one control, according to the following factors and levels: fat content (0.2, 1.7, and 3.2 g/100 g), κ-carrageenan (0 and 0.02 g/100 g), and caramel jam (10.0, 12.5, and 15.0 g/100 g). The samples were analysed at 0, 1, 2, 3, and 4 weeks of storage. The samples added with *cajeta* had lower luminosity (L*) and redness (a*) and greater soluble solids and moisture contents (MCs). L* increased as a function of fat content and the control was the brightest due to the absence of caramel jam. All samples exhibited shear thinning and plasticity, and similar pH within formulations. The yoghurts with added caramel showed similar flow response to natural and strawberry commercial yoghurts. The non-Newtonian behaviour of the yoghurt was fitted by the Herschel and Bulkley model, which indicated greater yield stress and consistency coefficient at higher fat contents and longer storage times. Scanning Electron Microscopy images showed that a higher fat content resulted in larger globular casein aggregates. In this work, the authors assessed the individual and interactive effect of the fat (A), gum (B), caramel jam (C), and time (D). However, they have not determined what the optimal formulation would be.

Moscatto et al. (2006) developed a functional chocolate containing inulin and yacon flour (ground dried yacon roots). The objective was to obtain the same quality characteristics as in chocolate cake made of wheat flour. An incomplete 3^3 factorial design was used to investigate the effect of yacon meal concentration (X_1), inulin concentration (X_2), and water volume (X_3) on hardness, cohesiveness, and specific volume of the final cakes. All formulations were compared to a control (no inulin, no yacon). The results of the experiment design (15 formulations plus the control) were analyzed by Analysis of Variance (ANOVA), Tukey's test, and Response Surface Methodology (RSM). In the optimization step, the Complex algorithm was employed. The optimal formulation (i.e. minimum hardness, maximum cohesiveness, and specific volume) corresponded to a formulation with 20% of wheat flour replaced with yacon meal, no inulin, and 153 ml of water. However, another formulation with technological characteristics similar to that of the control cake was also selected (40% of wheat flour was replaced with yacon meal and 6% replaced with inulin, with 126 ml water) for the higher concentration of fructooligosaccharide(s) (FOS) and inulin.

FCP is used to design food products with a required set of nutritional characteristics. For instance, from the set of protein-containing ingredients, it is possible to calculate the optimal food formulations, where the ratio of such ingredients provides a balanced content of amino acids in the product. Amino acid content for ingredients can be easily obtained from a nutrient database for a standard reference (eg. Food and Drug Administration (FDA), European Food Safety Authority (EFSA)). In order to obtain the desired nutritive features of the resulting food mixtures, FCP employs a set of mathematical equations that provide solutions for altering the composition of such mixtures, usually by manipulating the weights of the raw materials. Sets of mathematical equations can be expanded according to the production needs, and Musina et al. (2017) used FCP to develop a curd–grain blend with specific nutritional properties ascribed to the complementary macro- and micro-nutrients of milk and grain raw materials. Several formulations were prepared based on 14 different ingredients – curd, wheat germ flakes, bran, peas, wheat, rye, barley, oats, millet, buckwheat, rice, corn, beans, and lentils. Eight nutritional parameters related to essential amino acids were optimized.

Classic factorial designs can be time-consuming and often require extensive resources. In contrast, Taguchi methods, which are fractional–factorial designs, use a special set of arrays called orthogonal arrays that yields maximum information from a minimum number of experiments (Arteaga et al. 1994; Assadpour and Jafari 2017; Dimou et al. 2009). Dimou et al. (2009) investigated the effect of four blossom honey types – orange, chestnut, heather, and cotton – on the quality characteristics of thyme honey-based mixtures. The study aimed to optimize the mixture of honeys from different botanical origins to be added to thyme honey to yield the same quality characteristics as pure thyme honey. Mixture ratios were prepared at 0%, 50%, and 100% per blossom honey type and were blended with thyme honey at 1:1 (m/m). A three-level, four-factor orthogonal array (L9 or 3^4) was used according to the Taguchi method to design the experiment. Such a design is composed of nine experimental runs that can capture the effect of the four formulation factors on multiple quality characteristics of honey. The characteristics of each blend were determined, comprising microscopical – thyme-type pollen grains (TTPG), number of pollen grains per 10g of honey (NPG) – sensory (colour, aroma, and flavour) and physicochemical – electrical conductivity (EC, mS/cm) and fructose and glucose contents) properties. A pure orange-type blend yielded maximum aroma (and minimum electrical conductivity). The mixture of thyme honey with other types of honey did not influence significantly the original colour of thyme honey. Aroma and flavour were influenced by the addition of honey types other than thyme in a widely variable manner. Even small amounts of chestnut honey strongly affected the microscopical and physicochemical characteristics of the blends. Cotton honey and heather did not significantly influence the quality characteristics of the final product.

RSM has been extensively used in food formulation for decades (Chakraborty et al. 2011; Pardhi et al. 2019; Vaisey-Genser et al. 1987). The underlying principle of RSM is to relate product properties to input parameters (process conditions and/or formulation factors) from the experimental design through regression equations of polynomials of second order. The responses are plotted either in the 3D space or as contour plots that help visualize the shape of the response surface and explore for stationary points (maxima, minima, and saddle points). Extensive detail on the principles and applications of RSM can be found in Box and Draper (2007) and Myers et al. (2011).

Chakraborty et al. (2011) extruded different millet–legume formulations of varying blend ratios (BR) and MC. The extrusion parameters consisted of die head temperature (DHT), barrel temperature (BT), and screw speed (SS). Second-order polynomial models were fitted to the experimental data to assess the effect of BR and MC on density, sectional expansion index (SEI), water absorption index (WAI), and crispness of the extrudates. The RSM illustrated the large impact MC had on SEI, WAI, and crispness, and the clear effect of SS on density. The optimum formulation was determined as MC = 23.2% w.b. and BR = 19.9% legume when processed at DHT = 187 °C, BT = 121.1 °C, and SS = 123 rpm. Density, SEI, WAI, and crispness at these conditions were predicted as 0.52 kg/m^3, 5.1, 9.4, and 45, respectively. There were no differences ($p < 0.01$) between experimental and predicted results.

13.2.2 Mixture Designs

Very commonly, a food product consists of a mixture of different ingredients, whose fractions impact on the final product's characteristics. In mixture design, there are two main types of approaches: (i) when the response is a function of the fraction of each ingredient in the formulation; and (ii) when the final product's properties are also affected by the total amount of the mixture (Piepel and Cornell 1985; Schrevens and Portier 2005). For example, the viscosity of a juice is a function of its relative composition and not of the volume of juice. Mixture designs are especially useful in finding the single contribution of the individual components in a formulation, as well as the synergistic effects, for instance in food formulations favoured by the interaction between polymers or oils and proteins (Maher et al. 2011; Nikzade et al. 2012).

In mixture experiments, the independent variables represent proportionate amounts that must meet two conditions: (i) the fraction of each component must be between 0 and 1; and (ii) the fractions must sum to unity. Mixtures with only one component are called pure mixtures. In general, a simplex-centroid or D-optimal mixture design is performed to find a set of possible formulations in accordance with the problem constraints (e.g. the fraction of yeast in a loaf of bread must be greater than 0 and smaller than the fraction of flour). The researcher defines the key properties of the final product and their acceptable ranges based on technological issues and/or consumer preferences. Then, experiments are carried out in agreement with the number of replicates outlined in the experimental design. Instrumental determinations of technological properties and sensory tests are conducted and the results are fitted to mathematical models (i.e. linear, quadratic, cubic, etc.). This way, the effect of each property on consumer liking is determined to support the choice of the optimum formulation (Piepel and Cornell 1985; Schrevens and Portier 2005).

Sarteshnizi et al. (2015) found that the optimum formulation of a prebiotic sausage composed of resistant starch (RS), β-glucan (BG), and starch (ST) following the D-optimal mixture design approach. Their study aimed to determine the single effect of RS, BG, ST, and the ingredients' interactions on the product's properties and accordingly to obtain the best formulation. A D-optimal design was used with the constraints β-glucan (x1) + resistant starch (X2) + starch (X3) = 6% and component ranges as $1 < BG < 3$, $2 < RS < 3$, and $1 < ST < 3$. The experimental design, generated by the Design-Expert software (version 8.0.5), resulted in 13 formulations with a different number of replicates. The formulations were assessed for cooking yield, frying loss, texture profile analysis, and colour. Sensory tests were conducted with 15 trained panellists, who used a 9-point hedonic scale for each sensory attribute: flavour: 1 = extremely dislike to 9 = extremely like; firmness 1 = very soft to 9 = very hard;

juiciness 1 = very dry to 9 = very juicy; colour and odour 1 = extremely dislike to 9 = extremely like; chewiness 1 = not chewy to 9 = very chewy; texture acceptability 1 = extremely dislike to 9 = extremely like; and overall acceptability 1 = extremely dislike to 9 = extremely like. The responses were fitted to each sensory parameter using the linear, quadratic. and cubic models. The best models differed for each property: cubic for cooking yield and frying loss; cubic for texture (except resilience, quadratic model); linear for L* and cubic for a*and b*; quadratic for chewiness and odour; and cubic for the other sensory parameters. The optimum formulation was obtained based on maximum cooking yield, minimum frying loss, hardness in the range of 24.5–26.5 (according to the texture acceptability), and maximum overall acceptability. The optimum formulation corresponded to 2.216 starch, 1.328 β-glucan, and 2.456 starch, which had a desirability level of 0.878 (excellent acceptance is 0.8–1). Resistance starch had a negative effect on cooking yield and overall acceptability, which was lessened by the interaction with β-glucan.

Nikzade et al. (2012) optimized a formulation of low-cholesterol–low-fat mayonnaise containing soy milk as an egg yolk substitute (10%) and different fractions (0–0.36% each) of xanthan gum (XG), guar gum (GG), and mono- and diglycerides emulsifier (MDG). A simplex–centroid mixture design was applied and experiments were run in which stability, texture, and rheological properties were determined. Mayonnaise samples with one day of storage were presented to a trained panel of 10 people, who assessed them for appearance, colour, odour, texture, taste, and overall acceptance using a 5-point hedonic scale (1 = lowest; 5 = highest). The independent and dependent variables were fitted to linear, quadratic, special cubic, full cubic, and special quadratic models, and residuals plots were generated to assess the goodness of fit. The optimum formulation corresponded to 6.7% MDG, 36.7% GG, and 56.7% XG. XG had the greatest effect on all the properties, increasing stability, heat stability, consistency coefficient, viscosity, firmness, adhesiveness, adhesive force, and overall acceptance and lowering of the flow behaviour index. GG had the second-largest impact on the mayonnaise's properties. Soy milk had a good performance as the egg yolk substitute.

Maher et al. (2011) optimized the formulation of β-casein nanoemulsions using the mixture design approach. Formulations containing β-casein (2.5, 5, 7.5, and 10% w/w), lactose (0–20% w/w), and trehalose (0–20% w/w) were generated by D-optimal design through the software Design Expert (version 7.1.3). The problem constraints were defined as lactose + trehalose + β-casein = 20% w/w and β-casein ≤10% w/w. Glass transition temperature (Tg′), onset of ice-melting temperature, viscosity, particle size, and separation rate were determined for all formulations. The objective of the optimization was to minimize Tg′ and to maximize viscosity at fixed protein concentrations of 2.5, 5, 7.5, and 10% w/w. In that phase, the constraints were: lactose ≤17.5% w/w, trehalose ≤17.5% w/w, β-casein 2.5–10% w/w, and lactose + trehalose + β-casein = 20% w/w. Models were fitted to the experimental results and the best-performing ones at $p<0.05$ were used in the optimization procedure. β-casein had the greatest effect on the nanoemulsion separation rate due to increased viscosity at high protein levels. Important synergism was revealed between lactose and β-casein, and between trehalose and β-casein, with these mixtures maintaining a low separation rate. Lactose and trehalose fractions did not affect stability. Increasing levels of β-casein (2.5–10% w/w) yielded a significant increase in viscosity (5–156 mPa.s) and particle size (186–199 nm), and a significant decrease in Tg′ (−45 to −50 °C). Higher protein contents resulted in greater nanoemulsion stability.

13.3 Consumer-Based Food Formulation

Food companies design their products based on the nutritional needs, preferences, and convenience of a target group of consumers. As the ultimate goal of the food industry is to sell their products, pilot tests must be correlated to consumer studies. Failing to do so incurs in the risk of low market success rates. Consumer based-food formulations rely on scores of appearance (colour, shape, size, gloss, dullness, and transparency), flavour (odour, taste, and mouthfeel), aroma, texture (tenderness, softness consistency, thickness, fragility, chewiness, and size and shape of particles), and sound (e.g. effervescence of drinks, cracking of biscuits) given by consumers in relation to a specific product formulation (Sharif et al. 2017).

Consumer-driven food development employs methods that detect the consumer acceptability towards ingredients in the formulation or liking differences between formulations and a control. Consumer studies are commonly designed in such a way that a number of consumers get to rate a food product using labelled scales, for instance hedonic scales and Likert scales. Hedonic scales are used in sensory science research for capturing the linking data (i.e. 1 – dislike extremely, 2 – dislike slightly, 3 – neither like nor dislike, 4 – like slightly, and 5 – like extremely), whereas Likert scales are psychometric scales used in studies of purchase decisions and consumer attitudes where consumers reveal their level of agreement or disagreement with a statement or question (i.e. 1 – strongly disagree, 2 – disagree, 3 – neither agree nor disagree, 4 – agree, and 5 – strongly agree) (Joshi et al. 2015; Lim 2011). Most opinion rating scales contain 5–10 response categories. The literature suggests that five-point scales are less confusing and tend to increase the response rate (Cox 1980; Dawes 2008). In terms of human mind functioning (absolute judgement, memory, and attention), Miller (1956) concluded that there is no benefit in increasing the number of responses beyond six or seven. With regard to data processing, however, reliability is favoured when seven response scales are employed (Colman et al. 1997). It has also been reported that seven-point scales yield stronger correlations with t-test results (Lewis 1993). Johns (2010) concluded that data from Likert scales and similar rating scales becomes significantly less accurate at scale points lower than 5 and above 7. In practise, however, the traditional 5-point hedonic scale and the 9-point hedonic scale are the most widely used hedonic scales for investigating consumer acceptance and purchase intent of foods, respectively (Lawless and Heymann 2010).

Even though some companies rely exclusively on panels of trained experts in food formulation development, this practise is not encouraged. Consumers' (naïve panellists') perception of a formulation differs significantly from that of trained panellists, yielding inaccurate conclusions (Giménez et al. 2012). By using QDA, trained assessors can spot differences among formulations, processing conditions, and storage conditions, identify drivers of consumer acceptance, and investigate relationships between sensory and chemical attributes (Yang and Lee 2019). Importantly, consumer tests are more prone to result variability than trained panel studies, which requires extra attention on the impact of random effects (e.g. individual panellist agreement with the mean panellist scores). This impacts on the choice of the statistical methods used in the data process stage. Generally, data from sensory panels and consumer studies are correlated through statistical tools, for instance Preference Mapping (PM) and multinomial mixed effect models.

13.3.1 Preference Mapping

PM is a tool for relating sensory perception with consumer liking to identify the sensory attributes that drive liking in a product formulation. This method can be used in the development of new products, product improvement, and product adaptation (e.g. light, diet, lactose-free, and gluten-free). PM data (e.g. QDA) are usually processed by PCA, Multidimensional Scaling (MDS), and Unfolding (MacFie 2007). To correlate taste panel and consumer studies results, data are collected about the intensity of different attributes in different samples by a number of expert panellists. The samples are then presented to a group of consumers, who rank them according to liking or acceptability. Both sets of data (trained panel and consumer) are then processed by PCA and superimposed in loadings and factors graphs. These graphs yield a correlation between samples, their sensory descriptors, and the associated liking scores. In a PC graph (or map), consumers with vectors in the same direction have similar preferences, while products that are close together are liked similarly. Maps are especially useful for identifying segmentation between samples due to different treatments or ingredients in the formulations (Heyd and Danzart 1998; MacFie 2007).

PM has been used for decades in consumer preference studies on food (Bogue et al. 1999; Breternitz et al. 2017; Paulos et al. 2015; Viljoen et al. 2017). Paulos et al. (2015), for instance, used PM, Generalized Procrustes Analysis (GPA), and PCA to map the consumer preferences of goat and sheep fresh sausages, by correlating the results of a taste panel and a consumer panel. Sausage samples from 140 culled goats and 140 culled ewes were evaluated by a trained panel of 9 experts for raw meat odour intensity, off-odours, raw meat flavour intensity, off-flavours, hardness, juiciness, fibres presence, spiciness, and sweetness using QDA. Eight samples were evaluated in each of eight trained panel sessions using an unnumbered scale of 10 cm, with the extremes representing the minimum and maximum attribute intensity. Then, a consumer panel containing 82 people assessed the samples for taste liking, texture liking, spiciness liking, and overall acceptability using a 10 cm unstructured scale (from 0 cm – 'do not like' to 10 cm – 'like very much'). The data from the taste panel were analyzed by GPA, a multivariate technique extensively used in sensory evaluation. GPA uses translation, rotation, and isotropic scaling to minimize differences amongst panellists and summarizes the sets of 3D data (samples, sensory attributes, and assessors). To relate the consumers' preferences with the sausages' characteristics, preference maps were built for each of the following variables: taste, texture, spiciness, and overall acceptability. The rankings of samples by each consumer (or group of consumers) were modelled using the sensory characteristics as explanatory variables.

The five attributes of the taste panel were plotted against three PCs using GPA. The correlation between sensory parameters and the three PCs indicated that hardness, juiciness, and the presence of fibres were highly correlated with PC1; PC2 was highly correlated with the odour intensity, flavour intensity, off-flavour, spiciness, and sweetness; and very low correlations were found between the attributes and PC3. The sheep meat sausages were clustered in the PC1 positive quarter, while the goat meat sausages clustered in the negative part of the same factor, indicating that the sausages are separated by tenderness and juiciness. Sausages with paprika and no paprika were separated by PC2, indicating that the condiment influences taste, odour, and flavour. Species had a great influence on the sausage's texture with the hardness and fibrousness more associated with goat sausages, while juiciness was more associated with sheep sausage. In relation to taste (Figure 13.1A), consumers in class 1 (39 individuals) prefer tender

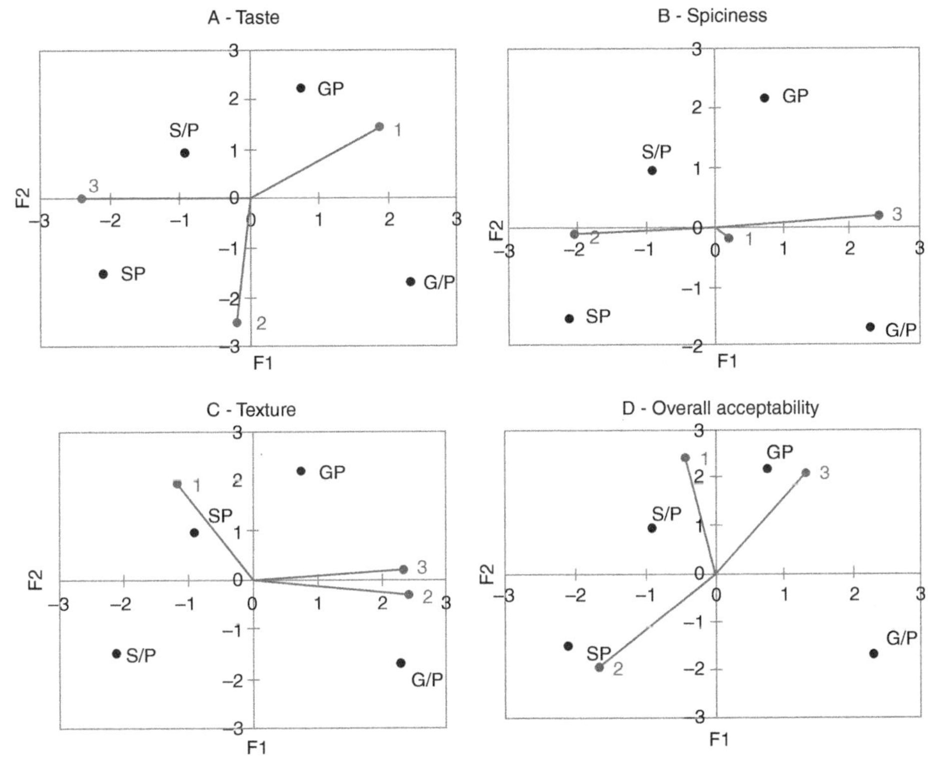

Figure 13.1 Preference mapping for sausages: (A) taste, (B) spiciness, (C) texture, and (D) overall acceptability. SP = sheep with paprika, S/P = sheep without paprika, GP = goat with paprika, and G/P = goat without paprika. Source: Paulos et al. (2015). Reproduced with permission of John Wiley & Sons.

and juicier sausages (i.e. sheep sausages without paprika); consumers in class 2 (33 individuals) prefer goat sausages with paprika characterized by being sweeter and with intense odour; and consumers in class 3 (10 individuals) prefer sausages without paprika, which have higher flavour intensity and are spicier. A similar analysis was done with spiciness (Figure 13.1B), texture (Figure 13.1C), and overall acceptability (Figure 13.1D). The combined methodology revealed the differences between fresh sausages of different species and formulations and the associated impact on consumer liking. There was no significant consumer preference for any of the fresh sausages, meaning there is a market for both of them.

13.3.2 Multinomial Regression

Most commonly, the result of a sensory study involves more than two responses, for example, several factors determining liking or disliking of a food product rather than only accept versus discard. In the case of binary responses, most models use either a probit or a logit specification. In turn, multinomial models (e.g. multinomial logit estimation) deal with a range of possible responses (Kohansal and Firoozzare 2013).

Multinomial logistic regression is also known as multiclass logistic regression, multinomial logit, and mlogit (Malouf 2002). Multinomial logit models (MLM) have a dependent variable to be predicted that comes from one of a limited set of items that cannot be meaningfully ordered, as well as a set of independent variables (or explanators). Multinomial logistic regression uses linear combinations and problem-specific parameters to estimate the probability of each particular outcome (i.e. each value of the dependent variable). MLM assume that data are case-specific, meaning that each independent variable has a single value for each case (Garson 2016). Whenever the multinomial logit is used to model choices, it relies on the assumption of independence of irrelevant alternatives (IIA). In the case of food choice, this means that the introduction of a third formulation (C) when there were previously only two formulations (A and B), would not affect the consumers' preference responses. When the IIA hypotheses do not hold true (i.e. the withdrawal of formulation C changes the consumer's responses), alternative models should be used, such as the nested logit or the multinomial probit (Baltas and Doyle 2001).

The basis of MLMs is to construct a linear predictor function that builds a score from a set of weights that are linearly combined with the explanatory variables of a given observation using a dot product, as shown in the equation below:

$$Score(x_i k) = \beta_k \cdot x_i \tag{13.2}$$

where x_i is the vector of explanatory variables describing observation i, β_k is a vector of weights (or regression coefficients) corresponding to outcome k, and Score (x_i, k) is the score associated with assigning observation i to category k.

In discrete choice theory, where observations represent people and outcomes represent choices, the score is considered the utility associated with person i choosing outcome k. The predicted outcome is the one with the highest score. The score can be converted to a probability value, indicating the probability of observation i choosing outcome k given the measured characteristics of the observation (Garson 2016).

In the mixed multinomial logit model (MIXLM), the vector of consumer utility weights on product attributes is assumed to have a multivariate normal distribution in the population. This means that preference is heterogeneous within the study's set of consumers, that is, some individuals give higher importance to some attributes than others, in this case, the IIA property of MLM (Goos and Hamidouche 2019; Kohansal and Firoozzare 2013).

Goos and Hamidouche (2019) modelled consumer choices between cocktail mixtures with different proportions of ingredients. Seven cocktails containing mango juice, cassis, and lime were prepared according to a simplex–centroid design by respecting some constraints regarding the lower proportions of the three main ingredients: 0.3 for mango, 0.15 for cassis, and 0.1 for lime. Sixty consumers made eight pair-wise comparisons each. For each cocktail pair, the respondents indicated which of the two tasted best and had the most attractive colour. To avoid bias related to the order of sample presentation, a linked-pair design was employed that ensured that every cocktail was tasted first as often as it was tasted second. The preference data were processed by three methods: multinomial logit model (MLM), mixed multinomial logit model (MIXLM), and latent class mixture-choice (LCM). The MLM did not fit the data appropriately when it was applied to the full data set due to heterogeneous preference between individuals. For example, cocktails with a small proportion of lime were preferred by the consumers according to MLM, MIXLM, and a subset of the data processed by the LCM. However, a second response segment processed by the LCM indicated a strong preference for cocktails with a larger proportion of lime. Therefore, the heterogeneity was modelled by using MIXLM, LCM, and MNM applied to the data from each individual separately. Based on the AIC (Akaike Information Criterion) and BIC (Bayesian Information Criterion) values for the three models, the MLM and the LCM out-performed MIXLM. To optimize the cocktail composition, the best mixture-choice model was identified for each segment/data subset, yielding 16 different models. Due to the heterogeneous characteristic of the respondents, several optimum points were found rather than one single optimal composition. Nevertheless, in general, the predicted optimal cocktail composition involved a minimum of lime ~0.1, a proportion of mango of 0.45–0.6, and cassis of 0.3–0.45 in terms of preference score. A similar prediction was made in terms of colour, which revealed that a minimum proportion of lime also resulted in a more attractive colour in most data subsets/segments. Then, an optimization of the two responses (taste and colour) simultaneously was performed. For a substantial number of segments, a formulation with a minimum proportion of lime, ~50% of mango, and ~40% of cassis. and another with ~20% of lime, 40% of mango. and 40% of cassis pleased 9 out of 14 segments in terms of both taste and colour.

13.4 Nutrikinetics and Nutridynamic Characteristics of Food Formulations

Nutridynamics and Nutrikinetics are fields within food and nutrition research that study the action mechanisms of a living organism and a food component or dietary supplement, respectively, within the human body (de Vos et al. 2006; van Duynhoven et al. 2012). Nutrikinetics, specifically, encompasses the ADME of food components in the host, which includes the interactions between the host metabolome and the gut microbiome (van Duynhoven et al. 2012).

13.4.1 Quantitative Structure–Activity Relationship (QSAR) Modelling

Quantitative structure–activity relationship (QSAR) modelling is widely used in Nutridynamics to predict the bioavailability of a food component by comparing its chemical structure with that of other substances with known biological activity (Lefebvre et al. 2015). Another common application of QSAR models is in the toxicology and safety assessment of foods (Blaauboer et al. 2016). QSAR modelling can involve simple regression methods when it comes to small sets of compounds from the same family to complex non-linear models generated by artificial intelligence (e.g. artificial neural network – ANN), in the case of large data sets comprising thousands of varying molecular structures (Cherkasov et al. 2014). Chemical descriptors employed in QSAR modelling range from 1D descriptors (molecular formula) to 2D (structural formula or topological representation), 3D (conformation-dependent formula), and even 4D (mutual orientation and time-dependent dynamics of molecules (Cherkasov et al. 2014).

Goel et al. (2018) used QSAR to predict the sweetness of nearly 500 compounds in relation to that of sucrose. Although this application is not related to food formulation optimization per se, finding the best performing compound is an example of how deductive models can support the earlier stages of food product design. Goel's study aimed at building a global model based on a comprehensive data set to predict the sweetness of any novel compound. A total of 487 molecules with relative sweetness varying from 0.2 to 220 000 were gathered and the concentration ratio between each of them and sucrose was calculated. Then, 564 molecular descriptors (e.g. structural, topological, thermodynamic, atomistic, etc.) were calculated for each compound. An earlier step of outlier detection was performed, in which compounds that received a poor estimation of the sweetness index by the QSAR model were disregarded in the next stages. The curated data set of 455 molecules was randomly split into a training set (70%) and test set (30%). The 319 molecules from the training set were used to build the model by Genetic Function Approximation (GFA) as a function of 61 molecular descriptors, structural, topological, thermodynamic, atomistic, etc. The remaining molecules were used to validate the model. GFA models were constructed using linear and quadratic fits of the responses as a function of 6 to 14 variables. Several statistical parameters for linear and quadratic fits were used to evaluate the goodness of fit, such as the Lack-Of-Fit (LOF) score, the correlation coefficient (R^2), the cross-validation coefficient (R_{cv}^2) obtained through the leave-one-out method, the Mean Absolute Error (MAE), the Mean Square Error (MSE), and the Mean Absolute Percentage Error (MAPE).

The quadratic fit performed better than the linear fit, which was improved with the addition of splines or high-order polynomial terms. For comparison purposes, another modelling technique was tested: ANN trained with the BFGS (Broyden-Fletcher-Goldfarb-Shanno) method using an S-shaped sigmoid function. The validation set again comprised 30% of the data set. The network was fed with the model-building data set and 61 descriptors, and worked iteratively (n = 4000) by comparing output values with the experimental ones towards error minimization. The two models had a satisfactory and comparative performance: $R_{training}^2 = 0.889$ and $R_{test}^2 = 0.831$ in ANN and $R_{training}^2 = 0.864$ and $R_{test}^2 = 0.832$ in GFA with linear spline. Both GFA with spline and ANN models were able to embody nonlinear effects in the data set. However, the authors concluded that the effects of a variable on the output were more evident in the GFA modelling, which enables the user to determine a more straightforward structure–property relationship.

13.4.2 Population-Based Non-Linear Mixed Effect Modelling

Non-linear Mixed Effect Models (NLME) are models that establish non-linear relationships between fixed factors and random effects and the studied responses from longitudinal data. Fixed effects are population parameters that are constant every time a data set is collected, while random effects are random variables associated with each experimental determination (Pinheiro and Bates 2000). As an example, in the context of food formulation, fixed effects could be the fixed amounts of ingredients in different formulations (samples), whereas random effects would be different liking scores assigned to the same sample by different panellists. NLME are often used in Nutrikinetics to analyse the ADME profile of a food compound. Importantly, this modelling approach enables the assessment of inter-individual variation in the effect of a compound in a food formulation on human health (van Duynhoven et al. 2012). Population-based NLME modelling is able to overcome the challenges related to longitudinal sampling of plasma and urine common in Nutrikinetics (van Velzen et al. 2014). Nutrikinetic MLME models differ from traditional pharmacokinetic models in that variation between individuals and background diet are taken into account (van Duynhoven et al. 2012).

Mertens-Talcott et al. (2008) assessed the antioxidant effects of anthocyanins on healthy individuals after consumption of acai juice and pulp. An acute crossover clinical trial was conducted that comprised 12 healthy individuals receiving samples at 7 mL/kg of body weight after a washout phase and overnight fast. Plasma and urine were sampled over 12 and 24 hours after consumption, respectively. Total anthocyanins were quantified as cyanidin-3-O-glucoside and yielded maximum concentration (Cmax) values of 2321 and 1138 ng/L at t_{max} 2.2 and 2.0 hours, and area under the curve (AUC) last values of 8568 and 3314 ng h L^{-1} for acai pulp and juice, respectively. The Cmax/AUC ratio, which is used to evaluate the equivalence of absorption rates, was calculated as 0.27 h^{-1} for acai pulp and 0.34 h^{-1} for acai juice. The NLME models indicated a negative nonlinear relationship for both acai pulp and acai juice between dose volume and oral bioavailability. Plasma antioxidant capacity was raised up to 2.3 and 3 times for acai juice and acai pulp, respectively.

Van Velzen et al. (2014) quantified polyphenol exposure by combining Nutrikinetics and population-based modelling. Experimental data were collected from a placebo-controlled crossover study comprising 20 male individuals who were given extracts of grapes/wine (GW) and black tea (BT). The urinary and plasma Nutrikinetics time–response curves and phylotypic analysis (gut microbial species) of stool samples were used to phenotype the gut microbial bioconversion capacity of the individuals. The participants consumed capsules containing GW (1410 mg), BT (2500 mg), or placebo (empty capsules) after fasting. Blood samples were collected at times 0, 1, 2, 4, 6, 9, 12, 16, 24, and 34 hours from intake, faecal samples were collected at the beginning and end of the study, and urine was sampled irregularly during the intervention period. Data on cumulative urinary excretion curves of sulphated pyrogallol and hippuric acid of each individual after a single oral dose of GW/BT/placebo, nutrikinetic signatures of two c-valerolactones (M2 and M6), and microbiota composition revealed inter-individual differences in the bioconverting capacity of the gut microbiota. In addition, *Clostridium orbiscindens* and *Eubacterium ramulus* were two of the few microorganisms that could be positively associated with the systemic exposure of c-valerolactone M6. The population model parameters, the individual model parameters, and the random effects were simultaneously estimated from all individual polyphenol concentration-time profiles. All individual model parameters were assumed to be log-normally distributed with mean zero and variance r2h, which indicate the level of inter-individual heterogeneity. The NLME model results indicated that the availability of polyphenols is highly variable between individuals and not as dependent on the product matrix as it is on gut microbial metabolism. This is a relevant conclusion around polyphenol exposure and related health benefits.

13.5 Conclusion

The combined use of experimental studies and various mathematical tools has been discussed. As food formulation design covers a wide range of stages, from experimental design to consumer studies to the assessment of nutrient bioavailability, it can involve a variety of different mathematical tools. This very nature of food formulation renders it difficult – if not impossible – to establish what are the best performing modelling techniques in this field. Ultimately, the choice of mathematical methods in food formulation design depends not only on the type of problem under analysis but also on the quality and size of the data set. Nevertheless, this chapter outlined various examples of studies having different objectives (e.g. ingredient effect on a formulation vs. consumer studies and panel studies coupling vs. nutrient delivery) and the mathematical tools utilized in each of them. Although this is not a systematic guide, this chapter is expected to help food technologists find the proper – or at least most used – mathematical tool(s) to aid different stages of food formulation development.

References

Abuhelwa, A.Y., Williams, D.B., Upton, R.N., and Foster, D.J.R. (2017). Food, gastrointestinal pH, and models of oral drug absorption. *European Journal of Pharmaceutics and Biopharmaceutics* 112: 234–248.

Antony, J. (2014a). Full factorial designs. In: *Design of Experiments for Engineers and Scientists*, 2e (ed. J. Antony), 63–85. Elsevier.

Antony, J. (2014b). Fraction factorial designs. In: *Design of Experiments for Engineers and Scientists*, 2e (ed. J. Antony), 87–112. Elsevier.

Ares, G. and Varela, P. (2017). Trained vs. consumer panels for analytical testing: fueling a long lasting debate in the field. *Food Quality and Preference* 61: 79–86.

Ares, G., Antúnez, L., Bruzzone, F. et al. (2015). Comparison of sensory product profiles generated by trained assessors and consumers using CATA questions: four case studies with complex and/or similar samples. *Food Quality and Preference* 45: 75–86.

Arteaga, G.E., Li-Chan, E., Vazquez-Arteaga, M.C., and Nakai, S. (1994). Systematic experimental designs for product formula optimization. *Trends in Food Science and Technology* 5 (8): 243–254.

Assadpour, E. and Jafari, S.-M. (2017). Spray drying of folic acid within nano-emulsions: optimization by Taguchi approach. *Drying Technology* 35 (9): 1152–1160.

Baltas, G. and Doyle, P. (2001). Random utility models in marketing research: a survey. *Journal of Business Research* 51 (2): 115–125.

Bao, C., Jiang, P., Chai, J. et al. (2019). The delivery of sensitive food bioactive ingredients: absorption mechanisms, influencing factors, encapsulation techniques and evaluation models. *Food Research International* 120: 130–140.

Berens, A.R. and Hopfenberg, H.B. (1978). Diffusion and relaxation in glassy polymer powders. *Polymer* 19 (5): 489–496.

Blaauboer, B.S., Boobis, A.L., Bradford, B. et al. (2016). Considering new methodologies in strategies for safety assessment of foods and food ingredients. *Food and Chemical Toxicology* 91: 19–35.

Bogue, J., Delahunty, C., Henry, M., and Murray, J. (1999). Market-oriented methodologies to optimise consumer acceptability of Cheddar-type cheeses. *British Food Journal* 101 (4): 301–317.

Box, G.E.P. and Draper, N. (2007). *Response Surfaces, Mixtures, and Ridge Analyses*. Hoboken, USA: John Wiley & Sons, Inc.

Breternitz, N.R., Bolini, H.M.A., and Hubinger, M.D. (2017). Sensory acceptance evaluation of a new food flavoring produced by microencapsulation of a mussel (*Perna perna*) protein hydrolysate. *LWT – Food Science and Technology* 83: 141–149.

Chakraborty, S.K., Singh, D.S., Kumbhar, B.K., and Chakraborty, S. (2011). Millet-legume blended extrudates characteristics and process optimization using RSM. *Food and Bioproducts Processing* 89 (4): 492–499.

Cherkasov, A., Muratov, E.N., Fourches, D. et al. (2014). QSAR modeling: where have you been? Where are you going to? *Journal of Medical Chemistry* 57 (12): 4977–5010.

Colman, A.M., Norris, C.E., and Preston, C.C. (1997). Comparing rating scales of different lengths: equivalence of scores from 5-point and 7-point scales. *Psychological Reports* 80: 355–362.

Cox, E.P. (1980). The optimal number of response alternate for a scale: a review. *Journal of Marketing Research* 17 (4): 407–422.

Crank, J. (1975). *The Mathematics of Diffusion*, 2e. New York: Oxford University Press.

da Cruz, M.C.R., Dagostin, J.L.A., Perussello, C.A., and Masson, M.L. (2019). Assessment of physicochemical characteristics, thermal stability and release profile of ascorbic acid microcapsules obtained by complex coacervation. *Food Hydrocolloids* 87: 71–82.

Dawes, J. (2008). Do data characteristics change according to the number of scale points used? An experiment using 5-point, 7-point and 10-point scales. *International Journal of Market Research* 50 (1): 61–77.

Dimou, M., Marnasidis, S., Antoniadou, I. et al. (2009). The application of Taguchi method to determine the optimum blend of unifloral honeys to most closely match thyme honey quality. *International Journal of Food Science and Technology* 44 (10): 1877–1886.

Garson, G.D. (2016). *Logistic Regression: Binary and Multinomial*. Asheboro, NC : Statistical Associates Publishers..

Giménez, A., Ares, F., and Ares, G. (2012). Sensory shelf-life estimation: a review of current methodological approaches. *Food Research International* 49 (1): 311–325.

Goel, A., Gajula, K., Gupta, R., and Rai, B. (2018). In-silico prediction of sweetness using structure-activity relationship models. *Food Chemistry* 253: 127–131.

Goos, P. and Hamidouche, H. (2019). Choice models with mixtures: an application to a cocktail experiment. *Food Quality and Preference* 77: 135–146.

Granato, D. and Calado, V.M.A. (2014). The use and importance of design of experiments (DOE) in process modelling in food science and technology. In: *Mathematical and Statistical Methods in Food Science and Technology* (eds. D. Granato and G. Ares), 1–18. Sussex, United Kingdom: John Wiley & Sons.

Heyd, B. and Danzart, M. (1998). Modelling consumers' preferences of coffees: evaluation of different methods. *LWT – Food Science and Technology* 31 (7–8): 607–611.

Hough, G., Sanchez, R.H., de Pablo, G.G. et al. (2002). Consumer acceptability versus trained sensory panel scores of powdered milk shelf-life defects. *Journal of Dairy Science* 85: 2075–2080.

Johns, R. (2010). Likert items and scales. Survey Question Bank: Methods Fact Sheet, 1. University of Strathclyde. Available at: https://www.sheffield.ac.uk/polopoly_fs/1.597637!/file/likertfactsheet.pdf. Accessed on 31 December 2020.

Joshi, A., Kale, S., Chandel, S., and Pal, D.K. (2015). Likert scale: explored and explained. *British Journal of Applied Science and Technology* 7 (4): 396–403.

Kohansal, M.R. and Firoozzare, A. (2013). Applying multinomial logit model for determining socio-economic factors affecting major choice of consumers in food purchasing: the case of Mashhad. *Journal of Agricultural Science and Technology* 15: 1307–1317.

Lawless, H.T. and Heymann, H. (2010). *Sensory Evaluation of Food: Principles and Practices*, 2e. New York: Springer.

Lefebvre, D.E., Venema, K., Gombau, L. et al. (2015). Utility of models of the gastrointestinal tract for assessment of the digestion and absorption of engineered nanomaterials released from food matrices. *Nanotoxicology* 9 (4): 523–542.

Lewis, J.R. (1993). Multipoint scales: mean and median differences and observed significance levels. *International Journal of Human Computer Interaction* 5 (4): 383–392.

Lim, J. (2011). Hedonic scaling: a review of methods and theory. *Food Quality and Preference* 22 (8): 733–747.

MacFie, H. (2007). Preference mapping and food product development. In: *Consumer-Led Food Product Development* (ed. H. MacFie), 551–592. Cambridge, UK: Woodhead Publishing.

Maher, P.G., Fenelon, M.A., Zhou, Y. et al. (2011). Optimization of β-casein stabilized nanoemulsions using experimental mixture design. *Journal of Food Science* 76 (8): C1108–C1117.

Malouf, R. (2002). A comparison of algorithms for maximum entropy parameter estimation. In: *Proceedings of the 6th Conference on Natural Language Learning (CoNLL)* (eds. D. Roth and A. van den Bosch), 49–55. New Brunswick, NJ, USA: ACL.

Mason, R.L., Gunst, R.F., and Hess, J.L. (2003). *Statistical Design and Analysis of Experiments: With Applications to Engineering and Science*, 2e. New York: Wiley.

Mertens-Talcott, S.U., Rios, J., Jilma-Stohlawetz, P. et al. (2008). Pharmacokinetics of anthocyanins and antioxidant effects after the consumption of anthocyanin-rich acai juice and pulp (*Euterpe oleracea* Mart.) in human healthy volunteers. *Journal of Agricultural and Food Chemistry* 56 (17): 7796–7802.

Miller, G. (1956). The magical number seven, plus or minus two: some limits on our capacity for processing information. *The Psychological Review* 63: 81–97.

Moscatto, J.A., Borsato, D., Bona, E. et al. (2006). The optimization of the formulation for a chocolate cake containing inulin and yacon meal. *International Journal of Food Science and Technology* 41: 181–188.

Moussaoui, K.A. and Varela, P. (2010). Exploring consumer product profiling techniques and their linkage to a quantitative descriptive analysis. *Food Quality and Preference* 21 (8): 1088–1099.

Musina, O., Putnik, P., Koubaa, M. et al. (2017). Application of modern computer algebra systems in food formulations and development: a case study. *Trends in Food Science and Technology* 64: 48–59.

Myers, R.H., Montgomery, D.C., and Anderson-Cook, C.M. (2011). *Response Surface Methodology: Process and Product Optimization using Designed Experiments*, Wiley Series in Probability and Statistics, 3e, vol. 705, 704 pp. New York: Wiley.

Nikzade, V., Tehrani, M.M., and Saadatmand-Tarzjan, M. (2012). Optimization of low-cholesterol–low-fat mayonnaise formulation: effect of using soy milk and some stabilizer by a mixture design approach. *Food Hydrocolloids* 28 (2): 344–352.

Pardhi, S.D., Singh, B., Nayik, G.A., and Dar, B.N. (2019). Evaluation of functional properties of extruded snacks developed from brown rice grits by using response surface methodology. *Journal of the Saudi Society of Agricultural Sciences* 18 (1): 7–16.

Paulos, K., Rodrigues, S., Oliveira, A.F. et al. (2015). Sensory characterization and consumer preference mapping of fresh sausages manufactured with goat and sheep meat. *Journal of Food Science* 80 (7): S1568–S1573.

Piepel, G.F. and Cornell, J.A. (1985). Models for mixture experiments when the response depends on the total amount. *Technometrics* 27 (3): 219–227.

Pinheiro, J.C. and Bates, D.M. (2000). Nonlinear mixed-effects models: basic concepts and motivating examples. In: *Mixed-Effects Models in S and S-PLUS. Statistics and Computing*, 273–304. New York: Springer.

Ramírez-Sucre, M.O. and Vélez-Ruiz, J.F. (2013). Physicochemical, rheological and stability characterization of a caramel flavored yogurt. *LWT – Food Science and Technology* 51 (1): 233–241.

Rødbotten, M., Tomic, O., Holtekjølen, A.K. et al. (2015). Barley bread with normal and low content of salt: sensory profile and consumer preference in five European countries. *Journal of Cereal Science* 64: 176–182.

Ronald, B.P. and Shahidi, F. (2007). Encapsulation, stabilization and controlled release of food ingredients and bioactives. In: *Handbook of Food Preservation*, 2e (ed. M.S. Rahman), 514–527. Boca Raton, FL: CRC Press,Taylor & Francis Group.

Roselli, L., Cicia, G., Cavallo, C. et al. (2018). Consumers' willingness to buy innovative traditional food products: the case of extra-virgin olive oil extracted by ultrasound. *Food Research International* 108: 482–490.

Sarteshnizi, R.A., Hosseini, H., Bondarianzadeh, D. et al. (2015). Optimization of prebiotic sausage formulation: effect of using β-glucan and resistant starch by D-optimal mixture design approach. *LWT – Food Science and Technology* 62 (1), Part 2: 704–710.

Schrevens, E. and Portier, K. (2005). Experimental design in product development and optimisation. In: *Handbook of Food Science, Technology, and Engineering* (ed. Y.H. Hui), vol. 3, 143-1–143-30. New York: Taylor & Francis.

Sharif, M.K., Butt, M.S., Sharif, H.R., and Nasir, M. (2017). Sensory evaluation and consumer acceptability. In: *Handbook of Food Science and Technology*. (eds. T. Zahoor and M.S. Butt), 361–386

Sidel, J.L. and Stone, H. (1993). The role of sensory evaluation in the food industry. *Food Quality and Preference* 4 (1–2): 65–73.

Tijskens, L.M.M., Hertog, M.L.A.T.M., and Nicolai, B.M. (2001). *Food Process Modelling*, vol. 59. Cambridge, UK: Woodhead Publishing.

Vaisey-Genser, M., Ylimaki, G., and Johnson, B. (1987). The selection of canola oil, water, and an emulsifier system in cake formulation by response surface methodology. *Cereal Chemistry* 64: 50–54.

van Duynhoven, J.P.M., van Velzen, E.J.J., Westerhuis, J.A. et al. (2012). Nutrikinetics: concept, technologies, applications, perspectives. *Trends in Food Science and Technology* 26 (1): 4–13.

Valentas, K.J., Rotstein, E., and Singh, R.P. (1997). *Handbook of Food Engineering Practice*. Boca Raton: CRC Press.

van Velzen, E.J.J., Westerhuis, J.A., Grün, C.H. et al. (2014). Population-based nutrikinetic modelling of polyphenol exposure. *Metabolomics* 10 (6): 1059–1073.

Verdenius, F. (2001). The power and pitfalls of inductive modelling. In: *Food Process Modelling*, vol. 59 (eds. L.M.M. Tijskens, M.L.A.T.M. Hertog and B.M. Nicolai), 510 pp. Cambridge, UK: Woodhead Publishing.

Viljoen, M., Muller, M., De Beer, D., and Joubert, E. (2017). Identification of broad-based sensory attributes driving consumer preference of ready-to-drink rooibos iced tea with increased aspalathin content. *South African Journal of Botany* 110: 177–183.

de Vos, W.M., Castenmiller, J.J.M., Hamer, R.J., and Brummer, R.J.M. (2006). Nutridynamics – studying the dynamics of food components in products and in the consumer. *Current Opinion in Biotechnology* 17 (2): 217–225.

Yang, J. and Lee, J. (2019). Application of sensory descriptive analysis and consumer studies to investigate traditional and authentic foods: a review. *Food* 8 (2): 1–17.

Yin, S., Wu, L., Du, L., and Chen, M. (2010). Consumers' purchase intention of organic food in China. *Journal of the Science of Food and Agriculture* 90 (8): 1361–1367.

Yu, P., Low, M.Y., and Zhou, W. (2018). Design of experiments and regression modelling in food flavour and sensory analysis: a review. *Trends in Food Science and Technology* 71: 202–215.

14
Regulatory and Legislative Framework for Novel Foods

Jessica Vapnek[1], Kai Purnhagen[2], and Ben Hillel[1]

[1]*University of California, Hastings College of the Law, San Francisco (UC Hastings), USA*
[2]*University of Bayreuth, Campus Kulmbach, Chair for Food Law, Bayreuth, Germany*

Summary

This chapter discusses how novel foods are regulated in the EU, US, and elsewhere. It reviews some of the reasons why novel foods might be regulated and how such regulatory goals can be accomplished. Regulation of novel foods is the policy maker's reaction to the unknown, although what is novel today may not be novel tomorrow. Indeed, the novel food industry has grown immensely over the past decade due to the need to feed a growing world population while using less land, water, and other limited natural resources. Ultimately, policy makers have the difficult task of balancing adequate government oversight over novel foods without stifling innovation and creativity.

14.1 Introduction

Scientific advances have made it possible to produce foods from unconventional sources or through new techniques. At the same time, globalization and integrated value chains have facilitated the movement of foods and food ingredients from regions where they have been in use for decades or longer, to other regions where they are unknown – and therefore perceived as 'novel'. Earlier chapters of this book provided an overview of novel food ingredients and processing techniques. They also explored the use of novel food ingredients in the formulation of novel and health-promoting food products. With more and more of these foods and food ingredients circulating in the market, interest in regulation is increasing.

The purpose of this chapter is to review how novel foods are currently regulated in the European Union (EU), United States (US), and elsewhere. We briefly review

Food Formulation: Novel Ingredients and Processing Techniques, First Edition.
Edited by Shivani Pathania and Brijesh K. Tiwari.

what novel foods are – and in later sections we look at special categories of foods such as foods derived from insects, 3D printed food, and food formulated for special population segments – and why they are regulated. We then discuss the various types of regulation of novel foods – primarily official review and approval (or rejection) of the introduction of these types of foods on the market, but also intellectual property law, marketing law, tort law, and criminal law.

This chapter builds on the scientific discussions in the earlier chapters and provides an overall picture of what public and private entities should take into account when regulating or trading in novel foods.

14.2 What is a Novel Food?

In most jurisdictions, new foods and ingredients introduced into the stream of commerce are generally referred to as 'novel foods'. In some ways it is a circular effort to define a novel food, because the definition varies depending on what the applicable legislation says it is. In turn, whether a food is 'novel' according to legislation may depend on geographic, temporal, or cultural variables, i.e. whether it is unfamiliar to a specific region, time period, or culture. The definition of a novel food may also differ, depending on the purpose for which the food is being regulated. For example, one piece of legislation might define a novel food for purposes of protecting public health, while another piece of legislation may propose a different definition for another purpose, such as securing intellectual property rights.

Some examples of what might be considered novel foods are newly developed and innovative foods, foods produced using new technologies and production processes, or foods traditionally eaten outside of a geographical region.[1] More than 30 years ago, the International Programme on Chemical Safety (IPCS), a joint UN body with representation from the World Health Organization, the International Labour Organization, and the United Nations Environment Programme, defined a novel food as 'a food or food ingredient produced from raw materials not normally used for human consumption or food that is severely modified by the introduction of new processes not previously used in the production of food'.[2] In the EU, the Novel Food Regulation (NFR),[3] which in its current form came into force on January 1, 2018, defines novel foods as foods that have not been used for human consumption to a significant degree within the EU before May 15, 1997 and that fall within one of 10 enumerated categories.[4] The European Commission provides the following specific examples:

> Examples of Novel Food include new sources of vitamin K (menaquinone) or extracts from existing food (Antarctic Krill oil rich in phospholipids from *Euphausia superba*), agricultural products from third countries (chia seeds, noni fruit juice), or food derived from new production processes (UV-treated food (milk, bread, mushrooms and yeast)).[5]

In the US, the Food and Drug Administration (FDA), which is one of the food regulatory agencies, has no formal definition and in fact does not recognize novel foods as a category of food. Instead, any new food (no matter its technological, temporal, or geographical origin) is regulated like any other food. Instead, a new substance or new food ingredient is regulated through the food additive process, unless it meets an exception under the Federal Food, Drug, and Cosmetic Act[6] or is considered Generally Recognized as Safe (GRAS).[7]

We examine the regulatory frameworks for new or novel foods in greater detail in the next section.

14.3 Overview of Regulatory Frameworks for Novel Foods

14.3.1 Why Regulate Novel Foods?

There are many reasons to regulate novel foods. At the most fundamental level, regulation of novel foods represents how society has chosen to deal with the unknown consequences of new technologies and materials. This decision-making process is implicated not only with respect to novel foods but also to all novel techniques and materials on the market. Scholars call this 'regulating the unknown'.[8] Policy makers weighing and choosing regulatory approaches must find a balance between, for example, protection and innovation, between financing of research and reaping of societal benefits, and between societal acceptance and necessary government interference. Depending on the culture and legal system, the balances may shift, and one factor may be privileged over others. In Ulrich Beck's work, which has guided EU regulation,[9] technologies create risks that trigger regulation.[10] Commentators in the economic arena have noted that this approach focuses on maintaining the status quo,[11] and critics charge that it therefore limits flexibility and stifles innovation. But at the least, regulation of technologies should preserve society's ability to understand potential future uses of the technology in order to evaluate whether risks are worth taking to reap potential future benefits.[12] Whatever the locale or justification, regulation in this context generally consists of testing, data, and scientific evaluation of potential adverse effects on health, safety, and the environment.

The problem arises when data on potential hazards is non-existent or inconclusive. Whereas the US' regulatory approach requires scientific proof of harm, Europe requires regulators to take action when, after assessment of available information, a possibility of harmful effects on health is identified but scientific uncertainty persists[13]: this is known as the precautionary principle, which applies under the General Food Law (GFL) (the preeminent EU Regulation on food)[14] and hence to novel foods.[15] Although this definition is widely shared, it is important to note that the exact wording and application of the precautionary principle in EU law depend on the legal sector and/or type of good, service or circumstance it applies to.

The European Commission describes the principle's justification as follows: 'decision-makers are constantly faced with the dilemma of balancing the freedom and rights of individuals, industry and organisations with the need to reduce the risk of adverse effects to the environment, human, animal or plant health'.[16] Equally, policy makers are divided over how to weigh and evaluate the relative importance of types of effects: should one prioritize conservation, innovation, or consumer health? These are political decisions which regulations cannot solve, but rather are left to the political realm. At least within the EU, the precautionary principle is the only tool available to regulate decisions under circumstances of uncertainty, and policy makers have determined that the precautionary principle applies to novel foods.[17]

14.3.2 How are Novel Foods Regulated?

Novel food regulations can be categorized in a variety of ways to facilitate understanding. The first division is between content-related versus information-related measures. Content-related measures, in this context, consist of measures such as approval procedures, where responsible authorities examine the type of food before approving or rejecting an application to commercialize it or before banning certain types of foods. Information-related measures, by contrast, are measures where regulators control information about the food through labelling or through information released to consumers to affect consumer choice.

Regulation of novel foods can also be categorized based on the phase of market introduction of the product. Here, measures can be conceptually divided into those regulating market access (e.g. pre-market measures such as approvals) and those regulating products already on the market (e.g. post-market measures such as liability).

Regulation can also be grouped into measures that are process-related (i.e. regulating the technique used to produce the food) and those that are product-related (i.e. regulating different features of the product). Which of these might be given more or less attention or weight in the regulation of novel foods at a particular time depends on the jurisdiction. It is worth noting, however, that most novel food regulations apply a product-related approach, which experts consider the favoured strategy in the face of uncertainty.[18] With this type of regulatory framework, whether a food is subject to special regulation as a novel food mainly depends on evaluation of the 'nutritional composition, presence of known toxins or anti-nutrients, and allergenicity of proteins as well as ... the potential nutritional impact of introducing the novel food into the human diet.'[19]

It is important to underline here that although some technology-related regulation exists, in some respects this is just another type of product-related measure. For instance, the EU applies the novel food regulatory regime to foods containing engineered nanomaterials, but in fact it is not the technique that is being regulated, but rather the risks stemming from the use of the technique.

Regardless of how one conceives of the regulation of novel foods, in all jurisdictions that regulate them we find most of these categories in use.

14.3.3 How Novel Foods Came to be Regulated Internationally

More than 30 years ago, the IPCS developed guidelines for the safety assessment of novel foods. The IPCS defined a novel food as 'a food or food ingredient produced from raw materials not normally used for human consumption or food that is severely modified by the introduction of new processes not previously used in the production of food.'[20] The definition applied to foods intended for direct consumption or for consumption after some alterations to make the product more acceptable.

The IPCS was particularly concerned about circumstances where novel foods are intended to replace a significant portion of traditional food in the diet, because of the likely nutritional impact. For this reason, in addition to recommending animal and human tests, the IPCS proposed examining the effect of the introduction of the new substance on the nutrient composition of the diet as a whole. It recommended particular care where the food is intended for groups such as 'children, the elderly, and "captive populations," e.g. hospital patients and school children'.[21]

As we saw in earlier chapters of this book, complete chemical identification of novel foods may not be feasible, but it is important that the regulatory system ensure microbiological purity and that levels of potentially hazardous contaminants be kept to a minimum. The IPCS pointed out that if the novel food is intended to be an alternative significant supply of protein, tests on its protein quality will be necessary.[22] Other studies should analyze: (i) the availability of vitamins and minerals in the novel food in comparison with the food it would replace; and (ii) any interaction the novel food might have with other items of the diet that would reduce the whole diet's nutritional value. IPCS also recommended testing for allergenicity.[23]

Currently, the Joint FAO/WHO Committee on Food Additives (JECFA), a scientific advisory body, evaluates the safety of food additives for purposes of advising the Codex Alimentarius Commission, which is the reference body for international food standards under the World Trade Organization (WTO)[24] that produces standards to guide national legislators. Although JECFA's safety determinations are not binding at national level (i.e. at the moment a particular jurisdiction is deciding whether to approve a particular new food additive), its recommendations are broadly accepted and so are likely to be taken into consideration.[25]

14.3.4 How Novel Foods are Regulated in the European Union

Before looking specifically at the regulation of novel foods in the EU, it may be useful to review the complex institutional and legislative framework underpinning the regulation of food within the Union.[26] Legislation issued at EU level in the form of a Regulation, such as the NFR, is paramount and directly applicable in member states, without any need for national adoption or enactment. Member state legislation that conflicts with an EU-level Regulation is invalidated in case of conflict, much as federal law pre-empts state law in the US. Since most food laws in the EU are in the form of Regulations, they are directly applicable to EU institutions, all member state authorities, and all persons within the EU. Enforcement responsibility rests with food business operators, although member states are also obligated to establish appropriate enforcement mechanisms.[27] The European Commission in turn carries out regular audits to identify and minimize variations among these member state controls.

The European Commission's integrated approach to food safety is set out in the White Paper on Food Safety,[28] now more than 20 years old, which has become binding law through the GFL. The regulatory system covers all stages 'from farm to fork', starting with feed production and primary production and continuing through food processing, storage, transport, and sale. The GFL sets out an overarching framework for the development of food legislation and regulation at EU and member state levels. It establishes the 'general principles, requirements, and procedures that underpin decision making in matters of food and feed safety'.[29] As noted, since the GFL is a Regulation, it is directly applicable and does not need to be transposed into member state legislation.

In the EU, pre-emption occurs only where EU law regulates the precise topic at issue. To take a simple example, if member state law says that strawberries must be green but EU law states that they must be red, then EU law pre-empts all national laws with regard to strawberries (but not to apples). What distinguishes the GFL from most EU law is that it applies a horizontal (systemic) approach, meaning that it covers not only legislation but all types of action by member states, businesses, and other private parties active on the internal EU market. If the GFL said that all fruits must have their natural colour, this would cover all laws, regulations, rules, and practices in a member state regarding strawberries, apples, and all other fruits – no matter the form of the rule or business practice in the member state. The GFL also introduces an ostensibly science-based approach, meaning that no decision on foods can be made without scientific evidence being available.[30]

Within the EU, the enactment of the NFR has provided the region with a comprehensive novel food definition and regulatory framework, which was substantially revised as of January 1, 2018. Under the NFR, novel foods are foods that have not been used for human consumption to a significant degree within the EU before May 15, 1997 and that fall within one of 10 enumerated categories:

1. food with a new or intentionally modified molecular structure;
2. food consisting of, isolated from, or produced from microorganisms, fungi, or algae;
3. food consisting of, isolated from, or produced from material of mineral origin;
4. food consisting of, isolated from, or produced from plants or their parts obtained by non-traditional propagating practices if significant changes in the composition or structure of the food affect its nutritional value, metabolism, or level of undesirable substances;
5. food consisting of, isolated from, or produced from animals or their parts obtained by non-traditional breeding techniques;
6. food consisting of, isolated from, or produced from cell culture or tissue culture derived from animals, plants, microorganisms, fungi, or algae;
7. food resulting from a new production process if significant changes in the composition or structure of the food affect its nutritional value, metabolism, or level of undesirable substances;
8. food consisting of engineered nanomaterials;
9. vitamins and minerals and other substances used in accordance with Food Supplements Directive 2002/46/EC obtained by a new food production process or containing engineered nanomaterials;
10. food used exclusively in food supplements within the EU before May 15, 1997, intended to be used in foods other than food supplements.[31]

The NFR's regulation of novel foods attempts a maximum harmonization approach,[32] under which EU law provides both floor and ceiling for member state regulation. What this means in practice is that member states, in general, cannot introduce higher protection measures in their national legislation than the EU standards. If, for example, EU member states would like to add additional labelling requirements for novel foods which are not foreseen in the NFR or accompanying EU texts, this would be against EU law. However, EU law provides some options to deviate from this strict regime if new scientific evidence comes to light.[33] In such circumstances, the member state has to notify the Commission, which decides whether the proposed member state measure is justified.[34] This accords with WTO rules, in that international standards are a floor, and member states may enact more rigorous standards so long as they are based on science and do not violate other WTO principles.[35]

The NFR applies to novel foods placed on the market within the EU.[36] Some foods, however, including those made with genetically-modified organisms (GMOs), food enzymes, food additives, food flavourings, and extraction solvents, are covered by other legislative measures and hence fall outside the scope of the NFR.[37] For example, if a food has been in continual use in a non-EU country, it is not considered novel if its safety has been confirmed with compositional data: continual use is defined here as use for at least 25 years in the customary diet of a significant number of people residing within in at least one non-EU country.[38] If these criteria are not met, non-EU or 'third-country foods' are considered novel.[39] Third-country foods are also considered novel if the food belongs to one of the categories that are excepted from the NFR, set out above.[40] The burden of establishing whether a foodstuff is novel lies with the food business operator that intends to place the food on the EU market,[41] although member states and the Commission have a duty to assist food business operators in making that assessment.[42]

Novel foods may only be placed on the market within the EU if they are authorized via inclusion on a publicly available list established by the Commission.[43] Article 10 of the NFR lists some generic requirements that must appear in all authorization applications, including:

1. the name and address of the applicant;
2. the name and description of the novel food;
3. the description of the production process(es);
4. the detailed composition of the novel food;
5. scientific evidence demonstrating that the novel food does not pose a safety risk to human health;
6. the analysis method(s) applied, where applicable;
7. a proposal for the conditions of intended use and for specific labelling requirements which do not mislead the consumer, or a verifiable justification why those elements are not necessary.[44]

In assessing the safety of a novel food under the NFR, the EU Commission may grant authorization only where:

1. the food does not pose any risk to human health on the basis of scientific evidence[45];
2. the food's intended use does not mislead the consumer, especially when the food is intended to replace another food and there is a significant change in the nutritional value[46];
3. where the food is intended to replace another food, it does not differ from that food such that its normal consumption would be nutritionally disadvantageous for the consumer.[47]

14.3.5 How Novel Foods are Regulated in the United States

As we did for the EU, in this section we briefly review the legislative framework for food in the US. The US system consists of federal law, which applies throughout the country and exists concurrently with the individual regulatory frameworks of the 50 states. Local legislatures (such as counties and cities) may also regulate aspects of food within their jurisdictions. Both state and local legislatures are free to regulate food so long as their regulation does not contradict federal law. However, certain topics are relegated to the federal government's purview exclusively, such as import, export, and interstate commerce.[48]

In the US, federal law is often, but not always, considered supreme to state law. Federal law operates within the federal government's constitutionally established bounds. In contrast, state law often concerns intrastate matters that fall outside of these boundaries or are not covered by federal law. State law can fill in the gaps that federal law, sometimes purposely, leaves behind for states to legislate. All US states have passed laws regulating foods and other products.

The Pure Food and Drug Act (PFDA), enacted in 1906, was the first federal law to regulate food in the United States; the Meat Inspection Act was passed the same year.[49] Both laws were adopted to respond to burgeoning public attention to food hygiene and safety issues. The PFDA hewed closely to the model of the state-level laws that had proliferated in the latter half of the nineteenth century, in that it mandated accurate product labelling, outlawed interstate trade in 'adulterated' and 'misbranded' foods, and required producers to indicate on food labels the presence of mixtures or impurities.[50] Congress later passed the 1938 Food, Drug, and Cosmetics Act (FDCA) to expand federal regulatory authority over drugs while also more generally strengthening the powers of the FDA.[51] Another piece of federal legislation, the Nutrition Labelling and Education Act enacted by Congress in 1990, required all packaged foods to contain standardized information on nutrition and on serving sizes.[52] Four years later, the Dietary Supplement Health and Education Act classified dietary supplements as foods rather than drugs.[53] In more recent years, Congress reformed major food safety laws via the 2011 Food Safety Modernization Act; this overhaul shifted the FDA's focus from reacting to food safety issues to preventing food-borne illnesses through expanded regulatory authority.[54]

The US does not specifically define or regulate novel foods; instead, they are regulated as food, regardless of their technological, temporal, or geographical origin. If not a food, a new substance would be regulated as a direct food additive or food contact substance, depending on its intended use. The FDA considers any new food ingredient either as a food additive (requiring a pre-market approval by the FDA) or GRAS for specific uses (as classified by a panel of experts and determined independently from the FDA).[55]

The FDCA defines 'food additive' as any substance that is intentionally added to food as a consumable component or will affect the characteristic of the final end product, unless it falls under a listed exception (such as a colour additive, new animal drug (NAD), pesticide chemical, or dietary supplement) or is considered a GRAS substance.[56] Food additives are considered unsafe until the manufacturer can prove in a Food Additive Petition (FAP) that the ingredient is food-grade pure and meets food safety regulations.[57] The FDA will review the scientific data and information submitted in the FAP to ensure it is 'safe', which is defined as a reasonable certainty of no harm to consumers.[58] After the FDA approves the FAP, the substance may enter interstate commerce.[59]

Most companies opt to list new food ingredients as GRAS where possible, to avoid the tedious FAP pre-market approval process.[60] The key difference between the GRAS and food additive classification process is who reviews the scientific data and information to make the final decision. As indicated above, the FDA approves food additive safety based on privately held data and information about the substance that the producer compiled in the FAP.[61] By contrast, a company relies on the opinion of 'experts qualified by scientific training and experience to evaluate ... safety' in making its own (not the FDA's) determination that a substance's use is GRAS.[62] This self-determination still requires both technical evidence of safety and a basis to conclude that this evidence is generally known and accepted.

There is no requirement for the FDA to corroborate the company's findings before the self-determined GRAS substance is released on the market; instead, the company may voluntarily submit a notification of its determination to the FDA for review before putting it on the market.[63] The notification includes a detailed description of the substance, the conditions of use, and the basis of the determination.[64] A 'no questions' response from FDA does not mean that the FDA has approved the substance; as noted, in the GRAS process the FDA relies on the determination made by the company submitting the substance for approval.

In the US, companies have already taken advantage of the GRAS pathway to get their vegetarian food products, which contain new food ingredients, to market quickly. For example, the FDA examined extensive test data about the Impossible Burger's key ingredient, a protein called 'soy leghaemoglobin', before affirming its GRAS status.[65] Approval for this protein was key to its success because it carries heme and makes the vegetarian patty 'bleed' like real meat.[66]

Alternatively, a substance can be classified as GRAS if it has been used in food before 1958 and has been commonly ingested by a significant number of consumers; however, this route is rarely utilized today.[67]

14.3.6 How Novel Foods are Regulated in Other Jurisdictions

Governments make different choices on how to design, enact, and enforce novel food regulations within their borders. In Australia and New Zealand, the establishment of a joint regulatory body has led to the creation of a set of food standards that apply in both countries. In Brazil, the country's own regulation exists alongside concurrent regulation across almost all of South America, through Mercosur, the South American Common Market. Novel foods are specifically defined and regulated in several countries including Australia/New Zealand, Brazil, Canada, and China; we look at each of these in turn. Although several of these are federal states, we only briefly touch on their central regulatory schemes for novel foods.

14.3.6.1 Australia/New Zealand

By agreement in July 1996, Australia and New Zealand authorized the predecessor to the bilateral agency 'Food Standards Australia New Zealand' (FSANZ) to establish and maintain food standards that apply in both countries.[68] The two countries have also harmonized labelling and compositional standards under the Australia New Zealand Food Standard Code (the Code).[69] Within Australia, enforcing compliance with the Code with respect to all foods is the responsibility of State/Territory Health Departments,[70] while for imported foods, enforcement is the responsibility of the Australian Quarantine Inspection Service.[71]

The Code regulates all aspects of food including labelling, food additives, processing aids, nutritive substances, contaminants, approval of new foods (novel foods, genetically-modified (GM) foods, and irradiated foods), and composition of standardized and special purpose foods, among others.[72] Other standards that are not part of the joint food standards setting system cover food safety, agricultural compounds, veterinary medicines, and primary food production and processing.[73]

Food ingredients are not defined under the Australia/New Zealand food regulations[74]; ingredients are generally considered to be either foods or substances added to food.[75] FSANZ defines a novel food as a non-traditional food with no history of safe use where the food requires an assessment of the public health and safety considerations with regard to:

1. the potential for adverse effects on humans;
2. the composition or structure of the food;
3. the process by which the food has been prepared;
4. the source from which it is derived;
5. patterns and levels of consumption of the food; or
6. any other relevant matters.[76]

Non-traditional food is defined as:

1. a food that does not have a history of human consumption in Australia/New Zealand;
2. a substance derived from a food where that substance does not have a history of human consumption in Australia/New Zealand other than as a component of that food;
3. any other substance, where that substance, or the source from which it is derived, does not have a history of human consumption as a food in Australia/New Zealand.[77]

An application to vary the Code is required to approve the use of a nutritive substance or to change the permissions for a currently used nutritive substance.[78]

14.3.6.2 Brazil

The main authority within Brazil for food regulation and safety assessment is the Ministry of Health (MOH) (Ministério da Saude) through its autonomous regulatory agency, Agencia Nacional de Vigilancia Sanitária (ANVISA – National Agency of Sanitary Surveillance).[79] ANVISA's function is to evaluate the safety of use of food additives and ingredients in foods. Specific working groups within ANVISA work on the approval of novel foods and novel food ingredients with functional health claims.[80] Novel foods and ingredients require a pre-market approval by ANVISA[81]: pre-market approvals are valid for five years from the date of their publication in the Brazilian Official Gazette and may be renewed.[82]

The following foods must be registered in the category of novel food, even if they do not bear functional and/or health claims[83]:

1. foods with no history of use in the country;
2. foods containing novel ingredients;
3. foods containing substances already consumed that may be added or used at levels much higher than those currently observed in the foods that constitute part of a regular diet; and
4. food offered in the form of capsules, pills, tablets, and the like.

Brazil is a member of the Southern Common Market, Mercosur, whose other members are Argentina, Paraguay, Uruguay, and Venezuela. Like the other member countries,[84] Brazil is gradually replacing many of its food standards with official Mercosur standards as they are developed. Mercosur standards are influenced by the EU, the *Codex Alimentarius*, and the FDA.[85] Mercosur does not have a definition of novel foods, although such foods are recognized in practice.[86]

14.3.6.3 Canada

In Canada, the Health Canada Food Directorate is responsible for establishing policies, setting standards, and providing advice and information on the safety and nutritional value of food.[87] It is also responsible for promoting the nutritional health and well-being of Canadians by defining, promoting, and implementing evidence-based nutrition policies and standards.[88] In addition, it administers the provisions of the Food and Drugs Act (1920) that relate to public health, safety, and nutrition.[89] The Canadian Food Inspection Agency is responsible for enforcing the food safety and nutritional quality standards established by Health Canada.[90]

Health Canada defines a novel food as:

1. a substance, including a microorganism, that does not have a history of safe use as a food;
2. a food that has been manufactured, prepared, preserved, or packaged by a process that has not been previously applied to that food and causes the food to undergo a major change; and
3. a food that is derived from a plant, animal, or microorganism that has been GM.[91]

Pre-market application is required for a new substance, for an extension of the use of a permitted food additive, for permission to change the maximum level of a permitted food additive, or for authorization to add a new organism to the list of permitted sources of enzymes used as a food additive.[92]

14.3.6.4 China

In China, the MOH is responsible for drafting health laws, regulations, policies, and standards.[93] It also supervises enforcement of these standards, unlike many other countries that assign enforcement authority to a different entity.[94]

Novel foods are referred to as new resource foods, which are defined as raw food materials or food ingredients that do not have a significant history of consumption in the country.[95] Chinese legislation under the MOH defines a new food additive as an additive that is not included in the national food safety standards, not included in the public announcement of permitted use issued by the MOH, and whose scope of use or dosage is increased.[96] Pre-market application is required for new food additives.[97]

14.4 Specific Regulatory Issues

14.4.1 Regulatory Issues Connected to Food Formulation

A food that results from a process not used for food production in the EU before May 15, 1997 but that gives rise to significant changes in the composition or structure of a food (affecting its nutritional value, metabolism, or level of undesirable substances) is considered a novel food under the NFR. The same applies if vitamins, minerals, and other substances are used in the production process and the process was not used for food production within the EU before May 15, 1997.

As discussed in the next section, more specific regulations apply if the novel food is being prepared for specifically defined vulnerable groups such as infants. In such circumstances, the rules of Regulation (EU) No 609/2013 on food intended for infants and young children, food for special medical purposes, and total diet replacement for weight control apply in addition to the rules of the NFR.[98]

In the US, federal agencies regulate food produced through unconventional methods within existing legal frameworks. Accordingly, the FDA focuses on the characteristics of the food product rather than the method by which the food is produced or developed.[99] It justifies this approach by reference to the agency's dual purpose: protecting the public's health by ensuring the safety of food that enters the US market, while also fostering innovation and technological advances in the food sector.[100]

Anticipating a boom in the application of the genomic editing technology called clustered regularly interspaced short palindromic repeats, also known as CRISPR-cas9 (CRISPR), the FDA released a draft guidance document in 2017. In it, the FDA proposes to expand the definition of 'new animal drug' (NAD) to include animals altered or developed through genome editing and genetic engineering (GE) technologies.[101] The agency justifies the expansion on the basis that altering an animal's genome to affect its bodily structure or function or to cure, treat, or prevent disease in the animal falls within the FDCA's definition of a drug.[102] In addition to CRISPR, the guidance document also identifies rDNA as an NAD, even though rDNA has been widely used in plants to create GMOs. The Guidance Document clarifies that the GM animal is not a drug, but rather the genome editing technology is an NAD.[103] Similar to food additives, NADs are subject to an expensive and time-consuming pre-market review process.[104] The FDA had not released a final guidance document by the end of 2020.[105]

14.4.2 Regulatory Issues Connected to Food Formulation for Special Populations

In the EU, Regulation (EU) No 609/2013 on food intended for infants and young children, food for special medical purposes, and total diet replacement for weight control (Food for Specific Groups [FSG])[106] stipulates a special regime for formulation of food for specific populations. According to FSG Recital 23, the rules in the FSG complement the rules of the NFR in the sense that no novel food shall be added to food covered by the FSG unless such substances fulfil the FSG's conditions for being placed on the market.

Where a food promises health benefits or nutritional benefits and the food business operator makes use of these benefits in its marketing communication, the EU's special regime of health claims applies. In this regard, Regulation (EC) No 1924/2006 on nutrition and health claims made on foods (NHCR)[107] introduces EU-wide quality requirements, which apply in addition to the novel foods regime. According to the NHCR, health and nutrition claims concerning foods need to be based on scientific evidence[108] and not be 'false, ambiguous or misleading'.[109] In addition, the NHCR introduces an authorization procedure for health claims.[110] Certain uses of health claims are also prohibited without exception, such as those that 'make reference to the rate or amount of weight loss'.[111] The EU Register of Nutrition and Health Claims lists all permitted nutrition claims and all authorized and non-authorized health claims.

14.4.3 Regulatory Issues Connected to Genetically Modified Food

In the EU, GM foods are not subject to the novel foods regime.[112] Rather, a special regulatory regime requires authorization and traceability of GMOs before they are released into the environment, introduced into the market, or used in food for human beings or animal feed.[113] Furthermore, EU law foresees certain liability requirements for damage caused by GMOs.[114]

In addition, specific labelling requirements exist for (authorized) GMOs. Regulation (EC) No 1829/2003 (GMO Labelling Regulation) concerns labelling of foods that contain or consist of GMOs or are produced from or contain ingredients produced from GMOs.[115] Article 13(1)(a) of this Regulation introduces an EU-wide labelling requirement, according to which each food containing or produced from GMOs must bear a label with mandatory information attached to it, indicating that the food or the respective ingredient contains GMOs. Article 12(2), however, exempts all 'foods containing material which contains, consists of or is produced from GMOs in a proportion no higher than 0.9 per cent of the food ingredients considered individually or food consisting of a single ingredient, provided that this presence is adventitious or technically unavoidable'.[116] It is also important to note that GMOs obtained by mutagenesis are not considered to be part of the special GMO regulation in the EU, with the exception of techniques developed after 2001 such as CRISPR.[117]

In the US during the 1970s and 1980s, there was increased application of innovative biotechnology in formulation of food products, specifically GMOs. The White House investigated how to regulate biotechnology products and who should be responsible for the review of their safety and for enforcement.[118] After 18 months of public comments, the White House finalized a federal policy entitled the Coordinated Framework for the Regulation of Biotechnology (Coordinated Framework) in 1986.[119] The Coordinated Framework outlines three basic regulatory principles.[120] First, agencies should focus on the end products, not the process used to create GE[121] food.[122] Second, GMOs are low risk and no more dangerous or risky than traditionally produced food.[123] Third, existing federal statutes are sufficient to ensure that GMOs and their by-products meet the same safety standards as their conventional counterparts; thus, GMOs do not require pre-market approval.[124] The Coordinated Framework tasks the Environmental Protection Agency, United States Department of Agriculture, and the FDA with enforcing biotechnology regulations according to the three principles just listed.[125] Under the Coordinated Framework, human and animal food derived from GE plants is held to the same food safety and labelling standards as all FDA-regulated food.[126]

The White House issued two updates to the Coordinated Framework, in 1992 and 2017. Both continued to emphasize a 'product-focused, not process-focused' regulatory approach to GMOs.[127] Accordingly, the Coordinated Framework continues to provide that GMOs should not be subject to excessive oversight compared to their non-modified counterparts, unless the risk posed by the GE product is unreasonable: this occurs when the amount of risk reduced by regulatory oversight exceeds the cost of regulation.[128]

As set out in the Coordinated Framework, GE plants are regulated under the relevant agency's existing framework, but additional regulatory programmes have been enacted to ensure GE food safety.[129] For example, while the FDA operates under the assumption that GE plants are not materially different from their conventional counterparts, the agency established a voluntary pre-market Plant Biotechnology Consultation Programme. To date, all GE plant developers have voluntarily participated in the Consultation Programme and waited for final FDA approval before selling their products.[130] The FDA evaluates factors such as whether the GE plant is as nutritious as its conventional counterpart and whether the genetic alteration may contain a new toxin or allergen.[131] After reviewing the sponsor's data, the agency issues a letter of its findings to the developer and publishes the consultation package online for the public.[132] As noted earlier, the FDA does not impose special GE labelling requirements.[133] The FDA treats GE products as GRAS, but GE developers must demonstrate that the modified crops are substantially equivalent to their non-modified version and are safe before placing them in interstate commerce.[134]

14.4.4 Regulatory Issues Connected to Food Containing Nanomaterials

In the EU, foods containing engineered nanomaterials are considered novel foods.[135] '[E]ngineered nanomaterial' means any intentionally produced material that has one or more dimensions of the order of 100 [nanometers] (nm) or less or that is composed of discrete functional parts, either internally or at the surface, many of which have one or more dimensions of the order of 100 nm or less, including structures, agglomerates or aggregates, which may have a size above the order of 100 nm but retain properties that are characteristic of the nanoscale. Properties that are characteristic of the nanoscale include:

i. those related to the large specific surface area of the materials considered; and/or
ii. specific physico-chemical properties that are different from those of the non-nanoform of the same material'.[136]

If a food contains such nanomaterials, the entire legal regime of the NFR applies.

Nanomaterials, or 'nanoparticles' are already present in many foods, both natural and processed. Some nanoparticles are specifically designed for incorporation into foods as a delivery system.[137] In the US, engineered nanomaterials are considered food additives and must go through the FAP pre-market approval process before entering the market. The FDA declared as policy in its guidance document that food ingredients created through nanotechnology and foods containing engineered nanomaterials are not eligible for GRAS status, although this guidance is not binding.[138]

14.4.5 Regulatory Issues Connected to 3D Printed Food Products

3D printing in the area of foods is still at an early stage. Potential applications include the production of food with special nutritional and dietary needs (e.g., for people suffering from illnesses) or substitution of food materials. In the EU there is no specific legal framework for the application of 3D printing techniques for foods. Therefore, in the absence of specific legislation, the applicable regulatory framework will depend on the product. It is unlikely that 3D foods will mimic foods which have been in circulation in the EU before May 15, 1997 and alter some of their structure, and so they are likely to fall under the regulatory scrutiny of the NFR. If the additional prerequisites of the NFR are met, i.e. the 3D printing is conducted such that it falls within the NFR's initial list,[139] most of these 3D foods would qualify as novel and hence require authorization in the EU. If, as expected, they are designed for specific groups, the rules of the FSG would also apply, as described above. Here, a regulatory obstacle with 3D printing is very likely to arise under FSG Article 4(2), which only allows such personalized foods to be sold prepacked in retail markets.

Although the US FDA has released a final guidance document for industry on 3D printed medical devices, there is no similar policy in place for 3D printed food.[140] The two main outstanding questions regarding 3D printed food are whether it should be considered 'imitation' food and be labelled as such, and whether 3D printed foods will lead to economic adulteration or 'food fraud'.[141] An example would be the sale of 3D printed food without proper labelling.[142]

14.4.6 Regulatory Issues Connected to Food Made from Insects

In the EU, foods made from insects are generally defined as novel foods. According to NFR Recital 8, if insects or their parts are used as food, they are also considered novel: this was subject to debate before the NFR came into force. Since many insects have been eaten in some way at some point in history in many places of the EU, it remains unclear if and how the NFR applies. Since insect food is a business of particular interest for start-ups, the compliance risks with the NFR can serve as a barrier to innovation for these companies.

In the US, there is no official FDA regulation or guidance document that prohibits or approves using insects as human food,[143] and novelty foods such as chocolate-covered ants have been on the market for years. The FDCA forbids the adulteration of food in interstate commerce, with adulteration defined as containing 'filthy ... substances, or if it is otherwise unfit for food'.[144] Courts have affirmed the FDA's interpretation of 'filthy' to include the *unintentional* addition of insects and insect parts,[145] but this would not apply to insects *intentionally* added to food. Recognizing the unavoidable presence of insects in certain food products, the FDA permits a certain amount of negligible insects in a food before it is deemed adulterated.[146] Again, where insects are the main feature of the food and are in more than negligible quantities, it is unlikely that the FDCA's prohibitions would apply.

To date, the FDA has given little attention to insects as an intentional component of food.[147] Many American consumers have a generally negative view of consuming insects, but there are a growing number of companies seeking to promote insects as an alternative form of protein.[148] Until specific guidelines appear regarding the use of insects, companies may find it risky to invest in this type of food.[149] It is also worth noting that companies that seek to sell insect-based human food in the interstate market will likely need to go through the pre-market food additive approval process.[150]

14.5 Other Legal Issues

To this point, this chapter has addressed and discussed regulation of novel foods – but only regulation in the sense of regulatory approval. Other regulatory (in the sense of legal and legislative) issues are implicated with respect to novel foods. For this reason, the remainder of this chapter turns more broadly to the constellation of other legal issues implicated by novel foods. In particular, we examine how producers of novel foods may fall afoul of legal requirements in different regulatory areas.

14.5.1 Intellectual Property Law

We saw in earlier chapters that scientists have become ever more creative in devising new food ingredients and processes. The question is at what point an inventor would seek intellectual property protection for a novel food.

Note first that 'novelty' for the purposes of patent law generally concerns prior art (e.g. whether the claimed invention has already been patented, described, in public use, on sale, or otherwise available to the public) rather than a history of consumption, which is key to the definition of novel foods.[151] Another feature of intellectual property law is that novelty alone does not make a product patentable. There is also a requirement of non-obviousness of the purported invention.[152] This means that a claimed invention is only patentable if, prior to the effective filing date, it would not have been obvious to a person having ordinary skill in the art to which the claimed invention pertains.[153] In terms of novel foods, this could mean that a food that merely substitutes one ingredient for another might not be considered patentable due to obviousness.

Alternatively, companies may seek to protect their investment by applying for patents on the technology that makes the novel food product possible. For example, as mentioned earlier, the US Patent Office has granted patents to the company Impossible Foods, Inc. for its technology which uses the key ingredient, soy leghaemoglobin, to make its plant-based Impossible Burger bleed like real meat.[154]

In the US and in the EU, the issues of intellectual property rights and GMOs are deeply intertwined. As members of the WTO, countries must comply with Trade-Related Aspects of Intellectual Property Rights (TRIPS) Article 27, 'Patentable Subject Matter'. TRIPS provides that 'patents shall be available for any inventions, whether products or processes, in all fields of technology', although member countries may exempt plants and animals from patentability.[155] Both the US and the EU have afforded patent protection for GM plants and animals.[156] For example, in *Diamond v. Chakrabarty*, the US Supreme Court held that 'anything under the sun made by man', such as genetically-engineered microorganisms, are patentable subject matters.[157]

14.5.2 Tort Law

Legal issues can also arise in the tort area. The seller of an unregulated food found harmful to consumers could be subject to a wide variety of potential litigation. In the EU, liability under tort law varies widely, as national tort law systems are generally not harmonized. One exception is the area of product liability, which according to Article 21 of the GFL also applies to foodstuffs. A comparable liability regime applies in the US, where a claim can be made under tortious product liability if the seller of a substance is found to have sold a product that caused the plaintiff injury due to its defective or unsafe nature.[158] In such a case, the seller could be liable for damages.

In addition, where a harm results from a substance which misleads the consumer, a cause of action might potentially be filed under the contract and consumer law-based principle of an implied warranty of fitness for a particular purpose, which is outlined in § 2-315 of the Uniform Commercial Code (UCC).[159] Under this claim, where a seller has reason to know any particular purpose for which the goods being furnished are required, and the buyer justifiably relies upon the seller's knowledge, skill, or judgement to furnish goods suitable for a particular purpose, the seller may be held liable if the goods do not fulfil that purpose.

14.5.3 Marketing Law

In the EU, the Food Information Regulation (FIR)[160] covers most food marketing and outlines a number of fair information practices.[161] The basic principles are that food information must be accurate, clear, and easy for consumers to understand. Specifically, food information must not be misleading:

- as to the characteristics of the food and, in particular, as to its nature, identity, properties, composition, quantity, durability, country of origin or place of provenance, method of manufacture or production;
- by attributing to the food effects or properties which it does not possess;
- by suggesting that the food possesses special characteristics (when in fact all similar foods possess such characteristics), in particular by specifically emphasizing the presence or absence of certain ingredients and/or nutrients;
- by suggesting, via the appearance, the description, or pictorial representations, the presence of a particular food or an ingredient when in reality a component naturally present or an ingredient normally used in that food has been substituted with a different component or ingredient.

Food information may not attribute to any food the property of preventing, treating, or curing a human disease, or refer to such properties. The prohibitions extend to advertising and the general presentation of foods, regarding their shape, appearance, or packaging; the packaging materials used; the way in which they are arranged; and the setting in which they are displayed. The FIR does not distinguish between foods and novel foods, and so all foods are treated the same, with no special requirements for marketing novel foods to consumers.

14.5.4 Criminal Law

Gene-editing technology, such as rDNA and CRISPR, holds great potential to revolutionize the food supply. Many laboratories are exploring the use of these technologies for the benefit of humankind, such as by formulating transgenic goat's milk to combat juvenile diarrhoea in developing countries.[162] However, just as food can be bioengineered to enhance or insert desirable characteristics, it could potentially be bioengineered for nefarious purposes. A bad actor could potentially target a nation's food supply chain to inflict serious damage on thousands of people.

In the EU, as with tort law, criminal law is generally a matter for individual member states. Hence, it would depend on the specific national legal system whether an act or an omission relating to novel food would be subject to criminal charges.

In the US, FDCA Chapter III explicitly prohibits certain actions and outlines potential civil and criminal liability.[163] For example, it prohibits introducing adulterated or misbranded food into the market,[164] and corporate officers have been found guilty of knowingly introducing adulterated food into interstate commerce.[165] Since criminal liability attached even where those corporate officers did not intentionally set out to create contaminated products, it is clear that criminal liability would also apply in the case of an intentional act.

14.6 Conclusion

This chapter discussed how novel foods are regulated within the EU, US, and elsewhere. We reviewed some of the reasons novel foods might be regulated and how such regulatory goals can be accomplished. Regulation of novel foods is the policy maker's reaction to the unknown, although what is novel today may not be novel tomorrow. Indeed, the novel food industry has grown immensely over the past decade due to the need to feed a growing world population while using less land, water, and other limited natural resources. Ultimately, policy makers have the difficult task of balancing adequate government oversight over novel foods without stifling innovation and creativity.

Acknowledgements

The authors would like to thank Jasmin Cohen, Dale Radford, Christen Sypolt, and Tiffany Wang for research assistance. We also thank Marsha Cohen and Hanna Schebesta for helpful comments on earlier versions of this chapter. Any mistakes remain our own.

Notes

1. European Commission. What is Novel Food? https://ec.europa.eu/food/safety/novel_food_en (accessed December 31, 2020).
2. World Health Organization (1987). International Programme on Chemical Safety [hereinafter IPCS], Environmental Health Criteria 70, Principles for the Safety Assessment of Food Additives and Contaminants in Food. http://www.inchem.org/documents/ehc/ehc/ehc70.htm#SubSectionNumber:6.2.4 (accessed December 31, 2020).
3. Regulation (EU) 2015/2283 of the European Parliament and of the Council of November 25, 2015 on novel foods (hereinafter NFR) Article 3(2) (a), amending Regulation (EU) No 1169/2011 of the European Parliament and of the Council and repealing Regulation (EC) No 258/97 of the European Parliament and of the Council and Commission Regulation (EC) No 1852/2001 (text with EEA relevance), OJ L 327, December 11, 2015, p. 1–22.
4. *See infra* text at n. 31.
5. European Commission, *supra* note 1.
6. Food, Drug, and Cosmetic Act 1938, 21 U.S.C. s 301 et seq.
7. Magnuson, B., Munro, I., Abbot, P. et al. (2013). Review of the regulation and safety assessment of food substances in various countries and jurisdictions. *Food Additives & Contaminants. Part A, Chemistry, Analysis, Control, Exposure & Risk Assessment*, 30(7): 1147–1220, at 1189. https://www.ncbi.nlm.nih.gov/pmc/articles/PMC3725665 (accessed December 31, 2020).
8. *See, e.g.* Finck, M. (2018). Blockchains: regulating the unknown. *German Law Journal*, 19:665–692 (blockchain technology); Åm, H. (2011). Regulating the Unknown: Governing Nanotechnologies by a Logic of Pre-emption, p. 11 ('nanotechnology brings to the fore the difficulties of regulatory policymaking under conditions of uncertainty'). Dissertation. University of Vienna. https://www.researchgate.net/profile/Heidrun_Am/publication/265276233_Regulating_the_Unknown_Governing_Nanotechnologies_by_a_Logic_of_Pre-emption/links/544fbe230cf24e8f7374a53a/

Regulating-the-Unknown-Governing-Nanotechnologies-by-a-Logic-of-Pre-emption.pdf (accessed December 31, 2020).

9. Opinion of AG Bobek delivered on March 30, 2017, Case C-111/16, Criminal proceedings against Giorgio Fidenato and Others, ECLI identifier: ECLI:EU:C:2017:248, para. 31 (stipulating that Ulrich Beck's work guided EU regulation).
10. Beck, U. (1992). *Risk Society: Towards a New Modernity*. London: Sage. *But see* Bergkamp, L. (2016). The concept of risk society as a model for risk regulation – its hidden and not so hidden ambitions, side effects, and risks. *Journal of Risk Research*, 20(10), 1275–1291 (criticizing the risk society approach, inter alia because it politicizes science and taints risk management decision-making).
11. In economic terms this is called the existence value. *See* Dana, D. (2004). Existence value and federal preservation regulation. *Harvard Environmental Law Review*, 28:343–399, at 345.
12. Cicchetti, C. and Wilde, L. (1992). Uniqueness, irreversibility, and the theory of non-use values, *American Journal of Agricultural Economics*. 74(5): 1121–1122.
13. General Food Law (hereinafter GFL), Regulation (EC) No 178/2002 of the European Parliament and of the Council of January 28, 2002 laying down the general principles and requirements of food law, establishing the European Food Safety Authority and laying down procedures in matters of food safety, OJ L 31, 1.2.2002, Art. 7(1).
14. GFL Arts. 1(3) and 4(1).
15. *Id.*
16. Commission of the European Communities (2000). Communication from the Commission on the precautionary principle, Document 52000DC0001, COM/2000/0001 final. https://eur-lex.europa.eu/legal-content/EN/ALL/?uri=celex:52000DC0001 (accessed December 31, 2020).
17. NFR Recital 20.
18. *See*, e.g. Finck, *supra* note 8, at 682–90. As one researcher pointed out, 'Uncertainty poses particular problems for governments, as decision-making is highly politicised when no recourse to "facts" is possible when faced with new technologies.' Åm, *supra* note 8, at 15. *But see* Cicchetti and Wilde, *supra* note 12 (an early example of the increasing literature in law and economics on the valuation of uncertainty, which can be used in assisting such decisions).
19. Magnuson, et al., *supra* note 7, at 1202.
20. IPCS, *supra* note 2.
21. *Id.*
22. *Id.*
23. *Id.*
24. FAO, *About Codex Alimentarius*, http://www.fao.org/fao-who-codexalimentarius/about-codex/en/#c453333 (accessed December 31, 2020). On the relationship between Codex standards and WTO law, *see* Dominique, S. and Purnhagen, K. (2016). Reversed harmonization or horizontalization of EU standards? *Wisconsin International Law Journal*, 34:102–104.
25. Magnuson et al., *supra* note 7, at 1148.
26. For more background, *see* Vos, E. and Wendler, F. (2007). Food safety regulation at the EU level. In: *Food Safety Regulation in Europe: A Comparative Institutional Analysis*, 65–138, at 74. Antwerp: Intersentia; Faure, M. (2018). The economics of harmonization of food law in the EU. In: *Regulating and Managing Food Safety in the EU* (ed. H. Bremmers and K. Purnhagen), 263–290, at 283. New York: Springer.
27. GFL Art. 17.
28. Commission of the European Communities (1999). White Paper on Food Safety. In 2020, the Commission adopted the Farm to Fork Strategy, which assigns food innova-

tion a key role to play in the food market of the future. European Commission, *Farm to Fork Strategy – for a fair, healthy and environmentally-friendly food system,* https://ec.europa.eu/food/farm2fork_en.
29. European Commission. General Food Law. https://ec.europa.eu/food/safety/general_food_law_en (accessed December 31, 2020).
30. There are exceptions when the precautionary principle applies. *See* GFL Art. 7.
31. NFR Art. 3 II (a).
32. *Id.* Recital 42.
33. Treaty on Functioning of the European Union Art. 114 (5).
34. *Id.* Art. 114 (5, 6).
35. World Trade Organization. Understanding the WTO: The Agreements, Standards and Safety. https://www.wto.org/english/thewto_e/whatis_e/tif_e/agrm4_e.htm (accessed December 31, 2020).
36. NFR Art. 2 I.
37. *Id.* Art. 2 II.
38. *Id.* Art. 3 II (b).
39. *Id.* Art. 3 II (c).
40. *See supra* text at n. 31.
41. *Id.* Art. 4 (1).
42. *Id.* Art. 4 (2, 3).
43. *Id.* Art. 6 (2).
44. *Id.* Art. 10.
45. *Id.* Art. 7 (a).
46. *Id.* Art. 7 (b).
47. *Id.* Art. 7 (c).
48. U.S. Constitution Art. I sec. 8.
49. Law, Mark T. History of Food and Drug Regulation in the United States. http://eh.net/encyclopedia/history-of-food-and-drug-regulation-in-the-united-states (accessed December 31, 2020).
50. *Id.*
51. *Id.*
52. *Id.*
53. *Id.*
54. *See* Strauss, D. (2011). An analysis of the FDA food safety modernization act: protection for consumers and boon for business. *Food and Drug Law Journal*, 66:354–355.
55. Intertek. Novel Food & Food Additive Services. http://www.intertek.com/food/consulting/novel-foods (accessed December 31, 2020).
56. 21 U.S.C. s 321(s); *see also* International Food Information Council and US Food and Drug Administration (2010). Overview of Food Ingredients, Additives & Colors. https://www.fda.gov/Food/IngredientsPackagingLabeling/FoodAdditivesIngredients/ucm094211.htm (accessed December 31, 2020).
57. Merrill, R. and Francer, J. (2000). Organizing federal food safety regulation. *Seton Hall Law Review*, 31:61–173, at 96; FDA. Determining the Regulatory Status of a Food Ingredient. https://www.fda.gov/food/ingredientspackaginglabeling/foodadditivesingredients/ucm228269.htm (accessed December 31, 2020).
58. FDA (2016). FAQs About GRAS for Substances Intended for Use in Human or Animal Food: Guidance for Industry. https://www.fda.gov/downloads/Food/GuidanceRegulation/GuidanceDocumentsRegulatoryInformation/UCM525233.pdf (accessed December 31, 2020); 21 C.F.R. § 170.3(i).
59. *See* 21 U.S.C. s 170.20(a) (establishing general principles for Food Additive Petition pre-market approvals); 21 U.S.C. s 171.1.

60. *See* Damewood, Kelly, Food Safety News (2014). The GRAS Process: How Companies Legally Add Ingredients to Food. https://www.foodsafetynews.com/2014/01/the-gras-process-how-companies-legally-add-ingredients-to-food (accessed December 31, 2020).
61. FDA, *supra* note 58.
62. 21 U.S.C. s 321(s); FDA (2005–2006). How US FDA's GRAS Notification Program Works. https://www.fda.gov/food/ingredientspackaginglabeling/gras/ucm083022.htm (accessed December 31, 2020).
63. Substances GRAS, 81 Fed. Reg. 54960 (August 17, 2016). *See also* 21 C.F.R. s 170 subpart E; FDA (2016). About the GRAS Notification Program https://www.fda.gov/Food/IngredientsPackagingLabeling/GRAS/ucm2006851.htm (accessed December 31, 2020).
64. Damewood, *supra* note 60.
65. Business Wire (2018). Impossible Foods Receives No-Questions Letter From US Food and Drug Administration. https://www.businesswire.com/news/home/20180723005786/en/Impossible-Foods-Receives-No-Questions-Letter-Food-Drug (accessed December 31, 2020).
66. *Id.*
67. Damewood, *supra* note 60. *See also* 21 C.F.R. s 170.30(c), 170.3(f).
68. Magnuson, et al., *supra* note 7, at 1150.
69. *Id.*
70. *Id.* at 1155.
71. *Id.*
72. *Id. See also* FSANZ 2011, www.foodstandards.gov.au (accessed December 31, 2020).
73. Magnuson, et al., *supra* note 7, at 1150.
74. *Id.* at 1156.
75. *Id.*
76. *Id.*
77. *Id.*
78. *Id.* at 1157.
79. For more information, *see* Ministry of Health, www.saude.gov.br (accessed December 31, 2020); *see also* National Agency of Sanitary Surveillance website, http://portal.anvisa.gov.br (accessed December 31, 2020).
80. Magnuson, et al., *supra* note 7, at 1161.
81. *See* Annex II of Resolution RDC 240/2018. *See also* http://portal.anvisa.gov.br/en/food (accessed December 31, 2020).
82. *Id.*
83. Magnuson, et al., *supra* note 7, at 1161.
84. *Id* at 1192.
85. *Id.* at 1161.
86. *Id.* at 1152.
87. *Id.* at 1164; *see also* Government of Canada (2019). Food and Nutrition https://www.canada.ca/en/health-canada/services/food-nutrition.html (accessed December 31, 2020).
88. Magnuson, et al., *supra* note 7, at 1164.
89. *Id.*
90. *Id.; see also* Government of Canada, Canadian Food Inspection Agency (2018). About the CFIA. http://www.inspection.gc.ca/about-the-cfia/eng/1299008020759/1299008778654 (accessed December 31, 2020).
91. Magnuson, et al., *supra* note 7, at 1199.
92. *Id.* at 1198.
93. *Id.* at 1170.
94. *Id.*

95. *Id.* at 1199.
96. *Id.* at 1198.
97. *Id.* at 1199.
98. Regulation (EU) No 609/2013 of the European Parliament and of the Council of June 12, 2013 on food intended for infants and young children, food for special medical purposes, and total diet replacement for weight control and repealing Council Directive 92/52/EEC, Commission Directives 96/8/EC, 1999/21/EC, 2006/125/EC and 2006/141/EC, Directive 2009/39/EC of the European Parliament and of the Council and Commission Regulations (EC) No 41/2009 and (EC) No 953/2009 Text with EEA relevance, OJ L 181, June 29, 2013, pp. 35–56 [hereinafter FIYC Regulation].
99. FDA (1992). Statement of Policy: Foods Derived From New Plant Varieties, 57 Fed. Reg. 22984. https://www.fda.gov/Food/GuidanceRegulation/GuidanceDocumentsRegulatoryInformation/Biotechnology/ucm096095.htm (accessed December 31, 2020).
100. FDA (2018). What Does FDA Do? https://www.fda.gov/about-fda/fda-basics/what-does-fda-do (accessed December 31, 2020); FDA (2018). Press Release, Statement from FDA Commissioner Scott Gottlieb, M.D. and FDA Deputy Commissioner Anna Abram on emerging food innovation, "cultured" food products. https://www.fda.gov/NewsEvents/Newsroom/PressAnnouncements/ucm610869.htm (accessed December 31, 2020).
101. Regulation of Intentionally Altered Genomic DNA in Animals: Notice of Availability of Draft Guidance for Industry #187 ('Notice of Availability: GFI #187'), 82 Fed. Reg. 6561, 6563 (January 19, 2017).
102. *Id.*; 21 U.S.C. s 321(g).
103. FDA (2017). Q&A on FDA Regulation of Intentionally Altered Genomic DNA in Animals. https://www.fda.gov/animal-veterinary/animals-intentional-genomic-alterations/qa-fda-regulation-intentionally-altered-genomic-dna-animals (accessed December 31, 2020).
104. FDA (2019). New Animal Drug Applications. https://www.fda.gov/animal-veterinary/development-approval-process/new-animal-drug-applications (accessed December 31, 2020).
105. FDA (2020). Guidances Under Development for 2020. https://www.fda.gov/animal-veterinary/guidance-industry/guidances-under-development-2020 (accessed December 31, 2020).
106. FIYC Regulation, *supra* note 98.
107. Regulation (EC) No 1924/2006 of the European Parliament and of the Council of December 20, 2006 on nutrition and health claims made on foods (OJ L 404, December, 30, 2006, p. 9), last amended by Commission Regulation (EU) No 1047/2012 of November 8, 2012 L 31036 November 9, 2012 [hereinafter NHCR].
108. *Id.* Art. 6 I.
109. *Id.* Art. 3(a).
110. *Id.* Art. 10.
111. *Id.* Art. 12(b).
112. NFR Art. 2(2)(a).
113. *See* Purnhagen, K. (2019). How to manage the Union's diversity: the regulation of new plant breeding technologies in confederation paysanne and others, *Common Market Law Review*, 56(5): 1379–1396.
114. Article 18(3)(b) Directive 2004/35/EC on Environmental liability with regard to the prevention and remedying of environmental damage, OJ 2004, L 143/56.
115. Regulation (EC) No 1829/2003 of the European Parliament and of the Council of September 22, 2003 on GM food and feed (Text with EEA relevance), OJ L 268, October 18, 2003, pp. 1–23.
116. NFR Art. 12(2).

117. *See* Purnhagen, K., Kok, E., Kleter, G. et al. (2018). EU Court casts new plant breeding techniques into regulatory limbo. *Nature Biotechnology*, 36(9): 799–800.
118. The National Academies of Science, Engineering, and Medicine et al. (2017). *Preparing for Future Products of Biotechnology*. National Academies Press. https://www.ncbi.nlm.nih.gov/books/NBK442204 (accessed December 31, 2020).
119. *Id.*
120. Coordinated Framework for the Regulation of Biotechnology, 51 Fed. Reg. 23302 (June 26, 1986) [hereinafter Coordinated Framework].
121. Organisms modified through recombinant DNA (rDNA) technology are known as GM, GMO, or GE, but the FDA prefers referencing the rDNA technology as 'genetic engineering'. FDA (2018). Questions & Answers on Food from Genetically Engineered Plants. https://wayback.archive-it.org/7993/20180908125544/https://www.fda.gov/Food/IngredientsPackagingLabeling/GEPlants/ucm346030.htm (accessed December 31, 2020).
122. *Id.*
123. *Id.*
124. Marden, E. (2003). Risk and regulation: US regulatory policy on genetically modified food and agriculture. *Boston College Law Review*, 44(734): 747; Coordinated Framework, *supra* note 120.
125. Marden, *supra* note 124, at 739.
126. FDA (2018). How FDA Regulates Food from Genetically Engineered Plants. https://www.fda.gov/food/food-new-plant-varieties/how-fda-regulates-food-genetically-engineered-plants (accessed December 31, 2020).
127. Exercise of Federal Oversight Within Scope of Statutory Authority: Planned Introductions of Biotechnology Products into the Environment, 57 Fed. Reg. 6753 (February 27, 1992) [hereinafter 1992 Coordinated Framework Update]; Executive Office of the President (2017). Modernizing the Regulatory System for Biotechnology Products: Final Version of the 2017 Update to the Coordinated Framework for the Regulation of Biotechnology. https://obamawhitehouse.archives.gov/sites/default/files/microsites/ostp/2017_coordinated_framework_update.pdf [hereinafter 2017 Coordinated Framework Update] (accessed December 31, 2020).
128. 1992 Coordinated Framework Update, *supra* note 127.
129. Coordinated Framework, *supra* note 120.
130. *Id.*
131. *Id.*
132. *Id.*
133. FDA (2019). Guidance for Industry: Voluntary Labeling Indicating Whether Foods Have or Have Not Been Derived from Genetically Engineered Plants. https://www.fda.gov/regulatory-information/search-fda-guidance-documents/guidance-industry-voluntary-labeling-indicating-whether-foods-have-or-have-not-been-derived (accessed December 31, 2020).
134. Consumer Reports (2001). Comment Letter from Consumers Union to FDA Regarding Docket No 00N-1396, Premarket Notice Concerning Bioengineered Foods. https://advocacy.consumerreports.org/press_release/cus-comments-on-premarket-notice-concerning-bioengineered-foods (accessed December 31, 2020).
135. NFR Art. 3(2) (a) (viii).
136. *Id.* Art. 3(2)(f).
137. Takhistova, K. (2009). Note, food nanotechnology – in search of a regulatory framework. *Rutgers Computer & Technology Law Journal*, 35:: 255–295, at 261.
138. Use of Nanomaterials in Food for Animals: Guidance for Industry #220, 80 Fed. Reg. 46587 (August 5, 2015).
139. *See supra* text at n. 31.

140. *See* Technical Considerations for Additive Manufactured Medical Devices, Final Guidance for Industry, 82 Fed. Reg. 57462 (December 5, 2017).
141. *Id.*
142. *Id.*
143. Boyd, M. (2017). Cricket soup: a critical examination of the regulation of insects as food. *Yale Law & Policy Review*, 36:17–81, at 66.
144. 21 U.S.C. s 342.
145. Boyd, *supra* note 143, at 40–41.
146. *Id.* at 43; 21 C.F.R. s 110.110(a).
147. Boyd, *supra* note 143, at 20.
148. *Id.* at 21.
149. *Id.* at 37.
150. *Id.* at 44–45 (discussing a federal case in which the FDA sought to enforce its regulation of intentionally added insects to animal-feed grain as a food additive).
151. *See, e.g.* 35 U.S.C. s 102 (US statute setting out rules for patentability).
152. Legal Information Institute. 35 U.S.C. §103; Conditions for patentability; non-obvious subject matter. https://www.law.cornell.edu/uscode/text/35/103 (accessed December 31, 2020).
153. *See, e.g.* 35 U.S.C. s 103 (US statute setting out the rules for non-obviousness).
154. US Patent # 9 700 067, Justia https://patents.justia.com/patent/9700067 (accessed December 31, 2020).
155. Torrance, A. (2007). Intellectual property as the third dimension of GMO regulation. Genetically Modified Organisms: Philosophy, Science, and Policy Symposium. *Kansas Journal of Law & Public Policy*. 16:257–285, at 279.
156. *Id.* at 280.
157. *Diamond v. Chakrabarty*, 447 U.S. 303, 309 (1980).
158. Legal Information Institute. Products Liability. https://www.law.cornell.edu/wex/products_liability (accessed December 31, 2020).
159. The UCC is not a federal law but rather a model commercial code for the states, governing commercial transactions to facilitate interstate commerce.
160. Regulation (EU) No 1169/2011 of the European Parliament and of the Council of October 25, 2011 on the provision of food information to consumers, amending Regulations (EC) No 1924/2006 and (EC) No 1925/2006 of the European Parliament and of the Council, and repealing Commission Directive 87/250/EEC, Council Directive 90/496/EEC, Commission Directive 1999/10/EC, Directive 2000/13/EC of the European Parliament and of the Council, Commission Directives 2002/67/EC and 2008/5/EC and Commission Regulation (EC) No 608/2004 Text with EEA relevance, OJ L 304, November 22, 2011, pp. 18–63 [hereinafter FIR].
161. FIR Art. 7.
162. Cooper, C., Klobas, L., Maga, E. et al. (2013). Consuming transgenic goats' milk containing the antimicrobial protein lysozyme helps resolve Diarrhea in young pigs. *PLoS One*, 8(3):e58409 https://journals.plos.org/plosone/article?id=10.1371/journal.pone.0058409 (accessed December 31, 2020).
163. 21 U.S.C. s 331 et seq.
164. *Id.* s 331(a).
165. Department of Justice (2015). Department of Justice, Former Peanut Company President Receives Largest Criminal Sentence in Food Safety Case; Two Others also Sentenced for Their Roles in Salmonella-Tainted Peanut Product Outbreak. https://www.justice.gov/opa/pr/former-peanut-company-president-receives-largest-criminal-sentence-food-safety-case-two (accessed December 31, 2020).

Index

Food Formulation: Novel Ingredients and Processing Techniques, First Edition.
Edited by Shivani Pathania and Brijesh K. Tiwari.